Linear Algebra & Matrix Theory

Robert R. Stoll

DOVER PUBLICATIONS, INC.
Mineola, New York

Bibliographical Note

Linear Algebra and Matrix Theory, published by Dover Publications, Inc., in 2012, is an unabridged republication of the work originally published by the McGraw-Hill Company, New York, in 1952 and republished by Dover Publications, Inc., in 1969.

International Standard Book Number

ISBN-13: 978-0-486-62318-4
ISBN-10: 0-486-62318-1

Manufactured in the United States by Courier Corporation
62318143 2013
www.doverpublications.com

To My Wife

PREFACE

This text is designed for use by advanced undergraduate or first-year graduate students, in a course on the theory of matrices viewed against the background of modern algebra. Chapters 1 to 5 treat topics in algebra and matrix theory that have risen to prominence in widely diversified fields: economics, psychology, statistics, the several engineering sciences, physics, and mathematics. The remaining chapters will be of more immediate utility to the latter half of the above-mentioned fields. Very little in the way of formal mathematical training is required. However, a certain amount of that elusive quality known as "mathematical maturity" is presupposed. For example, an understanding of "necessary and sufficient conditions" will be much more useful equipment than an understanding of "derivative." For a class whose membership is drawn from the social sciences, Chaps. 1 to 5, plus possibly a little time devoted to numerical methods in their special field of interest, should constitute a one-semester course. A coverage of the entire book, with the possible exception of elementary divisor theory (Secs. 7.6 to Sec. 7.9—an omission which will not interrupt the continuity), should be possible in approximately one and a half semesters, thereby leaving enough time in a two-semester course to take up one of such varied topics as (i) numerical methods in matrix theory, (ii) the Fredholm theory of integral equations, and (iii) applications of matrix methods to systems of differential equations, electrical networks, vibration problems, etc.

The original intention of the author was to write a book solely on the theory of matrices. In view of the widespread applications of this theory, as mentioned above, together with the scarcity of books on this subject that are suitable for texts, no apology is required for such a proposal. However, when it was realized that without much additional space it would be possible to discuss those topics in algebra which underlie most aspects of matrix theory and consistently interpret the results for matrices, the latter course was decided upon.

The advantages of studying matrices against the background of algebra are numerous. One sees questions about matrices arise in their natural setting and, consequently, is able to supply ample motivation for, and gain a full understanding of, the methods used to answer practical questions about matrices. In addition, the matter of what properties of a matrix are important in connection with a particular problem becomes clarified. When the matrix problem is identified with the underlying algebraic problem, the latter invariably suggests an equivalence relation for a class of matrices together with a guide as to what properties of a matrix are relevant to the equivalence class and, consequently, to the original problem. Finally, the reader has an opportunity to gain some understanding of the fundamentals of so-called "modern" algebra, a discipline of postulates and theorems.

Concerning the limitations on the coverage of matrix theory, numerical methods for the inversion of matrices of large orders, iterative schemes for approximating the characteristic values of a matrix, etc., are not discussed. However, many numerical methods are developed that are practical for matrices of small order and—what is more important—that are the starting point for devising computational methods adapted to computing machines.

To indicate the correlation of the algebra and matrix theory, as well as what are, in the author's opinion, the strong points of the book, a rapid survey of the contents is in order. In the first 20 pages of Chap. 1, the entire theory of systems of linear equations over a (commutative) field is developed "from scratch." Although it is assumed in the development that the linear equations have real coefficients, the analysis in Sec. 1.9 describes what properties of real numbers have been used, which leads in a natural way to the notion of a *field* and the conclusion that all earlier results for the real field extend to an arbitrary field.

Properties of the set of solutions of a homogeneous system of equations, and other similar examples, motivate and illustrate the notion of a vector space. The definition and basic properties of this type of algebraic system form the contents of Chap. 2. In Chap. 3, which is adequately described by the chapter heading, appears the first example of an equivalence relation for matrices. This is suggested by the algebraic problem of Chap. 1.

It is hoped that the reader has some knowledge of determinants before reading Chap. 4, if for no other reason than to put him in a better position to appreciate the elegance of the postulational approach to this topic. In Chap. 5, bilinear, quadratic, and Hermitian forms are studied in their natural setting, *viz.*, as representations of certain types of functions on a linear vector space to its field of scalars. In this chapter appear all of the standard theorems pertaining to real symmetric and Hermitian matrices with what are believed to be new and greatly simplified proofs, all of which stem from a single argument.

Chapter 6 begins with the definition of a linear transformation on a vector space followed by the correlation of this notion with matrices. The ensuing discussion leads to the introduction of two further types of algebraic systems—groups and rings.

Chapter 7 introduces and discusses the problem of finding canonical forms for matrices with respect to similarity via representations of linear transformations. Sections 7.6 to 7.9, on elementary divisor theory, are not needed for a study of Chap. 8. The latter portion of this chapter contains many proofs which are believed to be new.

One additional feature of this book is the inclusion of numerous examples. Their being ruled off from the text is solely for the purpose of distinguishing them from the main text.

The author is greatly indebted to several friends and former colleagues in the preparation of this book. Professor Everett Pitcher heads this group, having made many valuable suggestions and offered much constructive criticism while serving as a critical audience countless times. Professor Albert Wilansky was generous with his time and proposed useful ideas at difficult spots. Professor A. W. Tucker tendered several worthwhile suggestions for the plan of the book. A former student, Ward Cheney, was helpful with the exposition and correction of errors in the manuscript. Deep appreciation is expressed to Professor N. H. McCoy for his constructive criticism given after reading a near-final version of the manuscript.

Last, but not least, the author's wife deserves credit not only for typing the manuscript but also for her encouragement and understanding.

Robert R. Stoll

CONTENTS

Preface vii

Some Notation and Conventions Used in this Book xiii

Chapter 1 SYSTEMS OF LINEAR EQUATIONS

1.1 Examples 1
1.2 The consistency condition 3
1.3 The elementary operations and equivalent systems 6
1.4 Echelon systems 8
1.5 The general solution of (N) 9
1.6 The rank of a system of linear equations 12
1.7 An alternative form for the general solution of (N) 17
1.8 A new condition for solvability 20
1.9 Fields 22

Chapter 2 VECTOR SPACES

2.1 Vectors 30
2.2 Definition of a linear vector space 32
2.3 Subspaces 35
2.4 Dimension 38
2.5 Isomorphism in vector spaces 43
2.6 Equivalence relations 47
2.7 Direct sums and quotient spaces 50

Chapter 3 BASIC OPERATIONS FOR MATRICES

3.1 The matrix of a system of linear equations 55
3.2 Special matrices 56
3.3 Row equivalence of matrices 59
3.4 Rank of a matrix 61
3.5 Equivalent matrices 65
3.6 Matrix multiplication 67
3.7 Properties of matrix multiplication 73
3.8 Elementary matrices 77

Chapter 4 DETERMINANTS

4.1 Introduction 86
4.2 Postulates for a determinant function 87
4.3 Further properties of determinants 91
4.4 A proof of the existence of determinants 93
4.5 The adjoint 95
4.6 Matrices with determinant zero 98
4.7 Evaluation of determinants 100

Chapter 5 BILINEAR AND QUADRATIC FUNCTIONS AND FORMS
5.1 Linear functions . . . 105
5.2 Bilinear functions and forms . . . 107
5.3 Quadratic functions and forms . . . 111
5.4 Further properties of symmetric matrices . . . 118
5.5 Quadratic functions and forms over the real field . . . 119
5.6 Examples . . . 124
5.7 Hermitian functions, forms, and matrices . . . 128

Chapter 6 LINEAR TRANSFORMATIONS ON A VECTOR SPACE
6.1 Linear functions and matrices . . . 134
6.2 Non-singular linear transformations . . . 137
6.3 The group $L_n(F)$ . . . 139
6.4 Groups of transformations in general . . . 145
6.5 A reinterpretation of Chapter 5 . . . 148
6.6 Singular linear transformations . . . 151
6.7 The ring of all linear transformations on a vector space . . . 155
6.8 The polynomial ring $F[\lambda]$ . . . 157
6.9 Polynomials in a linear transformation . . . 163

Chapter 7 CANONICAL REPRESENTATIONS OF A LINEAR TRANSFORMATION
7.1 Introduction . . . 165
7.2 Invariant spaces . . . 166
7.3 Null spaces . . . 169
7.4 Similarity and canonical representations . . . 175
7.5 Diagonal and superdiagonal representations . . . 180
7.6 **A**-bases for a vector space . . . 187
7.7 Polynomial matrices . . . 191
7.8 Canonical representations for linear transformations . . . 199
7.9 The classical canonical representation . . . 208

Chapter 8 UNITARY AND EUCLIDEAN VECTOR SPACES
8.1 Introduction . . . 212
8.2 Unitary and Euclidean vector spaces . . . 214
8.3 Schwarz's inequality and its applications . . . 215
8.4 Orthogonality . . . 219
8.5 Orthonormal sets and bases . . . 221
8.6 Unitary transformations . . . 225
8.7 Normal transformations . . . 231
8.8 Normal, orthogonal, and symmetric transformations on Euclidean spaces . . . 240
8.9 Hermitian forms under the unitary group . . . 248
8.10 Reduction of quadric surfaces to principal axes . . . 252
8.11 Maximum properties of characteristic values . . . 255
8.12 The reduction of pairs of Hermitian forms . . . 259

References . . . 263

Index of Special Symbols . . . 264

Index . . . 267

SOME NOTATION AND CONVENTIONS USED IN THIS BOOK

The purpose of this section is to serve as a convenient reference for general mathematical vocabulary and conventions regarding notation.

I. Sets. Without exception, $\{a, b, c, \ldots\}$ is symbolism for the phrase "the set whose members (or elements), without regard to order, are $a, b, c, \ldots$." A set whose first member is a, whose second member is b, etc. will be designated by $(a, b, c, \ldots)$. If all elements of a set S are elements of a set T, we call S a *subset* of T and say that S is *contained in* T; in symbols

$$S \subseteq T$$

We also say that T *contains* S under these circumstances; in symbols

$$T \supseteq S$$

The *equality* of two sets S and T, written $S = T$, is synonymous with $S \subseteq T$ and $S \supseteq T$.

If T contains S and also elements not in S, we say that T contains S *properly*, written

$$T \supset S$$

or that S is a *proper subset* of T, written

$$S \subset T$$

The set R, consisting of all elements common to two given sets S and T is called the *intersection* of S and T; in symbols

$$R = S \cap T$$

If S and T have no elements in common, we say that $S \cap T$ is *empty*. We agree to speak of an *empty set* as one containing no elements and, moreover, that there is only one empty set. A set containing at least one element is called *nonempty*.

II. Subscripts and Superscripts. Use is made of both superscripts and subscripts to distinguish among sets of symbols. The sole disadvantage of superscripts occurs when a power of a symbol must be indicated; for example, the square of x^1 must be written as $(x^1)^2$. The advantage of using both superscripts and subscripts is this: With the criterion in mind that whenever possible summation should take place on an index which occurs once as a subscript and once as a superscript (for example, $\sum a_{ij}x^j$ should be summed with respect to j), an over-all harmonious notation suggests itself.

Whereas at the beginning of Chap. 1 it may seem awkward to use superscripts, it is felt that it is easier to get accustomed to new notation in familiar surroundings.

III. Summation Conventions. In an indicated sum, whenever the inclusion of the summation index and/or the range of the index are needed for the sake of clarity, they are included. However, one or both are often omitted when no misunderstanding can arise. For example, if the sum

$$\sum_{i=1}^{n} a_i x^i$$

appears in a discussion, it might be written simply as $\sum_1^n a_i x^i$ or $\sum a_i x^i$ at a later point.

Occasionally a sum over more than one index is used. The simplest example of this is of the form

$$\sum_{i=1}^{m}\left(\sum_{j=1}^{n} a_j^i\right)$$

As written, the addition takes place first with respect to j, and then i, and, consequently, amounts to the sum of all elements in the rectangular array

$$\begin{matrix} a_1^1 & a_2^1 & \cdots & a_n^1 \\ a_1^2 & a_2^2 & \cdots & a_n^2 \\ \cdot & \cdot & \cdots & \cdot \\ a_1^m & a_2^m & \cdots & a_n^m \end{matrix}$$

obtained by adding columns and then rows. Clearly the same sum results upon first adding the rows and then columns. In other words,

$$\sum_i\left(\sum_j a_j^i\right) = \sum_j\left(\sum_i a_j^i\right)$$

or the order of summation may be reversed.

IV. Functional Notations. Let $M = \{a, b, \ldots\}$ and $M' = \{a', b', \ldots\}$ denote two sets of elements. A *function* f *on* M *to* M' is a rule which associates with each element a of M a single element a' of M'. The element a' is called the *value of* f *at* a and designated by $f(a)$ or, simply, fa. The sets M and M' are called the *domain of existence* and *range* of f, respectively. If each element of M' occurs as a function value, which means that the equation $f(x) = a'$ has at least one solution in M for each a' in M', and it is desired to emphasize this, we shall refer to f as a *function on* M *onto* M'.

The notation $M \xrightarrow{f} M'$, or $f\colon M \to M'$, or $a \to a'$ may be used as symbolism for "a function on M to M'." On occasion it is suggestive to use the word "correspondence" or "transformation" in place of "function" and label $f(a)$, or fa, the *image* of a under f.

If f and g are functions on M to M', such that $fx = gx$ for all x in M, then we call f *equal to* g; in symbols, $f = g$.

Next we describe an important method for combining certain functions. Let f be a function on M to M' and let g be a function on M' to a third set M''. Then it is possible to define a function on M to M'' by the rule $x \to g(fx)$. This function is called the *product* of g and f and designated by gf.

In order to define the notion of a function of two variables, we introduce a further concept. If M and N are arbitrary sets, the *product set* $M \times N$ is defined to be the set of all pairs (x, y) with x in M and y in N. Then a function of two variables in M with values in M' may be defined as a function on $M \times M$ to M'. More generally, we can consider functions on $M_1 \times M_2$ to M'.

Of importance for our purposes are functions on $M \times M$ to M; these will be called *binary compositions* (or operations) in the set M. For example, ordinary addition and multiplication are binary compositions in the set of integers, in accordance with our definition. The terminology and notation used in this case are often employed for binary compositions in general. Thus, for a binary composition in M, the image in M of the (ordered) pair (a, b) in $M \times M$ is often called the *product* (in symbols $a \cdot b$, or simply ab) or the *sum* (in symbols $a + b$) of a and b.

CHAPTER 1

SYSTEMS OF LINEAR EQUATIONS

1.1. Examples. By a *linear equation* in $x^1, x^2, \ldots, x^n$† is meant an equation of the form

$$a_1x^1 + a_2x^2 + \cdots + a_nx^n = y \tag{1}$$

where $a_1, a_2, \ldots, a_n, y$ are (for the moment, at least) fixed real numbers. A *solution* of (1) is any set of n real numbers $\{x_0^1, x_0^2, \ldots, x_0^n\}$ such that

$$a_1x_0^1 + a_2x_0^2 + \cdots + a_nx_0^n = y$$

This chapter is devoted to the discussion of an elimination method for determining all simultaneous solutions of one or more linear equations in several unknowns. We begin with an examination of several simple examples which can be studied geometrically and which offer suggestions for handling the general case.

EXAMPLE 1.1

Find all solutions of the single equation

$$5x^1 - 3x^2 = -1 \tag{2}$$

We observe that this equation is *equivalent* to the equation

$$x^2 = \tfrac{5}{3}x^1 + \tfrac{1}{3} \tag{3}$$

in the sense that a solution of one is a solution of the other, so that we can study (3) in place of (2). In (3) it is clear that x^1 may be chosen arbitrarily and that for each choice of x^1, x^2 is uniquely determined. If we let t stand for any real number, our solution may be put in the form

$$x^1 = t$$

$$x^2 = \tfrac{5}{3}t + \tfrac{1}{3}$$

which shows that we have an endless number of solutions. If we interpret $\begin{pmatrix} x^1 \\ x^2 \end{pmatrix}$ as the rectangular coordinates of a point in the plane, the above

†The reader should examine the section Some Notation and Conventions Used in This Book (page xiii) in order to become acquainted with various conventions that are used consistently.

solutions of (2) are the points of a straight line l_1, called the *graph* of the equation.

EXAMPLE 1.2

Find all solutions of the system of linear equations

$$5x^1 - 3x^2 = -1 \tag{4}$$
$$10x^1 - 6x^2 = -2$$

Since the second equation is a nonzero multiple of the first, the system is equivalent to that consisting of the first equation alone. Hence the solutions of (4) are precisely those of (2). Interpreted geometrically, the graphs of the two equations of (4) coincide so that the graph of the first equation gives the solutions of the system.

EXAMPLE 1.3

Find all solutions of the system of equations

$$5x^1 - 3x^2 = -1$$
$$10x^1 - 6x^2 = 0$$

Since, for any solution $\begin{pmatrix} x_0^1 \\ x_0^2 \end{pmatrix}$ of the first equation, $5x_0^1 - 3x_0^2 = -1$, while, for any solution $\begin{pmatrix} x_1^1 \\ x_1^2 \end{pmatrix}$ of the second, $10x_1^1 - 6x_1^2 = 0$ or $5x_1^1 - 3x_1^2 = 0$, no solution of one is a solution of the other. Consequently, the system has no simultaneous solutions and because of this fact is called *inconsistent.* The graphs of the equations are parallel and distinct straight lines.

EXAMPLE 1.4

Find all solutions of the system of equations

$$5x^1 - 3x^2 = -1 \tag{5}$$
$$2x^1 + 5x^2 = 12$$

The first equation has the graph l_1 mentioned above, and the second has a straight-line graph l_2. Since these two lines have a single point $P\begin{pmatrix} 1 \\ 2 \end{pmatrix}$ in common, the system has a single solution

$$x^1 = 1$$
$$x^2 = 2$$

It cannot be emphasized too strongly that this solution is again a pair of simultaneous equations in x^1 and x^2; it is only because of its simplicity that we label this system the solution. Actually this pair of equations is one of infinitely many pairs which are equivalent to the original pair. Interpreted geometrically, the equations of any pair of distinct straight lines through P constitute a system equivalent to (5), since any such system has the same set of solutions as the original system. We have labeled as solution the pair of equations with graphs consisting of a vertical and a horizontal line through P.

Analytically the lines through P are obtained by assigning all possible values to k_1 and k_2 in the equation

$$k_1(5x^1 - 3x^2 + 1) + k_2(2x^1 + 5x^2 - 12) = 0$$

constructed from multiples of the equations of the system. Rewriting this equation in the form

$$(5k_1 + 2k_2)x^1 + (5k_2 - 3k_1)x^2 + k_1 - 12k_2 = 0$$

it is seen that a horizontal line is obtained for values of k_1 and k_2 such that $5k_1 + 2k_2 = 0$, while a vertical line results if k_1 and k_2 satisfy the equation $5k_2 - 3k_1 = 0$.

1.2. The Consistency Condition. Turning now to the general situation, the notation employed in the following system of linear equations has sufficient flexibility to represent any specific system:

$$\text{(N)}\qquad \begin{aligned} f^1(X) &= a^1_1x^1 + a^1_2x^2 + \cdots + a^1_nx^n = y^1 \\ f^2(X) &= a^2_1x^1 + a^2_2x^2 + \cdots + a^2_nx^n = y^2 \\ &\cdots\cdots\cdots\cdots\cdots\cdots\cdots\cdots\cdots \\ f^m(X) &= a^m_1x^1 + a^m_2x^2 + \cdots + a^m_nx^n = y^m \end{aligned}$$

Using summation notation, we can condense this array of equations to

$$\text{(N)}\qquad f^i(X) = \sum_{j=1}^{n} a^i_jx^j = y^i \qquad i = 1, 2, \ldots, m$$

A word of explanation about the notation convention is in order. First, X is an abbreviation for the array of variables

$$\begin{bmatrix} x^1 \\ x^2 \\ \cdot \\ x^n \end{bmatrix}$$

which henceforth will be written as

$$(x^1, x^2, \ldots, x^n)^T$$

where T (which stands for "transpose") indicates that the elements enclosed should be read as a column. Next, f^i denotes that function whose value at X is $f^i(X) = a_1^i x^1 + a_2^i x^2 + \cdots + a_n^i x^n$. A function whose value at X is computed with such a formula is called a *linear form*; hence the name linear equation for $f^i(X) = y^i$.

Our present point of view with regard to (N) is that the a_j^i's and y^i's are fixed real numbers and that we wish to determine all simultaneous solutions. With this interpretation, (N) is called a *system of m linear equations in the n unknowns* x^i. To justify the use of the word "unknown," observe that each equation of (N) is a constraint upon the variables x^i and that to solve (N) is to find all allowable values of these variables.

If the y^i's are all zero, the system (N) is called *homogeneous*; otherwise, it is called *nonhomogeneous*. The designation of the system by (N) is a reminder that, in general, it is a nonhomogeneous system. In keeping with the above definition of X we shall write a simultaneous solution as $(x_0^1, x_0^2, \ldots, x_0^n)^T$ or simply X_0.

In order to formulate a necessary condition for the existence of a solution of (N), it is appropriate to examine Example 1.3 since that system has no solution. The situation there is such that the relation

$$2(5x^1 - 3x^2) = 10x^1 - 6x^2 \tag{6}$$

holds between the values to the left of the equality sign for all choices of x^1 and x^2. Consequently if the number y^2 is not twice y^1 in the system

$$5x^1 - 3x^2 = y^1$$
$$10x^1 - 6x^2 = y^2$$

it is impossible to find a simultaneous solution.

The state of affairs indicated by (6) suggests that we turn to the functions f^1 and f^2, let us say, whose values at $X = (x^1, x^2)^T$ are $5x^1 - 3x^2$ and $10x^1 - 6x^2$, respectively, and call f^2 equal to twice f^1 since their values stand in this relation. Then we can state that since $2f^1 = f^2$, in order that the system

$$f^1(X) = y^1$$
$$f^2(X) = y^2$$

can have a solution, it is necessary that $2y^1 = y^2$. We now make several definitions suggested by the above in order to cope with the general case.

DEFINITION. *Let f and g denote functions of $X = (x^1, x^2, \ldots, x^n)^T$. Then* f equals g, *in symbols $f = g$, if $f(X) = g(X)$ for all X. The* sum *of f and g, in symbols $f + g$, is the function whose value at X is*

$$(f + g)(X) = f(X) + g(X)$$

The scalar multiple *of f by the scalar (real number) c, in symbols cf, is the function whose value at X is*

$$(cf)(X) = c \cdot f(X)$$

Thus the sum $f + g$ is the function whose value at X is the sum of the values of f and g, while cf is the function whose value at X is c times that of f. Observe that two unequal functions may have the same value at some X. For example, if f and g are the functions for which

$$f(X) = 2x^1 - 3x^2 \qquad g(X) = x^1 - 2x^2$$

then $f(X) = g(X)$ if $X = (1, 1)^T$, but $f \neq g$. In view of our distinction in notation between functions and function values it is clear that an equation $f(X) = g(X)$ expresses the equality of two numbers while an equation $f = g$ expresses the equality of two functions. One possible source of confusion in this connection is an equation in which the symbol 0 appears, since it is assigned the double role of indicating both the number zero and the *zero function* whose value at every X is 0. However it should be clear that $f = 0$ involves the zero function while $f(X) = 0$ involves the zero number.

Although the above definitions for functions of X are general, we are interested at present in their application to linear forms. It is clear that if f and g are linear then so are $f + g$ and cf since if

$$f(X) = \sum a_i x^i \qquad g(X) = \sum b_i x^i$$

then

$$(f + g)(X) = \sum (a_i + b_i)x^i \qquad (cf)(X) = \sum c a_i x^i$$

Hence if $f^1, f^2, \ldots, f^m$ are linear forms and $c_1, c_2, \ldots, c_m$ are real numbers, $\sum c_i f^i$, which is called a *linear combination* of the f^i's, is a linear form.

DEFINITION. *A set of linear forms $\{f^1, f^2, \ldots, f^m\}$ is said to be* linearly dependent *if there exist real numbers c_i, not all zero, such that the linear combination $\sum c_i f^i$ is the zero function.*† *Otherwise $\{f^1, f^2, \ldots, f^m\}$ is a* linearly independent set.

†Observe that the zero function is a linear form since its value at X can be written as $0x^1 + 0x^2 + \cdots + 0x^n$.

EXAMPLE 1.5

The set of linear forms $\{f^1, f^2, f^3\}$ defined by

$$f^1(X) = x^1 - x^2 + x^3 \qquad f^2(X) = x^1 + x^2 + 2x^3$$
$$f^3(X) = 3x^1 + x^2 + 5x^3$$

is linearly dependent since $f^1 + 2f^2 - f^3 = 0$. However, the set consisting of f^1 and f^2 is independent. The set consisting of the zero form alone is dependent; the set consisting of a single form $f \neq 0$ is independent. Any set of forms, one of which is 0, is dependent.

We now consider the following:

Consistency Condition for (N): For every set of real numbers $\{c_1, c_2, \ldots, c_m\}$ such that $\sum c_i f^i = 0$, it is true that $\sum c_i y^i = 0$. That is, every linear dependence among the forms in (N) persists among the corresponding right hand members y^i.

DEFINITION. *A system* (N) *for which the consistency condition holds is called* consistent.

Theorem 1.1. If (N) has a solution, then (N) is consistent.

Proof. Let X_0 denote a simultaneous solution of (N) and $\sum c_i f^i = 0$ be a dependence among the forms f^i. Then $\sum c_i f^i(X_0) = 0$ and since $f^i(X_0) = y^i$, $\sum c_i y^i = 0$.

1.3. The Elementary Operations and Equivalent Systems. We shall prove that the consistency condition is also sufficient for the solvability of (N) by describing an explicit method for determining all solutions in a finite number of steps whenever the system is known to be consistent. These steps consist merely of the operations used in reducing systems like those in Sec. 1.1 to solved form: multiplying an equation by a nonzero number, adding a multiple of one equation to another, and the addition or deletion of the vanishing equation $(0x^1 + 0x^2 + \cdots + 0x^n = 0)$. For convenience, we add to this list the operation of interchanging two equations and call these the *elementary operations* upon a linear system (N). For identification purposes we label these as follows:

Type I. The interchange of two equations of (N).

Type II. The multiplication of an equation of (N) by a nonzero number.

Type III. The addition to any equation of (N) of a constant multiple of another equation of (N).

Type IV. The addition to or deletion from (N) of the vanishing equation.

Since an operation of type I can be performed with a sequence of operations of types II and III,† any assertion made about the elementary operations need not be verified for this type.

Theorem 1.2. The application of a finite number of elementary operations upon (N) yields a *system equivalent to* (N), that is, one whose solutions coincide with those of (N).

Proof. Let (N)* denote a system obtained from (N) by the application of any one elementary operation. Then every solution of (N) is a solution of (N)*. We verify this for a typical case, *viz.*, that of a type III operation. If c times the jth equation of (N) is added to the ith, the ith equation of (N)* is

$$f^i(X) + cf^j(X) = y^i + cy^j$$

while the remaining equations agree with the corresponding numbered equations of (N). If X_0 is a solution of (N), substitution shows that it is a solution of (N)*.

Next, if (N)* results from (N) by a single elementary operation, then (N) can be obtained from (N)* by an elementary operation of the same type. For example, in the case of the type III operation above, if $(-c)$ times the jth equation of (N)* is added to the ith, the resulting system is (N). Hence every solution of (N)* is a solution of (N). Combining this with the reverse inclusion establishes the theorem for the case of one elementary operation, and consequently any finite number of elementary operations.

COROLLARY. If in the system (N) the ith equation is replaced by $\sum c_k f^k(X) = \sum c_k y^k$, where $c_i \neq 0$, the new system

$$\begin{aligned} f^1(X) &= y^1 \\ &\cdots\cdots \\ f^{i-1}(X) &= y^{i-1} \\ \textstyle\sum c_k f^k(X) &= \textstyle\sum c_k y^k \\ f^{i+1}(X) &= y^{i+1} \\ &\cdots\cdots \\ f^m(X) &= y^m \end{aligned}$$

is equivalent to the original.

†To interchange the ith and jth equations, add the jth to the ith, subtract the new ith equation from the jth, add the jth equation to the ith, and, finally, multiply the jth equation by (-1).

Proof. The new system can be obtained from (N) by one operation of type II [multiplication of $f^i(X) = y^i$ by c_i] and $m - 1$ operations of type III.

PROBLEM

1. Show that type I and type II operations can be effected by types III and IV.

1.4. Echelon Systems. Before applying the operations described above to reduce a system of linear equations, an additional notion is needed. It is clear that a linear form f, for which $f(X) = \sum a_i x^i$, determines an n-tuple $(a_1, a_2, \ldots, a_n)$. Indeed f determines this n-tuple uniquely since, if in addition $f(X) = \sum b_i x^i$, choosing $x^k = 1$ and $x^i = 0$, $i \neq k$, gives $a_k = b_k$, $k = 1, 2, \ldots, n$. Conversely it is clear that an n-tuple determines a linear form. We shall say that f *corresponds to* the unique n-tuple it determines. For example, the zero form corresponds to the n-tuple $(0, 0, \ldots, 0)$.

DEFINITION. *Let the linear form f correspond to the n-tuple $(a_1, a_2, \ldots, a_n)$. The* length *of f is 0 if $f = 0$, otherwise it is the index of the last nonzero a_i with respect to the natural order for the a_i's. If f has length $k \geqq 1$ and $a_k = 1$, f is said to be* normalized.

For example, the form corresponding to $(2, -1, 1, 0)$ is normalized and of length 3.

We can now describe the type of linear system that is basic in our exposition.

DEFINITION. *A linear system* (N) *is an* echelon system *if the following conditions are satisfied:*

(*i*) *Each linear form has positive length and is normalized.*

(*ii*) *If k_i is the length of f^i, then $k_1 < k_2 < \cdots < k_m$.*

(*iii*) *x^{k_i} does not appear (with nonzero coefficient) in any equation except the ith, $i = 1, 2, \ldots, m$.*

EXAMPLE 1.6

The following system is an echelon system:

$$\begin{aligned} 2x^1 + x^2 \qquad\qquad\qquad\qquad &= 1 \\ x^1 \qquad + x^3 + x^4 \qquad &= 2 \\ x^1 \qquad + 2x^3 \qquad + x^5 &= 3 \end{aligned}$$

Theorem 1.3. If the system (N) is consistent, it is equivalent to an echelon system of the following type:

$$(\text{N}_e)\quad \begin{aligned} g^1(X) &= b^1_1x^1 + \cdots + b^1_{k_1-1}x^{k_1-1} + x^{k_1} &&= z^1 \\ g^2(X) &= b^2_1x^1 + \cdots + b^2_{k_2-1}x^{k_2-1} \qquad + x^{k_2} &&= z^2 \\ &\cdots\cdots\cdots\cdots\cdots\cdots\cdots\cdots\cdots\cdots\cdots\cdots \\ g^r(X) &= b^r_1x^1 + \cdots \qquad\qquad\qquad\qquad + x^{k_r} &&= z^r \end{aligned}$$

where $1 \leqq k_1 < k_2 < \cdots < k_r \leqq n$.

Proof. If (N) contains a form of length n, by an interchange of equations, if necessary, we can assume that it is f^m. Multiplication of the equation $f^m(X) = y^m$ by $(a^m_n)^{-1}$ normalizes f^m. The addition to the ith equation for each $i \neq m$ of $-a^i_n$ times the last produces a new ith equation wherein the coefficient of x^n is zero. Thus the coefficient of x^n is zero in every equation preceding the last one. If, at the outset, every coefficient of x^n is zero, the above reduction is omitted.

Next the foregoing is repeated on the system obtained by ignoring the last equation. Continuing in this way, we obtain an equivalent system which, after all vanishing equations are removed, contains let us say r equations satisfying requirements (i) and (ii) for an echelon system. Clearly $k_r \leqq n$; that each linear form in the system has positive length (thus in particular $k_1 \geqq 1$) is a consequence of the consistency assumption.

Finally, requirement (iii) can be met by repeated use of type III operations to eliminate x^{k_1} in all but the first equation, etc. This completes the proof.

Observe that in every case (N) is equivalent to an echelon system of no more than n equations. If $m > n$, then at least $m - n$ equations vanish in the above reduction.

1.5. The General Solution of (N). It is now an easy matter to obtain from the echelon system of Theorem 1.3 a system equivalent to it, and hence equivalent to (N), which is called the *solved form* or the *general solution* of (N) since n-tuples which satisfy (N) can be read off at a glance. [Remember that we assume that (N) is consistent.] Namely, the echelon system can be solved for $x^{k_1}, x^{k_2}, \ldots, x^{k_r}$ as linear combinations of the remaining $n - r$ x^i's which we label $x^{k_{r+1}}, x^{k_{r+2}}, \ldots, x^{k_n}$ in some order. Then it is clear that $x^{k_{r+1}}, x^{k_{r+2}}, \ldots, x^{k_n}$ can be assigned arbitrary values and that for each choice of these letters, $x^{k_1}, x^{k_2}, \ldots, x^{k_r}$ are uniquely determined. We express this result as follows:

Theorem 1.4. If the linear system (N) is consistent, it is equivalent to one of the following type:

$$
\begin{aligned}
x^{k_1} &= c_1^1 u^1 + \cdots + c_{n-r}^1 u^{n-r} + z^1 \\
x^{k_2} &= c_1^2 u^1 + \cdots + c_{n-r}^2 u^{n-r} \qquad + z^2 \\
&\cdots\cdots\cdots\cdots\cdots\cdots\cdots\cdots \\
x^{k_r} &= c_1^r u^1 + \cdots + c_{n-r}^r u^{n-r} \qquad\qquad + z^r \\
x^{k_{r+1}} &= u^1 \\
x^{k_{r+2}} &= \qquad u^2 \\
&\cdots\cdots\cdots \\
x^{k_n} &= \qquad\qquad u^{n-r}
\end{aligned}
$$

where $u^1, u^2, \ldots, u^{n-r}$ (which appear only if $n - r > 0$) are parameters denoting arbitrary real numbers and $z^1, z^2, \ldots, z^r$ are fixed real numbers.

Thus the necessary consistency condition for the solvability of (N) is also sufficient. In practice, the question as to whether or not a given system (N) of linear equations is consistent causes no difficulty. Simply set out to reduce the system to echelon form. If no inconsistencies (*i.e.*, no contradictions) are encountered in the process, so that (N) is equivalent to an echelon system, then (N) is consistent since any echelon system is consistent by virtue of the independence of the forms in its make-up. To verify this last statement, observe that the linear form $\sum c_i g^i$ formed from the g^i's in (N_e) corresponds to the n-tuple whose k_ith component is c_i, $i = 1, 2, \ldots, r$. Hence $\sum c_i g^i = 0$ if and only if $c_i = 0$, $i = 1, 2, \ldots, r$, which proves that $\{g^1, g^2, \ldots, g^r\}$ is an independent set. This being true, the consistency condition is trivially satisfied.

EXAMPLE 1.7

Find the general solution of the following system:

$$
\begin{aligned}
x^1 + 2x^2 + x^3 + x^4 &= 6 \\
x^1 \qquad + x^3 - x^4 &= 0 \\
2x^2 + x^3 + x^4 &= 5 \\
-2x^1 + x^2 + x^3 \qquad &= 0
\end{aligned}
$$

We shall derive an echelon system equivalent to this by using the method of proof in Theorem 1.3. Subtracting the third equation from the first, adding the third to the second, and interchanging the third and fourth, we obtain the system

$$
\begin{aligned}
x^1 \qquad\qquad\qquad &= 1 \\
x^1 + 2x^2 + 2x^3 \qquad &= 5 \\
-2x^1 + x^2 + x^3 \qquad &= 0 \\
2x^2 + x^3 + x^4 &= 5
\end{aligned}
$$

Next, subtracting twice the third from the second yields $5x^1 = 5$ as a new second equation. Since this is a multiple of the first, we may discard it to obtain the system

$$\begin{array}{rcl} x^1 & & = 1 \\ -2x^1 + x^2 + x^3 & & = 0 \\ & 2x^2 + x^3 + x^4 & = 5 \end{array}$$

which satisfies conditions (i) and (ii) for an echelon system. To meet condition (iii) we eliminate x^1 from the second equation and x^3 from the third. The resulting echelon system is on the left below and the solved form is on the right.

$$\begin{array}{lr} x^1 & = 1 \\ x^2 + x^3 & = 2 \\ x^2 \qquad + x^4 & = 3 \end{array} \qquad \begin{array}{rcl} x^1 & = & 1 \\ x^3 & = & -u^1 + 2 \\ x^4 & = & -u^1 + 3 \\ x^2 & = & u^1 \end{array}$$

Next we present the results for a more elaborate example, which illustrates Theorem 1.4 in greater detail.

EXAMPLE 1.8

Find the general solution of the following system:

$$\begin{array}{rcl} 10x^1 + 3x^2 + 9x^3 + 4x^4 + x^5 + x^6 + 2x^7 & = & 9 \\ 6x^1 + x^2 + 10x^3 + 6x^4 + 2x^5 + 2x^6 + x^7 & = & 10 \\ 4x^1 - 2x^2 + 8x^3 + 5x^4 + 2x^5 + 2x^6 + x^7 & = & 8 \\ 2x^1 + 3x^2 + 2x^3 + x^4 & = & 2 \\ x^1 - 2x^2 + 10x^3 + 7x^4 + 3x^5 + 3x^6 & = & 10 \\ 4x^1 + 2x^3 + x^4 + x^7 & = & 2 \end{array}$$

An echelon system equivalent to the above is

$$\begin{array}{rcl} x^1 + 2x^2 + x^3 & = & 1 \\ -x^2 + x^4 & = & 0 \\ -3x^1 - 5x^2 + x^5 + x^6 & = & 0 \\ 2x^1 - 3x^2 + x^7 & = & 0 \end{array}$$

Hence the general solution is

$$\begin{aligned}
x^3 &= -u^1 - 2u^2 + 1 \\
x^4 &= u^2 \\
x^6 &= 3u^1 + 5u^2 - u^3 \\
x^7 &= -2u^1 + 3u^2 \\
x^1 &= u^1 \\
x^2 &= u^2 \\
x^5 &= u^3
\end{aligned}$$

PROBLEMS

In Probs. 1 to 5 find the general solution (via an equivalent echelon system) for those which are consistent.

1.
$$\begin{aligned}
x + y + z &= 0 \\
4x + y + 2z &= 0 \\
3x - 3y - z &= 0
\end{aligned}$$

2.
$$\begin{aligned}
2x^1 + 3x^2 + x^3 + x^4 &= 1 \\
x^1 - x^2 - x^3 + x^4 &= 1 \\
3x^1 + x^2 + x^3 + 2x^4 &= 0 \\
-x^1 + x^3 - x^4 &= -2
\end{aligned}$$

3.
$$\begin{aligned}
x^1 + 2x^2 + x^3 &= 2 \\
2x^1 - 2x^3 + x^4 &= 6 \\
4x^2 + 3x^3 + 2x^4 &= -1 \\
8x^2 + 7x^3 + x^4 &= 1
\end{aligned}$$

4.
$$\begin{aligned}
7y - 10z &= 4 \\
2x - 3y + 4z &= 4 \\
x + 2y - 3z &= 4
\end{aligned}$$

5.
$$\begin{aligned}
x^1 + x^2 + x^3 - x^4 &= 0 \\
2x^1 - 3x^2 - 2x^3 + 2x^4 &= 19 \\
-x^1 - 3x^2 - 2x^3 + 2x^4 &= 4 \\
3x^1 + x^2 - x^3 &= -3
\end{aligned}$$

6. Determine those values of k, l for which the following system is consistent. Find the general solution using these values.

$$\begin{aligned}
3x^1 + 2x^2 + 5x^3 + 4x^4 &= 0 \\
5x^1 + 3x^2 + 2x^3 + x^4 &= 1 \\
11x^1 + 7x^2 + 12x^3 + 9x^4 &= k \\
4x^1 + 3x^2 + 13x^3 + 11x^4 &= l
\end{aligned}$$

1.6. The Rank of a System of Linear Equations. Various questions present themselves in connection with an echelon system (N_e) equivalent to a given consistent system (N). For example, is the number of equations in (N_e) uniquely determined by (N)? Again, is (N_e) itself uniquely determined by (N)? In order to investigate such questions a preliminary result is needed.

Lemma 1.1. Let $\{f^1, f^2, \ldots, f^m\}$ denote a set of linear forms, not all of which are the zero form. Then there is an integer r, $1 \leqq r \leqq m$, such that at least one linearly independent subset contains r forms and no subset of more than r forms is linearly independent. A linearly independent set containing r forms is called *maximal*. Each remaining form has a unique representation as a linear combination of a maximal set of forms.

Proof. Among the subsets of $\{f^1, f^2, \ldots, f^m\}$ there is at least one that is linearly independent, since a nonzero form is the sole member of a linearly independent set. On the other hand, m is an upper bound on the number of forms in a linearly independent set. Since the set of integers corresponding to the number of forms in the various linearly independent subsets is bounded below by 1 and above by m, it contains a greatest positive integer r. Any independent set of r forms is maximal.

For the second assertion we may assume $r < m$. Let $\{f^1, f^2, \ldots, f^r\}$ be a maximal set, and consider $\{f^1, f^2, \ldots, f^r, f^k\}$ where $k > r$. It is linearly dependent so that constants c_i, not all zero, exist such that $c_1f^1 + c_2f^2 + \cdots + c_rf^r + c_kf^k = 0$. Here $c_k \neq 0$ owing to the independence of $\{f^1, f^2, \ldots, f^r\}$. Hence we can write $f^k = \sum_1^r c_i'f^i$. If also $f^k = \sum_1^r d_if^i$, then subtraction gives $0 = \sum_1^r (c_i' - d_i)f^i$, which can hold only if every coefficient is 0, that is, $c_i' = d_i$ for all i, since we have assumed that $\{f^1, f^2, \ldots, f^r\}$ is a linearly independent set. This proves the uniqueness.

DEFINITION. *The* rank *of a set of linear forms* $\{f^1, f^2, \ldots, f^m\}$, *not all of which are the zero form, is equal to the number of forms in a maximal linearly independent subset. If every* $f^i = 0$, *the rank of the set is* 0. *The* rank *of* (N) *is equal to the rank of the set of forms occurring in* (N).

Lemma 1.2. If the elementary operations in Sec. 1.3 are restated for linear forms (instead of equations), the rank of a set of linear forms $\{f^1, f^2, \ldots, f^m\}$ is unchanged by the application of any elementary operation to the set.

Proof. The assertion is trivial if the rank r is zero. The same is true when $r > 0$ for operations of types I, II, and IV. For one of type III it is sufficient to prove the lemma if f^i is replaced by $f^i + f^j$. For this in turn, it is sufficient to prove that the rank of the new set is greater than or equal to r, since this, together with the fact that the original system can be obtained from the new one by the same process, implies equality of rank. Suppose then that $\{f^1, f^2, \ldots, f^r\}$ is a maximal linearly independent set.

This is permissible since the invariance of rank under type I operations allows us to renumber the forms. If $i > r$, $\{f^1, f^2, \ldots, f^r\}$ is linearly independent in the new system; if $i, j < r$, $\{f^1, \ldots, f^{i-1}, f^i + f^j, f^{i+1}, \ldots, f^r\}$ is independent. If $i < r < j$, write f^j as a linear combination of $f^1, f^2, \ldots, f^r$ (Lemma 1.1): $f^j = \sum_1^r d_k f^k$. If $d_i = 0$, then $\{f^1, \ldots, f^{i-1}, f^i + f^j, f^{i+1}, \ldots, f^r\}$ is independent, while if $d_i \neq 0$, $\{f^1, \ldots, f^{i-1}, f^j, f^{i+1}, \ldots, f^r\}$ is independent. This completes the proof.

Since the reduction of (N) to (N_e) involves only elementary operations upon the forms f^i in (N), the rank of (N_e) is that of (N). But the former rank is r since the set $\{g^1, g^2, \ldots, g^r\}$ is linearly independent (proof at end of Sec. 1.5). Thus we have proved† the following result:

Theorem 1.5. The number of equations in an echelon system equivalent to (N) is equal to the rank of (N).

We investigate next the uniqueness of the echelon system (N_e). Suppose that, besides (N_e),

$$(N_e)^* \qquad \begin{aligned} h^1(X) &= c_1^1x^1 + c_2^1x^2 + \cdots + x^{l_1} && = w^1 \\ h^2(X) &= c_1^2x^1 + c_2^2x^2 + \cdots \qquad + x^{l_2} && = w^2 \\ &\cdots\cdots\cdots\cdots\cdots\cdots\cdots\cdots \\ h^r(X) &= c_1^rx^1 + c_2^rx^2 + \cdots \qquad\qquad + x^{l_r} && = w^r \end{aligned}$$

is an echelon system equivalent to (N). Since the reduction of (N) to (N_e) in Theorem 1.3 employs only elementary operations, and these are reversible, (N_e) can be "expanded" to (N) by elementary operations. Thus it is possible to get from the g^i's to the f^i's to the h^k's and hence from the g^i's to the h^k's (and conversely) by elementary operations.† It follows that each h^k is expressible as a linear combination of the g^i's, and conversely. We shall use this fact to show the identity of (N_e) and (N_e)*.

It is easily seen that the first equations in the two systems agree, since the left member of each can be characterized as the normalized form of least length (a positive number) that can be obtained from the f^i's by elementary operations. Since an echelon system is consistent, it follows that the first equations are identical.

Assume that the systems agree through the first $p - 1$ equations. We prove that the pth equations agree. There exists a linear combination of the g^i's—denote it by $\sum c_i g^i$—which is equal to h^p. We assert that in this sum each index exceeds $p - 1$ so that

$$h^p = c_p g^p + c_{p+1} g^{p+1} + \cdots$$

†A gap in the argument exists here since it has not been shown that every echelon system equivalent to (N) can be determined from (N) by elementary operations.

This is a consequence of the following three remarks: (i) $x^{l_1}, x^{l_2}, \ldots, x^{l_{p-1}}$ do not appear in h^p, since $(\mathrm{N}_e)^*$ is an echelon system, (ii) $x^{l_i} = x^{k_i}$, $i = 1, 2, \ldots, p-1$ by our induction assumption, and (iii) x^{k_i} appears in only g^i, $i = 1, 2, \ldots, p-1$.

With the above form of h^p established, it follows that the length of h^p is greater than, or equal to, that of g^p. Reversing the roles of the h^k's and g^i's establishes the reverse inequality, and hence the equality of lengths. Thus $h^p = c_p g^p$, and since both forms are normalized, $h^p = g^p$. Finally, since (N) is assumed to be consistent, it follows that the pth equations agree. This completes the proof of our next theorem.

Theorem 1.6. There is a uniquely determined echelon system equivalent to a given consistent system of linear equations.

Returning now to the notion of rank introduced earlier in this section, we list several consequences of Theorem 1.4, stated in terms of the rank of (N).

Theorem 1.7. Suppose that (N) is consistent and has rank r. Then $r \leqq n$. If $r = n$, the system has a unique solution; if $r < n$, there are infinitely many solutions.

Proof. If $r = n$ in the echelon system (N_e) of Theorem 1.3, it is of the type

$$\begin{aligned} x^1 &= z^1 \\ x^2 &= z^2 \\ &\cdots\cdots \\ x^n &= z^n \end{aligned}$$

which displays the unique solution. If $r < n$, the solved form of (N_e) in Theorem 1.4 involves $n - r > 0$ parameters $u^1, u^2, \ldots, u^{n-r}$. This completes the proof.

Every system of linear *homogeneous* equations obviously satisfies the consistency condition. This is in accordance with the observation that every homogeneous system has the solution $0 = (0, 0, \ldots, 0)^{\mathrm{T}}$. This is called the *trivial solution*; any other solution of a homogeneous system is *nontrivial*. The next result is a specialization of Theorem 1.7.

Theorem 1.8. Let r denote the rank of a system of linear homogeneous equations in n unknowns. If $r = n$, the only solution is the trivial solution; otherwise (that is, $r < n$) there is a nontrivial solution. In particular, a system of linear homogeneous equations has a nontrivial solution if the number of unknowns exceeds the number of equations.

The final assertion in this theorem is of great importance for our purposes; the next result is an immediate consequence of it.

Theorem 1.9. If $\{f^1, f^2, \ldots, f^k\}$ is a linearly independent set of forms, and $\{g^1, g^2, \ldots, g^l\}$ is any set of $l > k$ linear combinations of the f^i's, then the latter set is linearly dependent.

Proof. Let

$$g^i = \sum_{j=1}^{k} b_j^i f^j \qquad i = 1, 2, \ldots, l \tag{7}$$

We must demonstrate the existence of a set of numbers $\{c_1, c_2, \ldots, c_l\}$, not all zero, such that $\sum c_i g^i = 0$. To this end, it will be sufficient to choose the c_i's so as to satisfy the linear system

$$\sum_{i=1}^{l} b_j^i c_i = 0 \qquad j = 1, 2, \ldots, k \tag{8}$$

since these expressions will be the coefficients of the f^i's when in $\sum c_i g^i$ each g^i is replaced by its value in (7) and terms are collected. A nontrivial solution of (8) always exists by the preceding theorem, since the number l of unknowns exceeds the number k of equations. This completes the proof.

We leave it as an exercise for the reader to apply Theorem 1.9 to deduce the next result, which clarifies the notions of a maximal linearly independent set of forms and of the rank of a set of forms.

Theorem 1.10. Let $\{f^1, f^2, \ldots, f^m\}$ denote a set of linear forms. Then every linearly independent subset with the property that any set properly containing this subset is linearly dependent, is maximal (and conversely). In particular, if the rank of $\{f^1, f^2, \ldots, f^m\}$ is equal to $r > 0$, then every linearly independent subset with the above property contains r elements.

It is important to notice that in the proof of Theorem 1.9 no use was made of the assumption that the f^i's and g^i's are linear forms. Since Theorem 1.10 is a corollary of Theorem 1.9, the same remark applies. It follows, therefore, that *these theorems are valid for any set of elements for which the notions of linear independence, etc., are defined.*

PROBLEMS

1. Verify the statement in the proof of Theorem 1.6 that the forms g^1 and h^1 have least length as described there.

2. Prove the final statement in Theorem 1.8 by induction without making use of any results in the text.

3. Prove Theorem 1.10.

4. Verify that the left-hand members of (8) are the coefficients of the f^i's as stated.

1.7. An Alternative Form for the General Solution of (N). In the event a linear system (N) has infinitely many solutions, so that we cannot list them all, it is possible to do the next best thing, *viz.*, to list a finite number of solutions in terms of which all solutions can be simply described. In this connection we introduce definitions of equality, addition, and multiplication by real numbers for column n-tuples:

$$\begin{pmatrix} a^1 \\ a^2 \\ \cdot \\ a^n \end{pmatrix} = \begin{pmatrix} b^1 \\ b^2 \\ \cdot \\ b^n \end{pmatrix} \quad \text{if and only if } a^i = b^i,\ i = 1, 2, \ldots, n \tag{9}$$

$$\begin{pmatrix} a^1 \\ a^2 \\ \cdot \\ a^n \end{pmatrix} + \begin{pmatrix} b^1 \\ b^2 \\ \cdot \\ b^n \end{pmatrix} = \begin{pmatrix} a^1 + b^1 \\ a^2 + b^2 \\ \cdot \\ a^n + b^n \end{pmatrix} \qquad c\begin{pmatrix} a^1 \\ a^2 \\ \cdot \\ a^n \end{pmatrix} = \begin{pmatrix} a^1 \\ a^2 \\ \cdot \\ a^n \end{pmatrix} c = \begin{pmatrix} ca^1 \\ ca^2 \\ \cdot \\ ca^n \end{pmatrix}$$

The reader should notice the similarity of these rules with those for the n-tuples corresponding to linear forms. If the linear forms f and g correspond to $(a_1, a_2, \ldots, a_n)$ and $(b_1, b_2, \ldots, b_n)$, respectively, then $f + g$ and cf correspond to $(a_1 + b_1, a_2 + b_2, \ldots, a_n + b_n)$ and $(ca_1, ca_2, \ldots, ca_n)$, respectively, which are precisely the row analogues of the above definitions stated for column n-tuples.

Theorem 1.11. If (N) is solvable, the totality of solutions can be expressed in the form $X_0 + X_H$, where X_0 is any one fixed solution of (N) and X_H runs through the solutions of (H), the homogeneous system associated with (N):

$$\begin{aligned} f^1(X) &= a_1^1x^1 + a_2^1x^2 + \cdots + a_n^1x^n = 0 \\ f^2(X) &= a_1^2x^1 + a_2^2x^2 + \cdots + a_n^2x^n = 0 \\ &\cdots\cdots\cdots\cdots\cdots\cdots\cdots\cdots\cdots \\ f^m(X) &= a_1^mx^1 + a_2^mx^2 + \cdots + a_n^mx^n = 0 \end{aligned} \tag{H}$$

Proof. We have

$$f^i(X_0 + X_H) = f^i(X_0) + f^i(X_H) = y^i + 0 = y^i$$

so that $X_0 + X_H$ is a solution of (N). Conversely, if X_N is any solution of (N), then $f^i(X_N) = f^i(X_0) = y^i$, or $f^i(X_N) - f^i(X_0) = f^i(X_N - X_0) = 0$. Hence $X_N - X_0$ is a solution, X_H, of (H). Thus $X_N = X_0 + X_H$. This completes the proof.

DEFINITION. *Any one fixed solution of* (N) *is called a* particular solution.

The above result is trivial if the rank r of (N) is equal to n, since then both X_0 and X_H are uniquely determined with $X_H = 0$ according to Theorem 1.7 and 1.8. It is when $r < n$ that the theorem is useful. Then the echelon system equivalent to (H) is (N_e) with the z^i's replaced by 0's, and the solved form of (H) is

$$\begin{aligned}
x^{k_1} &= c_1^1 u^1 + c_2^1 u^2 + \cdots + c_{n-r}^1 u^{n-r} \\
x^{k_2} &= c_1^2 u^1 + c_2^2 u^2 + \cdots + c_{n-r}^2 u^{n-r} \\
&\cdots\cdots\cdots\cdots\cdots\cdots\cdots\cdots \\
x^{k_r} &= c_1^r u^1 + c_2^r u^2 + \cdots + c_{n-r}^r u^{n-r} \\
x^{k_{r+1}} &= u^1 \\
&\cdots\cdots \\
x^{k_n} &= \qquad\qquad\qquad\qquad u^{n-r}
\end{aligned} \tag{10}$$

Consider now the $n - r$ solutions obtained by choosing $u^1 = 1$, $u^2 = 0, \ldots, u^{n-r} = 0$; then $u^1 = 0$, $u^2 = 1$, $u^3 = 0, \ldots, u^{n-r} = 0$, etc. These are the n-tuples

$$X_1 = \begin{pmatrix} c_1^1 \\ c_1^2 \\ \cdot \\ c_1^r \\ 1 \\ 0 \\ 0 \\ \cdot \\ 0 \end{pmatrix}, X_2 = \begin{pmatrix} c_2^1 \\ c_2^2 \\ \cdot \\ c_2^r \\ 0 \\ 1 \\ 0 \\ \cdot \\ 0 \end{pmatrix}, \ldots, X_{n-r} = \begin{pmatrix} c_{n-r}^1 \\ c_{n-r}^2 \\ \cdot \\ c_{n-r}^r \\ 0 \\ 0 \\ 0 \\ \cdot \\ 1 \end{pmatrix} \tag{11}$$

if we adopt the order $x^{k_1}, x^{k_2}, \ldots, x^{k_n}$ for the x^i's. The usefulness of $X_1, X_2, \ldots, X_{n-r}$, by virtue of the definitions (9), is that the solutions of (H) can be described as the set of all linear combinations $u^1X_1 +$

$u^2X_2 + \cdots + u^{n-r}X_{n-r}$ of these n-tuples. For the equation

(12) $X_H = (x^{k_1}, x^{k_2}, \ldots, x^{k_n})^T = u^1X_1 + u^2X_2 + \cdots + u^{n-r}X_{n-r}$

is simply a compact way to write the n equations (10). Referring to the solved form of (N) in Theorem 1.4, $Z = (z^1, z^2, \ldots, z^r, 0, \ldots, 0)^T$ is a particular solution; hence by Theorem 1.11 the solutions of (N) may be written as

$$X_N = u^1X_1 + u^2X_2 + \cdots + u^{n-r}X_{n-r} + Z$$

where $u^1, u^2, \ldots, u^{n-r}$ are arbitrary real numbers. This is the form of the general solution that we set out to develop. As an illustration, the general solution of the system of Example 1.8 in the above notation is

$$X_N = \begin{pmatrix} x^3 \\ x^4 \\ x^6 \\ x^7 \\ x^1 \\ x^2 \\ x^5 \end{pmatrix} = u^1\begin{pmatrix} -1 \\ 0 \\ 3 \\ -2 \\ 1 \\ 0 \\ 0 \end{pmatrix} + u^2\begin{pmatrix} -2 \\ 1 \\ 5 \\ 3 \\ 0 \\ 1 \\ 0 \end{pmatrix} + u^3\begin{pmatrix} 0 \\ 0 \\ -1 \\ 0 \\ 0 \\ 0 \\ 1 \end{pmatrix} + \begin{pmatrix} 1 \\ 0 \\ 0 \\ 0 \\ 0 \\ 0 \\ 0 \end{pmatrix}$$

The above set of n-tuples $\{X_1, X_2, \ldots, X_{n-r}\}$ is seen to be (upon examination of the last $n - r$ components) linearly independent in the sense of our earlier definition for linear forms, *i.e.*, no linear combination which includes a nonzero coefficient is the zero n-tuple $0 = (0, 0, \ldots, 0)^T$.

DEFINITION. *A set of solutions of* (H) *which is linearly independent and has the property that the totality of linear combinations of its members is the totality of solutions of* (H) *is called a* basis *of the solution set of* (H).

The set of solutions $\{X_1, X_2, \ldots, X_{n-r}\}$ in (11) demonstrates the existence of a basis for (H) where $r < n$. When $r = n$, (H) has only the trivial solution and a basis does not exist.

Theorem 1.12. A basis for the solution set of (H) is a maximal (see Lemma 1.1) linearly independent set of solution n-tuples and conversely. If (H) has rank r, then the number of solutions in a maximal linearly independent set (and hence in every basis) is $n - r$.

Proof. Since the assertions apply only to the case $r < n$, we assume that this inequality holds. That a basis is a maximal linearly independent set follows from the definition of a basis, together with Theorem 1.9 (by virtue of the remark made there concerning the generality of that result).

The converse is left as an exercise. Next, since all maximal linearly independent subsets of the solution set of (H) have the same number of members, all bases have that same number of elements. This common number may be determined by counting the members of any one basis; according to the basis in (11) it is $n - r$.

PROBLEMS

1. Supply the omitted part of the proof of Theorem 1.12.

2. Write the solution of the system in Example 1.7 using a basis of the associated homogeneous system.

3. Determine explicitly the basis (11) for the solution set of the homogeneous system consisting of the single equation

$$x^1 + x^2 + \cdots + x^n = 0$$

Also show that the $n - 1$ n-tuples $(1, -1, 0, \ldots, 0)^T$, $(0, 1, -1, 0 \ldots, 0)^T \ldots$, $(0, \ldots, 0, 1, -1)^T$ form a basis for the solution set. Write each member of the first basis in terms of the second basis and conversely.

1.8. A New Condition for Solvability. In this section we shall discuss (N) from a slightly different viewpoint. We assume (as usual) that the forms f^i are fixed, but that $Y = (y^1, y^2, \ldots, y^m)^T$ is free to vary. The question that we raise is: How can we describe those Y's for which (N) is solvable? Of course, the validity of the consistency condition is necessary and sufficient; however, it is possible to devise a practical criterion. The observation that is needed in this connection is the following: $\sum c_i f^i = 0$ if and only if $C = (c_1, c_2, \ldots, c_m)$ is a solution of the *transposed homogeneous system* (H^T) associated with (N):

$$(\mathrm{H}^T) \qquad \begin{aligned} a_1^1 x_1 + a_1^2 x_2 + \cdots + a_1^m x_m &= 0 \\ a_2^1 x_1 + a_2^2 x_2 + \cdots + a_2^m x_m &= 0 \\ \cdots\cdots\cdots\cdots\cdots\cdots\cdots\cdots& \\ a_n^1 x_1 + a_n^2 x_2 + \cdots + a_n^m x_m &= 0 \end{aligned}$$

This new system has been obtained from (N) in an obvious way. The use of subscripts to distinguish among variables, and the designation of a solution of (H^T) by a row (rather than a column) of real numbers, is in harmony with the interchange of rows and columns of coefficients in (N) to obtain (H^T).

To verify the assertion in question, let us assume that $\sum c_i f^i = 0$. Then

$$0 = \sum_{i=1}^{m} c_i f^i(X) = \sum_{i=1}^{m} c_i \sum_{j=1}^{n} a_j^i x^j = \sum_{j=1}^{n} \left(\sum_{i=1}^{m} a_j^i c_i \right) x^j$$

which implies that

$$\sum_{i=1}^{m} a_j^i c_i = 0 \qquad j = 1, 2, \ldots, n$$

Hence $C = (c_1, c_2, \ldots, c_m)$ is a solution of (H^T) if $\sum c_i f^i = 0$. Since each of the foregoing steps is reversible, the converse also follows.

Defining the m-tuple $Y = (y^1, y^2, \ldots, y^m)^\mathrm{T}$ as *orthogonal* to the m-tuple $C = (c_1, c_2, \ldots, c_m)$, in symbols $Y \perp C$, if $\sum_1^m y^i c_i = 0$, we may state the above result as: (N) is solvable for those Y's and only those Y's which are orthogonal to all solutions of (H^T). From this in turn follows easily our final result in this connection which is stated next.

Theorem 1.13. Let r' denote the rank of (H^T). If $r' = m$, then (N) is solvable for every choice of Y. If $r' < m$, then (N) is solvable for those Y's and only those which are orthogonal to each member of a basis for the solution set of (H^T).

Proof. If (H^T) has rank m, its only solution is 0 and $Y \perp 0$ for every Y. If $r' < m$, the stated condition is obviously necessary, and the sufficiency is a consequence of the fact that if $Y \perp X^i$, $i = 1, 2, \ldots, m - r'$, then Y is orthogonal to every linear combination of these m-tuples.

It is easily shown that the rank, r', of (H^T) is equal to the rank, r, of (H) and (N). Consider first the case $r < m$. Then we may assume that $\{f^1, f^2, \ldots, f^r\}$ is linearly independent and conclude from Lemma 1.1 that constants b_i^j exist such that

$$f^{r+k} = b_1^k f^1 + b_2^k f^2 + \cdots + b_r^k f^r \qquad k = 1, 2, \ldots, m - r$$

In turn, each implied linear dependence among the f^i's determines a solution of (H^T):

$$\begin{aligned} X^1 &= (b_1^1, b_2^1, \ldots, b_r^1, -1, 0, \ldots, 0) \\ X^2 &= (b_1^2, b_2^2, \ldots, b_r^2, 0, -1, \ldots, 0) \\ &\cdots\cdots\cdots\cdots\cdots\cdots\cdots\cdots\cdots \\ X^{m-r} &= (b_1^{m-r}, b_2^{m-r}, \ldots, b_r^{m-r}, 0, 0, \ldots, -1) \end{aligned} \tag{13}$$

This is a linearly independent set of solutions of (H^T) and hence [applying Theorem 1.12 to (H^T)], $m - r \leqq m - r'$, or $r' \leqq r$. If $r = m$, the former, and hence the latter, inequality is trivially satisfied so that $r' \leqq r$ in all cases. Upon reversing the roles of (H) and (H^T), the reverse inequality is established, which proves that $r = r'$. Finally, it follows that the set of solutions (13) is a basis for the solution set of (H^T) in the event that $r < m$.

In the following theorem we collect all of our foregoing results for the important case where the number of equations agrees with the number of unknowns.

Theorem 1.14. For fixed a_j^i and $m = n$ there are the following possibilities for a system (N) of rank r:

(i) $r = n$, in which case there is a unique solution X for every choice of Y.

(ii) $r < n$, in which case the associated homogeneous system (H) has a set of $n - r > 0$ linearly independent solutions. The transposed homogeneous system likewise has a set of $n - r$ linearly independent solutions: $\{X^1, X^2, \ldots, X^{n-r}\}$. (N) has solutions for those and only those n-tuples Y orthogonal to X^i, $i = 1, 2, \ldots, n - r$. These solutions are determined by adding a fixed solution of (N) to those of (H).

PROBLEMS

1. Exhibit the general solution of the homogeneous system (H) associated with the system (N) below in terms of a basis. Write the general solution of (N) in terms of this basis. Find a basis for the solution set of (H^T), and verify that the column n-tuple $Y = (5, 8, 1, 3)^T$ is orthogonal to every member of this basis.

$$\begin{aligned}
4x^1 + x^2 + x^3 - 2x^4 + x^5 &= 5 \\
11x^1 + x^2 - 3x^3 - 4x^4 + 2x^5 &= 8 \\
2x^1 + 2x^2 - x^3 + 2x^4 - x^5 &= 1 \\
3x^1 \qquad\qquad - 2x^4 + x^5 &= 3
\end{aligned}$$

2. Show in detail that the set of solutions (13) of (H^T) is a basis for the solution set of (H^T).

1.9. Fields. In our discussion of systems of linear equations it has been explicitly assumed that all coefficients and constants are real numbers. This was done so that full attention could be devoted to the topic at hand. However if the material is examined, it is found that only a partial list of the properties of real numbers is used. These may be summarized by the statement that the real numbers are closed with respect to the "rational operations," *viz.*, addition, subtraction, multiplication, and division. There are many systems, called fields, that share these properties with the real numbers. Thus we may conclude that our discussion is valid for linear systems with coefficients in any field. This section is devoted to an elaboration of these statements for the benefit of those readers interested in systems of linear equations with coefficients other than real numbers.

In the reduction of a linear system to solved form we have used the two familiar compositions of addition and multiplication for combining the real numbers that enter as coefficients. These are binary compositions in the set of real numbers as defined in Part IV of the section on notation and conventions (page xiv), since each is a function on ordered pairs of real numbers to real numbers. At various times we have used the following properties of these compositions:

ADDITION

A_1. *Associative law:* For every triplet, (a, b, c), of real numbers, $a + (b + c) = (a + b) + c$.

A_2. *Commutative law:* For every pair, (a, b), of real numbers, $a + b = b + a$.

MULTIPLICATION

M_1. *Associative law:* For every triplet, (a, b, c), of real numbers, $a(bc) = (ab)c$.

M_2. *Commutative law:* For every pair, (a, b), of real numbers, $ab = ba$.

D. Distributive law: For every triplet, (a, b, c), of real numbers, $a(b + c) = ab + ac$.

Further, we have relied upon the numbers 0 and 1 with their following special properties: For each real number a, $a + 0 = a$, and there is a second real number, $-a$, such that $a + (-a) = 0$. Again, for each real number a, $a \cdot 1 = a$, and provided that $a \neq 0$, there is a second real number a^{-1} such that $a \cdot a^{-1} = 1$. The existence of an "inverse" (*i.e.*, negative and reciprocal, respectively) relative to addition and multiplication, respectively, guarantee that subtraction and division, except by 0, can be carried out, since $a - b = a + (-b)$ and $\frac{a}{b} = a(b^{-1})$. Phrased differently, it is possible to find a (unique) solution for each of the equations $a + x = b$ and $cx = d$, with $c \neq 0$ in the latter case.

This completes the list of those properties of real numbers used in the preceding sections. As mentioned above, systems with such properties are called fields. Thus a *field* is a set of elements $F = \{a, b, \ldots\}$ together with two binary compositions in F, which we call addition and multiplication and indicate by $+$ and $\cdot$, respectively, for which the postulates A_1, A_2, M_1, M_2, and D hold (restated, of course, with "elements of F" in place of "real numbers") together with the following two postulates:

Zero Element. The set F contains (at least) one element, called a zero element, and designated by 0, such that

(i) $a + 0 = a$ for all a in F.

(ii) For each a in F, there exists an element in F called the negative of a and designated by $-a$, such that $a + (-a) = 0$.

Unit Element. The set F contains (at least) one element, called a unit element, and designated by 1, such that

(i) $a \cdot 1 = a$ for all a in F.

(ii) For each a in F, $a \neq 0$, there exists an element in F called the inverse of a, and designated by a^{-1}, such that $a \cdot a^{-1} = 1$.

It can be shown that F contains a single zero element and a single unit element and that the negative and inverse elements are unique in every case. As indicated for the real numbers, the existence of a negative and inverse is a guarantee that the equations $a + x = b$ and $cx = d(c \neq 0)$ have solutions in F; that is, that subtraction and division (and hence *all* rational operations) can be carried out in F.

The reader will observe that the description of a field includes the definition of a set, together with two compositions in that set. Upon the adoption of symbols for each of these, for example, the symbols F, $+$, and $\cdot$ used above, a field can be indicated by a symbol of the type $F, +, \cdot$. Frequently the symbol used to designate the set of the field will be used for the field as well. For example, R^* will always be used to designate both the set of real numbers and the field (*i.e.*, the set plus compositions) of real numbers.

In order to obtain further examples of fields we turn first to various subsets of R^*. The reader is familiar with the fact that the ordinary sum and product of two rational numbers are again rational numbers. Regarding the set R of rational numbers as a subset of R^*, this result may be stated as follows: When both arguments of the function called addition in R^* are restricted to R, the function values are in R; this also holds true for multiplication. We describe this state of affairs by saying that R is closed under the compositions in R^*. In general, a subset S of a set M is said to be *closed* under a binary composition $\cdot$ in M if a and b in S imply that $a \cdot b$ is in S. The closure of R under addition and multiplication in R^* means that these compositions induce two compositions in R. It is easily verified that R, together with these induced compositions, is a field. This field is a subfield of R^* in the following sense:

DEFINITION. *Let $F, +, \cdot$ be a field and S a subset of F which is closed under $+$ and $\cdot$. If S, together with the compositions induced in S by $+$ and $\cdot$, is a field, this field is called a* subfield *of $F, +, \cdot$.*

This precise definition can be paraphrased with no essential sacrifice of accuracy as: A subfield of a field F is a subset which is closed under the compositions in F and which, together with the compositions in F, is a field. In order that a subset S of a field F determine a subfield, it is necessary and sufficient that S (i) be closed under addition and multiplication, (ii) contain 1, and (iii) contain the negative and inverse of each of its nonzero members. To prove the sufficiency of this statement, observe that since the properties required of addition and multiplication hold throughout F, a fortiori they hold in S. Finally, 1 together with $1 - 1 = 0$ are in S.

Let us use the above conditions to verify that the set of all real numbers of the form $a + b\sqrt{3}$, where a and b are rational, together with ordinary addition and multiplication, is a subfield of $R^{\#}$. Let us first verify that the set is closed under these compositions. We have

$$(a + b\sqrt{3}) + (c + d\sqrt{3}) = (a + c) + (b + d)\sqrt{3}$$

$$(a + b\sqrt{3}) \cdot (c + d\sqrt{3}) = (ac + 3bd) + (ad + bc)\sqrt{3}$$

Since R is a field, $a + c$, etc., are rational numbers; hence this sum and product have the desired form. Next, a negative $-(a + b\sqrt{3})$ of $a + b\sqrt{3}$ is present, since $-(a + b\sqrt{3}) = -a - b\sqrt{3}$ and R (as a field) contains $-a$ and $-b$ if it contains a, b. Finally, the set in question contains $1 = 1 + 0 \cdot \sqrt{3}$ and an inverse $(a + b\sqrt{3})^{-1}$ of any nonzero number since

$$\frac{1}{a + b\sqrt{3}} = \frac{1}{a + b\sqrt{3}} \cdot \frac{a - b\sqrt{3}}{a - b\sqrt{3}} = \frac{a}{a^2 - 3b^2} + \frac{-b}{a^2 - 3b^2}\sqrt{3}$$

has the required property. Notice that if $a + b\sqrt{3} \neq 0$, not both of a, b are zero, from which it follows that $a^2 - 3b^2 \neq 0$, since $\sqrt{3}$ is irrational.

The set C of complex numbers $a + bi$(a and b in $R^{\#}$) is easily shown to be a field when compositions are defined in the familiar way:

$$(a + bi) + (c + di) = (a + c) + (b + d)i$$

$$(a + bi) \cdot (c + di) = (ac - bd) + (ad + bc)i$$

Like the real field, C has many subfields.

Our final example is that of a *finite field*, *i.e.*, one containing a finite number of elements. For this we use the set J of integers: $\{0, \pm 1, \pm 2, \ldots\}$. Of course J itself is not a field since an integer different from 1 has no inverse in J. We propose to partition J into classes, define compositions in this collection of classes, and verify that the field postulates are satisfied. Let p denote a fixed prime, and define

$$a \equiv b(\text{mod } p)$$

read "a is congruent to b, modulo p," for integers a, b as follows: $a - b$ is divisible by p. It is easily verified that this relation between pairs of integers has the following three properties:

Reflexivity: $a \equiv a(\text{mod } p)$,

Symmetry: $a \equiv b(\text{mod } p)$ implies $b \equiv a(\text{mod } p)$,

Transitivity: $a \equiv b(\text{mod } p)$ and $b \equiv c(\text{mod } p)$ imply $a \equiv c(\text{mod } p)$.

Hence it is possible to decompose J into classes (*i.e.*, subsets) such that each integer appears in exactly one class, *i.e.*, the congruence relation *partitions* J. To prove this, let $\bar{a}$ denote the class of all integers x which are congruent to a. Then a is in $\bar{a}$ since $a \equiv a$, so that every integer is in at least one class. But a is in no more than one class, since a in $\bar{b}$ and $\bar{c}$ implies $a \equiv b$ (hence $b \equiv a$) and $a \equiv c$ so that $b \equiv c$ or $\bar{b} = \bar{c}$. This completes the proof.

Next we show that there are exactly p distinct classes. Since no two of $0, 1, 2, \ldots, p-1$ are congruent modulo p, there are at least p distinct classes. On the other hand, any integer a can be written uniquely in the form

$$a = qp + r \qquad 0 \leqq r < p$$

and hence belongs to exactly one of the classes

$$\bar{0}, \bar{1}, \bar{2}, \ldots, \overline{p-1}$$

since $a \equiv r (\text{mod } p)$. These classes of congruent elements are called *residue classes modulo p*. The set whose elements are these residue classes will be denoted by J_p. We emphasize that a residue class is defined by any one of its members. For example, the class $\bar{1}$ which consists of the numbers

$$\ldots, -2p + 1, -p + 1, 1, p + 1, 2p + 1, \ldots$$

i.e., all numbers x such that $x \equiv 1 (\text{mod } p)$, is the same class as $\overline{p+1}$ or $\overline{-2p+1}$, because $x \equiv 1(\text{mod } p)$ and $p + 1 \equiv 1(\text{mod } p)$ imply $x \equiv p + 1(\text{mod } p)$, and conversely. The reader should dwell upon the foregoing long enough to convince himself that the essence of the transition from J to J_p is the introduction of a new equality relation in J: two integers are "equal" if they differ by a multiple of p.

We now define compositions in J_p in terms of those for J by the equations

$$\bar{a} + \bar{b} = \overline{a + b} \qquad \bar{a} \cdot \bar{b} = \overline{ab}$$

Having pointed out that a residue class is defined by any one of its constituents, we must verify that the sum and the product class are independent of the representatives a and b used to define them. Since

$$a \equiv a' \qquad \text{and} \qquad b \equiv b'$$

imply

$$a + b \equiv a' + b' \qquad \text{and} \qquad ab \equiv a'b'$$

this independence follows. The validity of the properties required of addition and multiplication for a field is an immediate consequence of the validity of the corresponding rules for these operations in J. We illustrate this by verifying the associative law for addition.

$$\begin{aligned} \bar{a} + (\bar{b} + \bar{c}) &= \bar{a} + \overline{b + c} = \overline{a + (b + c)} \\ &= \overline{(a + b) + c} = \overline{a + b} + \bar{c} = (\bar{a} + \bar{b}) + \bar{c} \end{aligned}$$

The elements $\bar{0}, \bar{1}$ are zero and unit element, respectively, and $-\bar{a} = \overline{p - a}$

is the negative of $\bar{a}$. It remains to show that each $\bar{a} \neq \bar{0}$ has an inverse, *i.e.*, that the equation $\bar{a}\bar{x} = \bar{1}$ has a solution in J_p for $\bar{a} \neq \bar{0}$. Translated into J, this equation becomes the congruence $ax \equiv 1(\text{mod } p)$ for $a \not\equiv 0(\text{mod } p)$. The condition $a \not\equiv 0(\text{mod } p)$ means that a is not divisible by p; hence since p is a prime (and this is the first time we have used this assumption), a is relatively prime to p. This implies the existence of integers x, y such that $ax + py = 1$.† In turn this equation implies $ax \equiv 1(\text{mod } p)$ so that $\bar{a}$ has an inverse.

EXAMPLE 1.9

The field J_5 of residue classes modulo 5 consists of five elements $\{\bar{0}, \bar{1}, \bar{2}, \bar{3}, \bar{4}\}$. Below are an addition and a multiplication table for these elements. In the first, the sum $\bar{a} + \bar{b} = \overline{a + b}$ appears at the intersection of the $\bar{a}$ row and the $\bar{b}$ column. In the second, the product $\bar{a}\bar{b} = \overline{ab}$ appears at the same location.

$+$	$\bar{0}$	$\bar{1}$	$\bar{2}$	$\bar{3}$	$\bar{4}$
$\bar{0}$	$\bar{0}$	$\bar{1}$	$\bar{2}$	$\bar{3}$	$\bar{4}$
$\bar{1}$	$\bar{1}$	$\bar{2}$	$\bar{3}$	$\bar{4}$	$\bar{0}$
$\bar{2}$	$\bar{2}$	$\bar{3}$	$\bar{4}$	$\bar{0}$	$\bar{1}$
$\bar{3}$	$\bar{3}$	$\bar{4}$	$\bar{0}$	$\bar{1}$	$\bar{2}$
$\bar{4}$	$\bar{4}$	$\bar{0}$	$\bar{1}$	$\bar{2}$	$\bar{3}$

$\cdot$	$\bar{0}$	$\bar{1}$	$\bar{2}$	$\bar{3}$	$\bar{4}$
$\bar{0}$	$\bar{0}$	$\bar{0}$	$\bar{0}$	$\bar{0}$	$\bar{0}$
$\bar{1}$	$\bar{0}$	$\bar{1}$	$\bar{2}$	$\bar{3}$	$\bar{4}$
$\bar{2}$	$\bar{0}$	$\bar{2}$	$\bar{4}$	$\bar{1}$	$\bar{3}$
$\bar{3}$	$\bar{0}$	$\bar{3}$	$\bar{1}$	$\bar{4}$	$\bar{2}$
$\bar{4}$	$\bar{0}$	$\bar{4}$	$\bar{3}$	$\bar{2}$	$\bar{1}$

This completes our introduction to the important algebraic concept of field. In spite of its incompleteness a newcomer to the notion should realize that our discussion of systems of linear equations is valid for a great variety of coefficient domains. Moreover, since only the rational operations are used in passing to a solved form of a system having solutions, a solution exists in every field that contains the coefficients. For example, if every coefficient of a solvable system of linear equations is of the form $a+b\sqrt{3}$, where a and b are rational numbers, then some solution has this form. This state of affairs stands in contrast with that encountered in solving equations of degree greater than one where, among those fields that contain all coefficients, some may contain solutions and others may not.

†For proof, consider the set S of all integers $ax + py$, where x, y range over J. Then S contains a positive integer, hence a least positive integer d. We leave it as an exercise to show that S consists precisely of all multiples of d. Then, in particular, d divides a and p, since $a = a \cdot 1 + p \cdot 0$ and $p = a \cdot 0 + p \cdot 1$ are in S. It follows that $d = 1$, that is, $1 = ax + by$ for suitable integers x and y.

EXAMPLE 1.10

To illustrate the force of the validity of our elimination method for systems of equations with coefficients in any field, consider the problem of finding all integral solutions x^1, x^2, x^3 of the following system of simultaneous congruences:

$$\begin{aligned} 2x^1 - 2x^2 + 3x^3 &\equiv 7 \pmod 5 \\ x^1 + x^2 \phantom{{}+3x^3} &\equiv 3 \pmod 5 \\ 2x^1 - x^2 + 2x^3 &\equiv 1 \pmod 5 \end{aligned}$$

That this problem falls within the scope of our discussion is seen upon restating it in the field J_5, where it becomes

$$\begin{aligned} \bar{2}\bar{x}^1 - \bar{2}\bar{x}^2 + \bar{3}\bar{x}^3 &= \bar{2} \\ \bar{x}^1 + \bar{x}^2 \phantom{{}+\bar{3}\bar{x}^3} &= \bar{3} \\ \bar{2}\bar{x}^1 - \bar{x}^2 + \bar{2}\bar{x}^3 &= \bar{1} \end{aligned}$$

Using the tables for J_5 in Example 1.8, this system can be reduced to echelon form as follows:

(i) Multiply the third equation by $\bar{3}$ to get the system

$$\begin{aligned} \bar{2}\bar{x}^1 - \bar{2}\bar{x}^2 + \bar{3}\bar{x}^3 &= \bar{2} \\ \bar{x}^1 + \bar{x}^2 \phantom{{}+\bar{3}\bar{x}^3} &= \bar{3} \\ \bar{x}^1 - 3\bar{x}^2 + \bar{x}^3 &= \bar{3} \end{aligned}$$

(ii) Add $\bar{2}$ times the third equation to the first to get

$$\begin{aligned} \bar{4}\bar{x}^1 + \bar{2}\bar{x}^2 \phantom{{}+\bar{x}^3} &= \bar{3} \\ \bar{x}^1 + \bar{x}^2 \phantom{{}+\bar{x}^3} &= \bar{3} \\ \bar{x}^1 - \bar{3}\bar{x}^2 + \bar{x}^3 &= \bar{3} \end{aligned}$$

(iii) Add $\bar{3}$ times the second equation to the first to get

$$\begin{aligned} \bar{2}\bar{x}^1 \phantom{{}+\bar{x}^2+\bar{x}^3} &= \bar{2} \\ \bar{x}^1 + \bar{x}^2 \phantom{{}+\bar{x}^3} &= \bar{3} \\ \bar{x}^1 - \bar{3}\bar{x}^2 + \bar{x}^3 &= \bar{3} \end{aligned}$$

(iv) Multiply the first equation by $\bar{3}$, etc., to get

$$\begin{aligned} \bar{x}^1 &= \bar{1} \\ \bar{x}^2 &= \bar{2} \\ \bar{x}^3 &= \bar{3} \end{aligned}$$

Thus the original equation has the solution

$$x^1 \equiv 1,\ x^2 \equiv 2,\ x^3 \equiv 3 (\text{mod } 5)$$

Although in the future the word field will be used frequently, a reader who is interested in the concepts discussed solely relative to the real or the complex field may always think in terms of these particular fields. Regarding the usage of the field concept, we shall make one mild restriction in order to avoid a special argument occasionally. Namely, we rule out fields such as J_2, wherein $1 + 1 = 0$. The reason is that in such fields $a + a = 0$ for all a so that every element is its own negative and consequently one cannot conclude from the equation $a = -a$ that $a = 0$.

PROBLEMS

1. Show by an example (say $p = 6$) that the assumption that p is a prime is necessary in order to conclude that J_p is a field.

2. Find the general solution of the following system of simultaneous congruences:

$$\begin{aligned} x^1 - 3x^2 + 6x^3 + 8x^4 &\equiv 5 (\text{mod } 11) \\ -3x^1 + 2x^2 - x^3 + 3x^4 &\equiv 5 (\text{mod } 11) \\ 4x^1 + x^2 + x^3 - x^4 &\equiv 0 (\text{mod } 11) \\ -6x^1 + x^2 - 2x^3 + 3x^4 &\equiv 4 (\text{mod } 11) \end{aligned}$$

CHAPTER 2

VECTOR SPACES

2.1. Vectors. Frequently in the study of concrete mathematical systems, which may be quite different from each other superficially, resemblances will be noticed in the central features of the theories accompanying these systems. The observer may then try to bring such diverse systems under a single heading by extracting all properties common to these systems, listing them as postulates for an otherwise unrestricted set, and making deductions from these postulates.† This postulational development has been very fruitful in algebra in bringing a great variety of algebraic systems under a very few general classifications. One example of this was indicated in the preceding chapter by the concept of field, which is a powerful unifying concept in algebra. In Chap. 1 also occur two other algebraic systems whose similarity might attract the attention of someone interested in the postulational approach. We refer to the system of linear forms in $x^1, x^2, \ldots, x^n$ and the system of column n-tuples. In fact, the rules for equality, addition, etc., of linear forms when described by n-tuples are precisely the same as those we found it convenient to introduce for column n-tuples of solutions. Moreover, the notions of linear combination, maximal linearly independent subset, etc., are valuable in both types of systems. We mention next a third system that, though superficially different, makes extensive use of the same basic principles.

†It is perhaps not out of place to elaborate on this statement. We mention first that by a postulate we do not mean a self-evident truth or a statement that we are unable to prove. A postulate is rather an assumed property. The postulates P appearing in a postulational system S are the assumptions made about the elements of S. The system S itself consists of elements, compositions, assumptions about both, and finally, theorems which are the logical consequences of the assumptions. Whenever a specific system C is encountered whose elements and operations satisfy the postulates P, the theorems of S may be applied to C. As a second remark, it is important to realize that one and the same system can stem from various sets of postulates; in one approach a statement will appear as a postulate, while in another approach it will appear as a theorem. If one's interest is solely in the "body of facts" accompanying a system, he may have no preference for one system of postulates over another. On the other hand a mathematician very likely is sensitive to the aesthetic qualities of a postulational system and will have a preference based upon elegance, simplicity, ease with which the theory can be deduced, etc.

The reader is familiar with the concept of a vector or vector quantity in one sense or another. For example, in elementary physics such concepts as displacement, velocity, and acceleration are classified as vectors. In any event, statements to which everyone will subscribe are, first, that a vector is or can be represented graphically by a directed line segment or arrow, α; second, that the following rules are valid for these arrows:

V_1. Two arrows are equal if they have the same magnitude and direction (Fig. 1).

V_2. Two arrows α and β can be combined by the parallelogram rule to produce a third called the sum, $\alpha + \beta$, of α and β (Fig. 2).

V_3. An arrow can be multiplied by a real number (scalar) c to indicate a magnification of the magnitude in the ratio $|c| : 1$, and if the multiplier is negative, a reversal of direction is performed (Fig. 3).

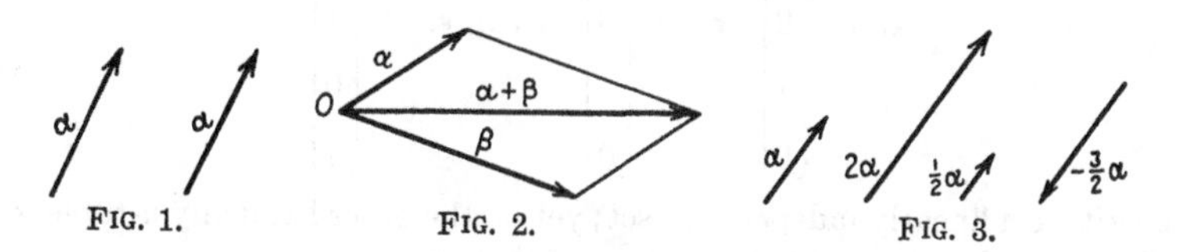

FIG. 1. FIG. 2. FIG. 3.

These rules, which form the basis for so-called "vector algebra," are easily seen to have the following properties upon drawing appropriate diagrams. Here vectors are indicated by Greek letters and scalars by Latin letters.

$$\alpha + \beta = \beta + \alpha$$

$$(\alpha + \beta) + \gamma = \alpha + (\beta + \gamma)$$

$$c(\alpha + \beta) = c\alpha + c\beta \qquad (a + b)\gamma = a\gamma + b\gamma$$

$$(ab)\gamma = a(b\gamma)$$

$$1\alpha = \alpha$$

Using analytic geometry, it is possible to reduce the study of planar vectors to a study of pairs $(a^1, a^2)^T$ of real numbers. For by V_1 we can restrict our attention to vectors issuing from the origin, and such vectors can be uniquely identified with the rectangular coordinates of their respective endpoints. For these number pairs, the rules V_1, V_2, and V_3 can be restated as follows:

V_1'. *Equality*: $(a^1, a^2)^T = (b^1, b^2)^T$ if and only if $a^i = b^i$, $i = 1, 2$

V_2'. *Addition*: $(a^1, a^2)^T + (b^1, b^2)^T = (a^1 + b^1, a^2 + b^2)^T$

V_3'. *Scalar multiplication*: $c(a^1, a^2)^T = (ca^1, ca^2)^T$

These of course are in agreement with our previous definitions.

We could supply other examples which, taken together with those discussed, justify the study of an abstract system whose defining properties are those underlying these examples. The justification rests with the economy of effort and standardization of notation and methods of proof.

A system that would embrace those mentioned so far is that of the set of all n-tuples $\alpha = (a^1, a^2, \ldots, a^n)^T$ with components a^i in a field F, together with compositions defined by V_2' and V_3' when extended from 2 to n components. However such a system has an inherent limitation which follows in part from Theorem 1.9, *viz.*, that n is the maximum number of linearly independent elements in any collection of n-tuples. For on one hand the *unit n-tuples*

$$\epsilon_1 = \begin{pmatrix} 1 \\ 0 \\ 0 \\ \cdot \\ 0 \end{pmatrix}, \quad \epsilon_2 = \begin{pmatrix} 0 \\ 1 \\ 0 \\ \cdot \\ 0 \end{pmatrix}, \quad \ldots, \quad \epsilon_n = \begin{pmatrix} 0 \\ 0 \\ \cdot \\ 0 \\ 1 \end{pmatrix}$$

constitute a linearly independent set; yet on the other hand any n-tuple is a linear combination of the members of this set:

$$(a^1, a^2, \ldots, a^n)^T = \sum_{i=1}^{n} a^i \epsilon_i$$

The existence of systems which include an addition and scalar multiplication, yet which contain an infinite number of linearly independent elements, is demonstrated by the set of all real-valued continuous functions on the closed interval (0, 1), with addition of functions and multiplication by a real number defined in the usual way. Indeed, the set of functions $\{f_0, f_1, \ldots, f_n\}$, where $f_k(x) = x^k$, is linearly independent for every positive integer n.

To carry out the postulational method which we have discussed, we should determine the algebraic properties of these systems and take them as our starting point. This is the origin of the definition in the next section.

2.2. Definition of a Linear Vector Space. The underlying properties of the systems mentioned so far are collected in the following definition:

DEFINITION. *A* linear vector space *over a field F (with elements $a, b, \ldots$, which are called* scalars) *is a set of elements $V = \{\alpha, \beta, \ldots\}$ called* vectors, *together with two compositions called* addition *and* scalar multiplication, *such that*

ADDITION

Addition, which is symbolized by $+$, is a binary composition in V satisfying the following conditions:

A_1. *Associative law:* $\alpha + (\beta + \gamma) = (\alpha + \beta) + \gamma$ for all α, β, γ in V

A_2. *Commutative law:* $\alpha + \beta = \beta + \alpha$ for all α, β in V

A_3. Existence of a zero element 0 with the property $\alpha + 0 = \alpha$ for all α in V

A_4. Existence of a negative, $-\alpha$, or, in other words, the solvability of the equation $\alpha + \xi = 0$ for all α in V and each zero element

SCALAR MULTIPLICATION

Scalar multiplication, which is symbolized by $\cdot$, is a symmetric function on $F \times V$ to V satisfying the following conditions:

M_1. *Distributive law with respect to addition in* V: $a \cdot (\alpha + \beta) = a \cdot \alpha + a \cdot \beta$

M_2. *Distributive law with respect to addition in* F: $(a + b) \cdot \alpha = a \cdot \alpha + b \cdot \alpha$

M_3. *Associative law*: $a \cdot (b \cdot \alpha) = (ab) \cdot \alpha$

M_4. $1 \cdot \alpha = \alpha$, where 1 is the unit element of F

We shall often use the term "vector space V" for the set part of the vector space. The designation of scalar multiplication as a symmetric function means that $c \cdot \alpha = \alpha \cdot c$ for a scalar c and vector α.

EXAMPLE 2.1

The set of all n-tuples $(a^1, a^2, \ldots, a^n)^T$ with components a^i in a field F, together with the compositions introduced earlier for n-tuples, is a vector space over F. This space will always be denoted by $V_n(F)$. The verification of the axioms for addition and multiplication are left as an exercise. We wish to point out, however, that (i) axioms A_1, A_2, M_1, ..., M_4 are consequences of properties of addition and multiplication in a field, (ii) the zero element 0 in F determines a zero element $(0, 0, \ldots, 0)^T$ for $V_n(F)$, and (iii) the existence of a negative for each field element guarantees the existence of a negative $(-a^1, -a^2, \ldots, -a^n)^T$ for the n-tuple $(a^1, a^2, \ldots, a^n)^T$.

Sometimes we shall write our n-tuples horizontally, $(a_1, a_2, \ldots, a_n)$, and shall distinguish between the row and column n-tuples by calling them row and column vectors, respectively.

If F is taken to be R^*, the field of real numbers, the resulting vector space $V_n(R^*)$ is frequently called the *n-dimensional Euclidean space*, since for $n = 2$ and 3 the elements can be thought of as the rectangular coordinates of points in the Euclidean plane and space, respectively. Since points and their coordinates are usually identified in analytic geometry, we shall also refer to the n-tuples of $V_n(R^*)$ as the *points* of the space. On the other hand, since a point P in Euclidean space is described by its position vector (the line segment from the origin to P), an n-tuple may be regarded as a position vector. Sometimes this is a more natural interpretation of an n-tuple in so far as the operations for n-tuples are concerned.

EXAMPLE 2.2

The set of all solution vectors (*i.e.*, column n-tuples) of a system of linear homogeneous equations in n unknowns and with coefficients in a field F, together with the compositions for $V_n(F)$, is a vector space over F. As a subset of $V_n(F)$ the validity of $A_1, A_2, M_1, \ldots, M_4$ is assured, so that only closure under addition and multiplication together with the existence of a zero element and negative must be verified. It is obvious that the sum of two solutions and a scalar multiple of a solution are solutions. Since the trivial solution is always present, the existence of a zero element is assured. The existence of a negative causes no difficulty, since if α is a solution vector, its negative is the scalar multiple $(-1)\alpha$.

If the homogeneous system has rank n, the vector space of solutions consists of the single zero element, *i.e.*, the trivial solution.

EXAMPLE 2.3

The set $P(C)$ of all polynomials with complex coefficients in a real variable t, together with the compositions of ordinary addition of polynomials and multiplication of a polynomial by a complex number, is a vector space over the complex field C. The same compositions and the set $P_n(C)$ of all polynomials with complex coefficients, in a real variable t, and of degree less than a fixed positive integer n, is a vector space over C.

We now derive some immediate consequences of the axioms for a vector space. These are primarily computational rules.

Rule 1. The generalized associative law for addition is valid, *i.e.*, the terms in any finite sum may be arbitrarily regrouped.

The axiom A_1 states just this for the case of three elements. The general result follows by induction.

Rule 2. A linear space V has a *unique* zero element, the *zero*, or *null*, *vector*.

If $0'$ is a second zero element, $0 = 0 + 0' = 0' + 0 = 0'$.

Rule 3. The cancellation law for addition holds: $\alpha + \beta = \alpha + \gamma$ implies $\beta = \gamma$.

Since V has a unique zero element, it is unambiguous to denote a negative of α by $-\alpha$. Adding $-\alpha$ to each side of the given equation and using A_1 and A_2, we obtain in succession

$$-\alpha + (\alpha + \beta) = -\alpha + (\alpha + \gamma), (-\alpha + \alpha) + \beta = (-\alpha + \alpha) + \gamma,$$
$$(\alpha + (-\alpha)) + \beta = (\alpha + (-\alpha)) + \gamma, 0 + \beta = 0 + \gamma,$$
$$\beta + 0 = \gamma + 0, \beta = \gamma$$

Rule 4. Subtraction is possible and unique, that is, the equation $\alpha + \xi = \beta$ has a unique solution.

Upon substitution, $\xi = \beta + (-\alpha)$ is found to be a solution. If ξ_1 and ξ_2 are both solutions, so that $\alpha + \xi_1 = \beta = \alpha + \xi_2$, the cancellation rule gives $\xi_1 = \xi_2$. The solution $\beta + (-\alpha)$ will be written $\beta - \alpha$.

Rule 5. Each element α has a *unique* negative, $-\alpha$.

A negative is a solution of the equation $\alpha + \xi = 0$.

It should be observed that there are two zero elements present in our discussion, the zero vector and the zero scalar. We denote these elements, for the moment, by 0_V and 0_F, respectively. They are uniquely defined by the equations

$$\xi + 0_V = \xi \qquad \text{for all } \xi \text{ in } V$$

$$x + 0_F = x \qquad \text{for all } x \text{ in } F$$

We apply the distributive laws M_1 and M_2 to obtain

$$x\xi + x0_V = x(\xi + 0_V) = x\xi = x\xi + 0_V$$

$$x\xi + 0_F\xi = (x + 0_F)\xi = x\xi = x\xi + 0_V$$

Canceling $x\xi$ in these two equations gives the following rule:

Rule 6. The zero vector 0_V and the zero scalar 0_F obey the laws

$$0_F\xi = 0_V \qquad \text{for all } \xi$$

$$x0_V = 0_V \qquad \text{for all } x$$

Since the likelihood of confusing these two elements in an equation is slight, the subscripts will be omitted.

Rule 7. The negative, $-\xi$, and the scalar multiple $(-1)\xi$ of a vector ξ are one and the same.

This follows from the equation (which uses M_4 for the first time)

$$\xi + (-1)\xi = 1\xi + (-1)\xi = (1 + (-1))\xi = 0\xi = 0$$

since it implies that $(-1)\xi$ is the (unique) negative of ξ.

2.3. Subspaces. In the development of a postulational system one relies upon concrete examples to suggest objects for study. For example, in three-dimensional Euclidean space, along with points, lines and planes are of interest. The points in a line or in a plane through the origin constitute a set which is closed under the compositions of addition and scalar multiplication. Moreover, such a set, together with the corresponding induced compositions, is a vector space. It is the analogue of lines and planes through the origin that we shall introduce for vector spaces in general. Parenthetically, we remark that closure under addition of a subset S of a vector space V over F means, by definition, that $\alpha + \beta$ is in S for α and β in S; closure under scalar multiplication means that $x\alpha$ is in S for α in S and all x in F.

Theorem 2.1. Let S be a nonempty subset of a vector space V over F which is closed under addition and scalar multiplication, so that

(i) for all α, β in S, $\alpha + \beta$ is in S, and
(ii) for all α in S and all x in F, $x\alpha$ is in S,

or what amounts to the same,

(iii) for all α, β in S and all x, y in F, $x\alpha + y\beta$ is in S.

Then S, together with the compositions induced by those of V, is a vector space which is called a *subspace* of V.

Proof. The equivalence of (i) and (ii) with (iii) is left as an exercise. To demonstrate that a subset S with the stated properties determines a subspace, observe that we need concern ourselves only with postulates pertaining to existence (*viz.*, A_3 and A_4) since the other requirements (for example, A_2) hold throughout V, hence a fortiori in a subset. If S consists of the zero vector alone, these are satisfied; if S contains $\alpha \neq 0$, it contains $(-1)\alpha = -\alpha$ and $\alpha - \alpha = 0$. This completes the proof.

The subset O, consisting of the zero vector alone (see Example 2.2), as well as that consisting of all elements of V, are obviously subspaces of V. These extremes are called *improper* subspaces, while all others are called *proper* subspaces of V.

EXAMPLE 2.4

Let us determine the proper subspaces of $V_3(R^*)$. If α_1 is a nonzero vector in a subspace S, then S contains all multiples $x^1\alpha_1$ for x^1 in R^*. These vectors form a subspace; if they exhaust S, it is simply the set of vectors on a line through the origin. Otherwise S contains a vector $\alpha_2 \neq x^1\alpha_1$ for all x^1. Hence S contains all vectors $x^1\alpha_1 + x^2\alpha_2$, x^1 and x^2 in R^*; these fill out the plane through the origin determined by α_1 and α_2. We assert that these vectors exhaust S; otherwise, $S = V_3(R^*)$. For if S contains $\alpha_3 \neq x^1\alpha_1 + x^2\alpha_2$ for all x^1 and x^2, hence all vectors $x^1\alpha_1 + x^2\alpha_2 + x^3\alpha_3$, we can show that any vector ξ in $V_3(R^*)$ is some linear combination of the α_i's. In fact, our assumptions imply that $\sum x^i\alpha_i = 0$ if and only if every $x^i = 0$. If $\alpha_j = (a_j^1, a_j^2, a_j^3)^T$, $j = 1, 2, 3$, this means that the homogeneous system

$$\sum_{j=1}^{3} a_j^i x^j = 0 \qquad i = 1, 2, 3$$

has only the trivial solution, hence has rank 3. Consequently the corresponding nonhomogeneous system in the unknowns x^i, deduced from the equation $\xi = \sum x^i \alpha_i$, has a unique solution for all choices of ξ (Theorem 1.14). Thus the only proper subspaces are the lines and planes through the origin.

EXAMPLE 2.5

Let m and n be two positive integers with $m < n$. The set of all row vectors $(a_1, a_2, \ldots, a_n)$, with every a_i in F, for which $a_1 = a_2 = \cdots = a_m = 0$ determines a subspace of $V_n(F)$. Again, referring back to Example 2.2, the solution space of a system of linear homogeneous equations of rank r is a proper subspace of $V_n(F)$ if $0 < r < n$.

We mention next several general methods for forming subspaces of a given vector space. The first is based on the notion used in Example 2.4.

Theorem 2.2. The set of all *linear combinations* $x^1\alpha_1 + x^2\alpha_2 + \cdots + x^m\alpha_m$ (x_i in F) of a fixed set of vectors $\{\alpha_1, \alpha_2, \ldots, \alpha_m\}$ in a vector space V over F is a subspace which will be denoted by $[\alpha_1, \alpha_2, \ldots, \alpha_m]$.

Proof. Since $x\sum x^i\alpha_i + y\sum y^i\alpha_i = \sum xx^i\alpha_i + \sum yy^i\alpha_i = \sum(xx^i + yy^i)\alpha_i$, the assertion follows from Theorem 2.1.

The subspace $[\alpha_1, \alpha_2, \ldots, \alpha_m]$ is evidently the smallest subspace containing all the vectors α_i. More generally, starting with an arbitrary set of vectors, the set of all linear combinations of all finite subsets is obviously the smallest subspace which contains the original set. If this technique is applied to form the smallest subspace containing two given subspaces S and T, it is seen that a linear combination of elements of S and T reduces to an element $\sigma + \tau$ with σ in S and τ in T. This proves the following theorem:

Theorem 2.3. If S and T are subspaces of V, the set of all sums $\sigma + \tau$ with σ in S and τ in T is a subspace called the *linear sum of* S *and* T and written $S + T$.

DEFINITION. *A set of vectors* $\{\alpha_1, \alpha_2, \ldots, \alpha_m\}$ *in a vector space* V spans *or* generates V *if* $V = [\alpha_1, \alpha_2, \ldots, \alpha_m]$.

For example, the unit vectors ϵ_i span $V_n(F)$ since $(a^1, a^2, \ldots, a^n)^T = \sum a^i\epsilon_i$. Again, $V_3(R^\#)$ is spanned by any three vectors which do not lie in a plane.

Theorem 2.4. The set of all vectors which belong to both of two subspaces S and T of a vector space V is a subspace called the *intersection* $S \cap T$ of S and T.

Proof. If α, β are in $S \cap T$, then, as members of S, $x\alpha + y\beta$ is in S according to (iii) of Theorem 2.1. Similarly, $x\alpha + y\beta$ is in T and hence in $S \cap T$.

The intersection of two subspaces is never empty since they always have the zero vector in common. Whereas $S + T$ is the smallest space containing both S and T, $S \cap T$ is the largest space contained in both S and T.

EXAMPLE 2.6

Let S be the subspace of $V_3(R)$, where R is the field of rational numbers, consisting of all vectors of the form $(0, x^2, x^3)^T$ and T the subspace spanned by $\alpha = (1, 1, 1)^T$ and $\beta = (2, 3, 0)^T$. Describe $S \cap T$ and $S + T$.

A vector $x\alpha + y\beta$ of T has the form of a vector in S if and only if $x + 2y = 0$. Thus $S \cap T$ is the space generated by $(0, 1, -2)^T$. Since S contains the unit vectors ϵ_2 and ϵ_3, $S + T$ contains $\alpha - \epsilon_2 - \epsilon_3 = \epsilon_1$. Thus $S + T = V_3(R)$ since it contains $\epsilon_1, \epsilon_2, \epsilon_3$.

PROBLEMS

1. Determine which of the following subsets of $V_n(R)$, where R is the field of rational numbers and $n \geqq 2$, constitute subspaces

(*a*) All $\xi = (x^1, x^2, \ldots, x^n)^T$ with x^1 an even integer

(*b*) All ξ with $x^1 = x^2 = 0$

(*c*) All ξ with $x^2 = x^1 - 1$

(*d*) All ξ with $x^1 = x^2$

(*e*) All ξ orthogonal to $(1, 1, \ldots, 1)$

(*f*) All ξ with $x^k = x^{n-k+1}$

2. In $V_3(R)$ let S be the subspace spanned by the vectors $(1, 2, -1)$, $(1, 0, 2)$, and $(-1, 4, -8)$ and T the subspace consisting of all vectors of the form $(x_1, 0, x_3)$. Describe $S \cap T$ and $S + T$.

3. Construct $V_3(J_2)$ and list all of its subspaces.

4. Show that every (row) vector in $V_3(R^\#)$ can be written as a linear combination of the set $\{(1, 1, 1), (1, 0, 1), (1, -1, -1)\}$.

2.4. Dimension. The motivation for this section is to be found in the preceding chapter. The various definitions and theorems stated in connection with sets of linear forms and the set of solution vectors of a system of homogeneous equation are useful in all vector spaces and will now be developed in full generality.

DEFINITION. *The set of vectors* $\{\alpha_1, \alpha_2, \ldots, \alpha_m\}$ *of a vector space* V *over a field* F *is* linearly dependent over F *if there exist scalars* c_i *in* F, *not all* 0, *such that* $\sum_1^m c^i\alpha_i = 0$. *Otherwise the set is* linearly independent. *An infinite set of vectors is linearly independent if every finite subset is linearly independent.*

To justify the inclusion of the phrase "over F" in the definition of linear dependence, we consider an example. The set C of complex numbers with ordinary addition is a vector space over C; in addition, C, with the same operation, is a vector space over the real field $R^{\#}$. Clearly the pair of vectors $\{1, i\}$ is linearly dependent over C but linearly independent over $R^{\#}$.

The set consisting of the zero vector alone is linearly dependent, while that consisting of one nonzero vector is linearly independent. Any set of vectors containing the zero vector is dependent. The same equation which demonstrates that the n unit vectors ϵ_i generate $V_n(F)$ verifies the independence of this set, since $\sum_1^n c^i\epsilon_i = (c^1, c^2, \ldots, c^n)^T = 0$ if and only if every $c^i = 0$. The essence of the linear dependence of a set of vectors is expressed in the next theorem.

Theorem 2.5. The set of nonzero vectors $\{\alpha_1, \alpha_2, \ldots, \alpha_m\}$ of V is linearly dependent if and only if some one of the vectors is a linear combination of the preceding ones.

Proof. If α_k is a linear combination of its predecessors, for example, $\alpha_k = \sum_1^{k-1} c^i\alpha_i$, then the linear dependence of $\{\alpha_1, \alpha_2, \ldots, \alpha_m\}$ is exhibited by the equation

$$c^1\alpha_1 + c^2\alpha_2 + \cdots + c^{k-1}\alpha_{k-1} + (-1)\alpha_k + 0\alpha_{k+1} + \cdots + 0\alpha_m = 0$$

Conversely, if $\{\alpha_1, \alpha_2, \ldots, \alpha_m\}$ is linearly dependent, suppose α_k is the last α with a nonzero coefficient in an equation exhibiting the dependence. Then it is possible to solve for α_k as a combination of $\alpha_1, \alpha_2, \ldots, \alpha_{k-1}$ unless $k = 1$. However, this cannot occur since an equation $c^1\alpha_1 = 0$ with $c^1 \neq 0$ implies $\alpha_1 = 0$, contrary to assumption.

COROLLARY. Any finite set of vectors, not all of which are the zero vector, contains a linearly independent subset which generates the same space.

Proof. Clearly any zero vectors may be discarded. Then, starting at the left of the given sequence $\alpha_1, \alpha_2, \ldots, \alpha_m$, discard the vector α_k if and only if it is a combination of its predecessors. The resulting subset spans the same space, and since no vector is a combination of those preceding it, the subset is independent.

DEFINITION: *Let S be a nonempty set of vectors in a vector space over F. If there exists an integer $m \geqq 0$ such that: (i) S contains a set of m vectors linearly independent over F, and (ii) any set of more than m vectors is linearly dependent over F, then m is called the* dimension (*or* rank) *of S over F; in symbols, $m = d[S]$. If no such m exists, S is said to have* dimension infinity. *If the dimension of S is m, any linearly independent set of m vectors in S is called* maximal.

It should be noticed that the first concept defined above is an extension of the notion of rank as defined prior to Lemma 1.2 (p. 13) for a finite set of linear forms. We prefer the term dimension to rank since our principal application will be to the case of a vector space for which the terminology is well established. The dimensions of several of our examples of vector spaces are easily determined. For example, that of the improper subspace O of any vector space is 0. The space $P(C)$ of Example 2.3 has dimensions infinity, since each of the sets $\{1, t, t^2, \ldots, t^r\}$, $r = 0, 1, 2, \ldots$, is linearly independent. In the same example, the space $P_n(C)$ has dimension n since $\{1, t, t^2, \ldots, t^{n-1}\}$ is linear independent and any set of more than n polynomials of $P_n(C)$ is easily seen to be linearly dependent using Theorem 1.8.

Theorem 2.6. If a vector space V is spanned by a finite set of vectors then V has finite dimension and $d[V]$ is equal to the dimensions of any generating system.

Proof. Let $S = \{\alpha_1, \alpha_2, \ldots, \alpha_m\}$ be a generating system for V. Clearly the dimension of S is finite. If $d[S] = r$, then S contains a maximal linear independent set of r vectors which, without loss of generality, we may take as $S' = \{\alpha_1, \alpha_2, \ldots, \alpha_r\}$. Then S' is a generating system for V. Indeed, $\{\alpha_1, \alpha_2, \ldots, \alpha_r, \alpha_{r+j}\}$, $j = 1, 2, \ldots, m - r$, is a linearly dependent set and in the relation

$$c^1\alpha_1 + c^2\alpha_2 + \cdots + c^r\alpha_r + c^{r+j}\alpha_{r+j} = 0$$

expressing this fact, $c^{r+j} \neq 0$, since the contrary would assert the linear dependence of S'. It follows that a representation of an element of V as a linear combination of vectors of S can be replaced by one involving only vectors of S'. Thus S' is a generating system for V.

Next, by virtue of Theorem 1.9 and the observation made upon the completion of the proof, *viz*, that the proof does not depend upon n-tuples, it follows that any set of more than r linear combinations of vectors of S' is linearly dependent. In other words, any set of more than r vectors of V is linearly dependent. Thus $d[V]$ exists and is less than or equal to r. But clearly $d[V] \geqq d[S] = r$. Thus $d[V] = r = d[S]$.

COROLLARY. If $\{\alpha_1, \alpha_2, \ldots, \alpha_n\}$ is a linearly independent set of vectors and spans V, then $d[V] = n$.

Theorem 2.7. The space $V_n(F)$ has dimension n over F.

Proof. This follows from the corollary to Theorem 2.6 using the unit vectors ϵ_i.

Thus our notion of dimension is in agreement with that for the familiar Euclidean spaces.

DEFINITION. *A set of vectors in a vector space V which (i) is linearly independent and (ii) spans V is called a* basis *of V.*

The corollary to Theorem 2.6 implies the following result:

Theorem 2.8. The bases of a finite dimensional vector space V coincide with the maximal linearly independent sets of vectors in V. If $d[V] = n$, then every basis of V contains n members.

Proof. If $\{\alpha_1, \alpha_2, \ldots, \alpha_n\}$ is a basis of V, then using the corollary to Theorem 2.6, $d[V] = n$. It follows that $\{\alpha_1, \alpha_2, \ldots, \alpha_n\}$ is a maximal linearly independent set. The converse is left as an exercise.

Next we state two further useful theorems concerning bases.

Theorem 2.9. If $d[V] = n$ and $\{\alpha_1, \alpha_2, \ldots, \alpha_n\}$ spans V, then $\{\alpha_1, \alpha_2, \ldots, \alpha_n\}$ is a basis for V.

Proof. By the corollary to Theorem 2.5, the α_i's contain a linearly independent subset which spans V and hence is a basis. According to the previous theorem, this subset must contain n elements and hence is the full set.

Theorem 2.10. Every vector ξ in a vector space V is expressible in a unique way as a linear combination of the members of a given basis $\{\alpha_1, \alpha_2, \ldots, \alpha_n\}$; in symbols, $\xi = \sum x^i \alpha_i$. The scalars x^i are called the *coordinates of ξ relative to the given basis.*

Proof. If $\xi = \sum x^i \alpha_i = \sum y^i \alpha_i$, then subtraction gives $\sum (x^i - y^i)\alpha_i = 0$, which holds if and only if $x^i = y^i$ for all i.

Most computations in three-dimensional vector algebra are carried out using the unit vectors ϵ_1, ϵ_2, and ϵ_3. However for many purposes any basis will serve equally well. Geometrically a basis can be described as any set of three vectors not lying in a plane. In Example 2.4 can be found the analytic condition that a set of three vectors $\alpha_j = (a_j^1, a_j^2, a_j^3)^T$, $j = 1, 2, 3$, is a basis, *viz.*, that the homogeneous system

$$\sum_{j=1}^{3} a_j^i x^j = 0 \qquad i = 1, 2, 3$$

has rank 3. Then the corresponding nonhomogeneous system

$$\sum_{j=1}^{3} a_j^i x^j = y^i \qquad i = 1, 2, 3$$

has a unique solution for an arbitrary vector $\eta = (y^1, y^2, y^3)^T$. This is a restatement of Theorem 2.10 for the case at hand, that *every* vector has a *unique* representation as a linear combination of basis elements.

EXAMPLE 2.7

Show that the set $\{\alpha_1 = (1, i, 0)^T, \alpha_2 = (2i, 1, 1)^T, \alpha_3 = (0, 1 + i, 1 - i)^T\}$ is a basis for $V_3(C)$. Find the coordinates of ϵ_1 relative to this basis. The homogeneous system determined by the vector equation $x^1\alpha_1 + x^2\alpha_2 + x^3\alpha_3 = 0$ has only the trivial solution; consequently the set is independent. The nonhomogeneous system determined by the equation $x^1\alpha_1 + x^2\alpha_2 + x^3\alpha_3 = \epsilon_1$ has the unique solution $\frac{1}{10}(4 - 2i, 1 - 3i, -2 + i)^T$.

As for further examples, an obvious basis for the vector space $P(C)$ of Example 2.3 is $\{1, t, \ldots, t^n, \ldots\}$, while one for $P_n(C)$ is $\{1, t, \ldots, t^{n-1}\}$. Again, the field C may be regarded as a vector space of dimension 2 over $R^\#$ since $\{1, i\}$ is a basis for C over $R^\#$. Of course this interpretation of C ignores the possibility of multiplying one complex number by another. In Chap. 1 the definition of basis made in connection with the solutions of a homogeneous system is an instance of our general definition. As a final example, observe that the set of linear forms g^i in the echelon system (N_e) is a basis for the space spanned by the set of linear forms f^i in the original system (N). *Hence the reduction of* (N) *to* (N_e) *is simply the replacement of a given generating set by a particular basis for the space.*

The variety of bases that exist for a vector space is indicated by the next theorem.

Theorem 2.11. Any linearly independent set $\{\beta_1, \beta_2, \ldots, \beta_m\}$ of vectors in a vector space V is part of a basis.

Proof. Let $\{\alpha_1, \alpha_2, \ldots, \alpha_n\}$ be a basis for V, and apply the method used in the proof of the corollary to Theorem 2.5 to the set $\{\beta_1, \beta_2, \ldots, \beta_m, \alpha_1, \ldots, \alpha_n\}$.

Theorem 2.12. If S and T are any two subspaces of V, their dimensions satisfy the equation

$$d[S] + d[T] = d[S + T] + d[S \cap T]$$

Proof. Let

$\{\alpha_1, \alpha_2, \ldots, \alpha_n\}$ be a basis for $S \cap T$

$\{\alpha_1, \alpha_2, \ldots, \alpha_n, \beta_1, \ldots, \beta_r\}$ be a basis for S

$\{\alpha_1, \alpha_2, \ldots, \alpha_n, \gamma_1, \ldots, \gamma_s\}$ be a basis for T

That bases for S and T can be written as indicated is a consequence of Theorem 2.11. We assert that the collective set $\{\alpha_i, \beta_j, \gamma_k\}$ is a basis for $S + T$. It spans $S + T$ by Theorem 2.3. To verify its independence, consider the equation

$$\sum x^i \alpha_i + \sum y^j \beta_j + \sum z^k \gamma_k = 0$$

This implies that $\sum y^j \beta_j = -\sum x^i \alpha_i - \sum z^k \gamma_k$, whence $\sum y^j \beta_j$ is in T. But as a linear combination of the β_j's, this vector is also in S, hence in $S \cap T$, which means that we can write $\sum y^j \beta_j = \sum u^i \alpha_i$ for scalars u^i. The independence of the α_i's together with the β_j's implies that every coefficient in this equation is zero; thus every $y^j = 0$. Hence $\sum x^i \alpha_i + \sum z^k \gamma_k = 0$, and now the independence of the α_i's together with the γ_k's implies that all x^i's and z^k's vanish. Our conclusion follows from the equation

$$(n + r) + (n + s) = (n + r + s) + n$$

PROBLEMS

1. For what values of x and y is the set of two vectors $\{(x, y, 3), (2, x - y, 1)\}$ linearly independent?

2. Show that the set of row vectors $\{(2, 3, 1, 0), (0, 1, 1, -2), (1, 0, -1, 3), (5, 3, -2, 9)\}$ is linearly dependent, and extract a linearly independent subset which generates the same subspace of $V_4(R^\#)$. Write each remaining vector as a linear combination of the members of this subset.

3. Verify that the set of all vectors (x_1, x_2, x_3, x_4) in $V_4(R)$ for which $x_1 = x_2 - x_3$ and $x_3 = x_4$ is a subspace S. Find a basis for S.

4. Complete the linearly independent set of vectors $\{(2, 3, 1, 0), (1, 0, -1, 3)\}$ to a basis for $V_4(R)$.

5. Show that the set $\{(1, 2 + i, 3), (2 - i, i, 1), (i, 2 + 3i, 2)\}$ is a basis for $V_3(C)$, and determine the coordinates of the unit vectors ϵ_1, ϵ_2, and ϵ_3 with respect to this basis.

6. Find all bases for $V_3(J_2)$.

7. In $V_4(R^\#)$ let S be the subspace spanned by $\{(1, 2, -1, -2), (2, -3, -2, 3), (-3, 4, 3, -4)\}$ and T the subspace spanned by $\{(-1, 2, 1, 2), (-2, -3, 2, -3), (3, 4, -3, 4)\}$. Determine the dimensions of S, T, $S \cap T$, $S + T$.

8. In $V_3(R^\#)$ consider the following four vectors: $\alpha_1 = (2, 2, -1)$, $\alpha_2 = (-1, 2, 2)$, $\alpha_3 = (2, -1, 2)$, $\alpha_4 = (3, 0, -3)$.

(*a*) Is $\{\alpha_1, \alpha_2, \alpha_3\}$ a basis for $V_3(R^\#)$? If possible, express α_4 as a linear combination of this set.

(*b*) Determine the dimension of the space S spanned by α_1 and α_2 as well as that of the space T spanned by α_3 and α_4. Determine $d[S \cap T]$, and specify a basis for this intersection.

9. For each of the following spaces of polynomials with rational coefficients show that the sets listed are bases:

(*a*) All polynomials in $2^{\frac{1}{3}}$: $\{1, 2^{\frac{1}{3}}, 2^{\frac{2}{3}}\}$.

(*b*) All polynomials in $2^{\frac{1}{2}}(1 + i)$: $\{1, 2^{\frac{1}{2}}, i, i2^{\frac{1}{2}}\}$.

2.5. Isomorphism in Vector Spaces. Theorem 2.10 has an important consequence which we shall discuss now. According to this result, once a basis, let us say $\{\alpha_1, \alpha_2, \ldots, \alpha_n\}$, is chosen in a vector space V over F, every vector is uniquely described by listing the n-tuple of its coordinates relative to this basis. Since V consists of all linear combinations of the α_i's, we have, conversely, that each n-tuple of scalars from F determines a unique vector in V. We shall always write these n-tuples of coordinates as column vectors and for the moment indicate this association of vectors and n-tuples with an arrow. Thus the notation

$$\xi \to (x^1, x^2, \ldots, x^n)^T_\alpha$$

implies that ξ has the coordinates $x^1, x^2, \ldots, x^n$ relative to $\{\alpha_1, \alpha_2, \ldots, \alpha_n\}$, that is, $\xi = \sum x^i\alpha_i$. If there is no ambiguity about the basis, the subscript α can be omitted. This correspondence of vectors and n-tuples is an example of a one-one correspondence as described in the following definition. First, let us remind the reader that the remainder of this section makes use of the conventions concerning functional notation discussed in Part IV of the section titled Some Notation and Conventions Used in This Book (page xiv).

DEFINITION. *A function φ on a set $M = \{a, b, \ldots\}$ to a set $M' = \{a', b', \ldots\}$ is said to be a* one-one correspondence *on M to M' if*

(i) Distinct elements in M have distinct images in M' (that is, $a \neq b$ implies $\varphi a \neq \varphi b$ or, what amounts to the same, $\varphi a = \varphi b$ implies $a = b$).

(ii) Every element of M' occurs as an image [that is, $\varphi x = a'$ has a solution x in M; by (i) the solution is unique].

It is to be observed that a one-one correspondence φ on M to M' exhibits a pairing of the elements of M with those of M', and consequently defines a correspondence on M' to M. Indeed, by (ii) above, if a' is any element of M', there exists a uniquely determined element a in M such that $\varphi a = a'$. Thus, if we associate with a' this element a, we obtain a function on M' to M which we call the *inverse function* of φ; in symbols φ^{-1}. Clearly φ^{-1} is one-one.

The symmetry that is present when a one-one correspondence on M to M' exists motivates the terminology "there exists a one-one correspondence between M and M'." There exists a one-one correspondence between two finite sets if (and only if) they have the same number of elements. Once an order is adopted in each set, the two first elements can be paired, the two second elements, etc. The familiar identification of points in the plane by their rectangular coordinates is a one-one correspondence. It is because of the one-oneness that in analytic geometry we speak of a point (x, y) as an abbreviation for the (unique) point with coordinates (x, y). Although there exists no one-one correspondence on a finite set to a proper part of itself, there always exists a one-one correspondence on an infinite set to a proper subset, as the correspondence $n \to 2n$ of the positive integers with the positive even integers illustrates.

In passing we note that the product of two one-one correspondences is a one-one correspondence. That is, if φ on M to M' is one-one, and ψ on M' to M'' is one-one, then $\psi\varphi$, the correspondence on M to M'' such that $a \rightarrow \psi(\varphi a)$, is one-one.

Returning to the one-one correspondence between vectors of V and n-tuples over F, if one decides to represent all vectors by their n-tuples of coordinates, the question arises as to what n-tuples accompany the sum of two vectors and the scalar multiple of a vector, respectively. If

$$\xi \rightarrow (x^1, x^2, \ldots, x^n)^T \qquad \text{and} \qquad \eta \rightarrow (y^1, y^2, \ldots, y^n)^T$$

the following rules are easily established:

$$\xi + \eta \rightarrow (x^1 + y^1, x^2 + y^2, \ldots, x^n + y^n)^T$$

and

$$c\xi \rightarrow (cx^1, cx^2, \ldots, cx^n)^T$$

Thus the n-tuples combine according to the rules already established for vectors in $V_n(F)$. To describe this feature of the correspondence we introduce a further definition.

DEFINITION. *A function φ on V to V', both vector spaces over F, is called an* isomorphism *between V and V' if*

(i) φ is a one-one correspondence

(ii) $\varphi(\alpha + \beta) = \varphi(\alpha) + \varphi(\beta)$

(iii) $\varphi(x\alpha) = x\varphi(\alpha)$ for all x in F

The spaces V and V' are called isomorphic *if there exists such a correspondence; in symbols, $V \cong V'$.*

Properties (ii) and (iii), which are equivalent to the single relation

$$\varphi(x\alpha + y\beta) = x\varphi(\alpha) + y\varphi(\beta)$$

constitute the definition of a *linear function* on a vector space. The linear forms in Chap. 1 are linear functions in the sense of this definition.

In contrast to the above definition of one-one correspondence, which was stated for arbitrary sets, that of isomorphism has been restricted to vector spaces simply because these are the systems we have in mind at the moment. Actually, the notion of isomorphism is applicable to all types of algebraic systems. For example, applied to fields, it reads: Two fields F and F' are isomorphic if there exists a one-one correspondence $a \rightarrow a'$ between the elements such that, if $a \rightarrow a'$ and $b \rightarrow b'$, then $a + b \rightarrow a' + b'$ and $ab \rightarrow a'b'$. The essence of the concept in every case is that there can be found some one-one correspondence between two systems relative to which the systems behave alike with respect to the operations defined for the systems, or *the correspondence is preserved by the operations defined for the respective systems.* In the light of our remarks in Sec. 2.1 about the postulational approach, we can describe isomorphic systems as those which not only are indistinguishable from the axiomatic point of view but also are indistinguishable in every detail except possibly in the notation for, or interpretation of, the elements. For example, the vector spaces $P_2(R^\#)$ and C, of dimension 2 over $R^\#$, are isomorphic under the correspondence $a + bt \rightarrow a + bi$.

Returning to the representation of vectors by their coordinates, the preservation of the operations for n-tuples by the correspondence we have exhibited may be expressed as follows:

Theorem 2.13. A vector space of dimension n over F is isomorphic to $V_n(F)$.

Since two algebraic systems isomorphic to a common third system are isomorphic to each other, it follows that any two vector spaces V and V', both of dimension n over F, are isomorphic to each other. With this result known, it is an easy matter to see how to establish all such isomorphisms.

Theorem 2.14. If V and V' are vector spaces of dimension n over F, each one-one correspondence between a basis for V and one for V' defines an isomorphism between V and V'. Moreover all isomorphisms on V to V' are obtainable in this way.

Proof. If $\{\alpha_1, \alpha_2, \ldots, \alpha_n\}$ and $\{\alpha_1', \alpha_2', \ldots, \alpha_n'\}$ are bases for V and V', respectively, let φ denote a one-one correspondence between them. For simplicity of notation let

$$\varphi\alpha_i = \alpha_i' \qquad i = 1, 2, \ldots, n$$

We extend this function to a one-one correspondence on V to V' by defining

$$\varphi(\sum x^i\alpha_i) = \sum x^i\varphi(\alpha_i) = \sum x^i\alpha_i'$$

Clearly this is an isomorphism. Conversely, if φ is a given isomorphism on V to V' and $\{\alpha_1, \alpha_2, \ldots, \alpha_n\}$ is a basis for V, set $\varphi\alpha_i = \alpha_i'$, $i = 1, 2, \ldots, n$. Then $\{\alpha_1', \alpha_2', \ldots, \alpha_n'\}$ is a basis for V', and since φ is linear, $\varphi(\sum x^i\alpha_i) = \sum x^i\varphi(\alpha_i) = \sum x^i\alpha_i'$, that is, φ is an isomorphism of the type constructed above. This completes the proof.

Theorem 2.13 can also be described as stating that the space $V_n(F)$ is a model for every vector space of dimension n over F. One might infer from this, together with the previous statement that isomorphic systems differ only in superficial details, that henceforth we can confine our attention to the spaces $V_n(F)$. Actually, however, the most important properties of vectors and vector spaces are those which are independent of coordinate systems, in other words, those which are invariant under (unchanged by) isomorphism. The correspondence between V and $V_n(F)$ in Theorem 2.13 was established, however, by choosing a basis for V. Thus if we were always to study $V_n(F)$, we would be restricted to a particular coordinate system, with the alternative of showing that our definitions and results are independent of the coordinate system for which they were stated. Moreover, many individual vector spaces would lose much of their intuitive content if regarded as spaces of n-tuples. It is with the aid of various examples of vector spaces that we gain insight into the structure of the theory as a whole. Thus, in effect, we shall ignore Theorem 2.13 and continue to discuss vector spaces without being dependent upon one fixed base.

PROBLEMS

1. An isomorphism between $V_3(R)$ and the space of polynomials in Prob. 9*a*, Sec. 2.4, is to be constructed. If we set $(1, 0, 1) \rightarrow 1$ and $(2, 3, -1) \rightarrow 2^{\frac{1}{3}}$ what choices are there for vectors to match with $2^{\frac{2}{3}}$?

2. The set of all pairs of complex numbers is a vector space V of dimension 4 over the real field. As such it is isomorphic to $V_4(R^\#)$. If possible, construct an isomorphism between these spaces so that the subspace $S = [(0, 1, 0, -1), (1, 0, -1, 0)]$ of $V_4(R^\#)$ is associated with the subspace $T = [(1, 0), (0, 1)]$ of V.

3. Let φ denote a function on a finite set to itself. Show that either of the two requirements for a one-one correspondence implies that φ is one-one and hence that each implies the other.

2.6. Equivalence Relations. We interrupt our study of vector spaces to present in this section the concept of an equivalence relation. It is introduced at this time, not only for future applications, but also to interpret results thus far obtained. This notion is a prerequisite for a careful phrasing, as well as a full understanding, of a great variety of statements in mathematics, particularly in algebra. It is a product of the postulational approach; its repeated occurrence in different connections warrants its examination divorced from any particular example.

As a preliminary, we formulate the notion of a *binary relation* $\mathcal{R}$ over a set $S = \{a, b, c, \ldots\}$ as any function on the product set $S \times S$ to the set consisting of two elements: {true, false}. In the event $\mathcal{R}(a, b) =$ true, we write

$$a\mathcal{R}b$$

and in the remaining case

$$a\mathcal{R}'b$$

EXAMPLE 2.8

Instances of binary relations are:

(i) T_α similar to T_β, in the set of all plane triangles,

(ii) A is obtainable from B by a rotation and translation in the set of all proper conic sections,

(iii) a divides b, or $a < b$, or a is relatively prime to b, or $a \equiv b(\text{mod } m)$ in the set of integers.

DEFINITION. *A binary relation* $\mathcal{R}$ *over a set* S *is called an* equivalence relation *if it satisfies the following postulates:*

E_1. *Reflexivity*: $a\mathcal{R}a$ *for every* a *in* S

E_2. *Symmetry*: $a\mathcal{R}b$ *implies* $b\mathcal{R}a$

E_3. *Transitivity*: $a\mathcal{R}b$ *and* $b\mathcal{R}c$ *imply* $a\mathcal{R}c$

These are the three properties that were listed in Sec. 1.9 for the binary relation $a \equiv b(\text{mod } p)$; hence in the set of integers, congruence modulo p (or, more generally, modulo m for any integer m) is an equivalence relation. Every equivalence relation over S defines a partition of S (just as congruence modulo m defines a partition of the integers), where a *partition* P of S is a collection of nonempty subsets (classes) $\{A, B, \ldots\}$ such that each element of S belongs to one and only one subset.

Theorem 2.15. An equivalence relation $\mathcal{R}$ over a set S defines a partition of S into classes $\bar{a}$, where the class $\bar{a}$ determined by a consists of all x in S for which $x\mathcal{R}a$. Conversely, a partition of S defines an equivalence relation $\mathcal{R}$, where $a\mathcal{R}b$ if and only if a and b belong to the same class.

Proof. By E_1, a is a member of $\bar{a}$, so that every element is in at least one class. But if a is in both $\bar{b}$ and $\bar{c}$ so that $a\mathcal{R}b$ and $a\mathcal{R}c$, E_2 implies $b\mathcal{R}a$ and then, using E_3, $b\mathcal{R}c$. Hence $\bar{b} = \bar{c}$, and a is in only one class. For the converse, it is clear that E_1, E_2, E_3 are satisfied.

In Example 2.8, (i) and (ii) are equivalence relations, while in (iii) only congruence is. It is well to point out here that some specific instances of equivalence relations have well-established names, *e.g.*, similarity of triangles, congruence modulo m; but in general an equivalence relation is called just that. As another example, consider the set of all finite dimensional vector spaces over a field F. The relation of isomorphism is an equivalence relation over this set. According to a remark just after Theorem 2.13, an equivalence class can be described simply as the set of all vector spaces having a given dimension.

Finally, in the set of all systems of linear equations in $x^1, x^2, \ldots, x^n$ with coefficients in a fixed field F, our earlier definition of the equivalence of two systems is seen to be an actual equivalence relation. An equivalence class consists of all systems with the same set of solutions.

It should be clear from these examples that an equivalence relation $\mathcal{R}$ over a set S is the formulation of a certain likeness exhibited by various elements in S and that an equivalence class consists of all elements of S indiscernible with respect to $\mathcal{R}$. An equivalence relation is frequently called an equality relation since it does equate elements that, although originally unequal, are equal with respect to $\mathcal{R}$. The adoption of $\mathcal{R}$ as an equality relation in S amounts to the transition to the set, which we suggestively denote by $S/\mathcal{R}$, whose elements are the various equivalence classes. Recall, for example, the construction of the rational numbers from the integers where we start with pairs (a, b) of integers with $b \neq 0$, and define an equality relation by the rule $(a, b) = (c, d)$ if and only if $ad = bc$. Then a rational number is defined as a class of equal pairs. In practice we actually work not with equivalence classes but with representatives of these classes. Since an equivalence class is defined by any one of its members a representative may be chosen at will. A set constructed by choosing in any manner a single element from each equivalence class is called a *canonical set* under $\mathcal{R}$. A member a of a canonical set is called a *canonical* (or *normal*) *form* under $\mathcal{R}$ of each member of the equivalence class determined by a.

Frequently the natural definition for an equivalence relation $\mathcal{R}$ over a set S does not place in evidence a *useful* algorithm for deciding whether or not $a\mathcal{R}b$ for a specified pair of elements $\{a, b\}$. This state of affairs creates one of the two important problems in connection with $\mathcal{R}$, *viz.*, that of deriving a characterization (in the form of a necessary and sufficient condition that $a\mathcal{R}b$) that will meet this need. The other important problem is that of obtaining a canonical set under $\mathcal{R}$ whose members are the simplest possible class representatives. Here, simplicity is a relative notion that is gauged in terms of the study that motivated the introduction of $\mathcal{R}$.

A valuable technique for treating these two major problems for an equivalence relation $\mathcal{R}$ over S is based upon the notion of an *$\mathcal{R}$-invariant* (or invariant under $\mathcal{R}$) which we define as a function f on S such that $f(a) = f(b)$ if $a\mathcal{R}b$. According to this definition, a function f on S such that $f(a)$ is independent of a is an $\mathcal{R}$-invariant, indeed, for every relation $\mathcal{R}$†; since such invariants do not discriminate among the elements of S, they have no value in this discussion and we shall ignore them. A set of $\mathcal{R}$-invariants $\{f_1, f_2, \ldots, f_k\}$ is said to be *complete* if the (ordered) set of values $(f_1(a), f_2(a), \ldots, f_k(a))$ coincides with the set of values $(f_1(b), f_2(b), \ldots, f_k(b))$ only if $a\mathcal{R}b$. Thus, by definition, a complete set of $\mathcal{R}$-

†For example, if we write all elements of S using a red pencil, then the function f, such that $f(a)$ is the color of a, is an invariant under any $\mathcal{R}$, since $f(a)$ = red for all a in S.

invariants characterizes $\mathcal{R}$. Often it is possible to discover a complete set of $\mathcal{R}$-invariants which describes, without further computations, a uniquely determined member in each equivalence class, and, consequently, a canonical set under $\mathcal{R}$. In this event both of the major problems are solved simultaneously.

Beginning in Chap. 3, numerous examples of the foregoing will be encountered; at the moment we may mention the equivalence relation of isomorphism over the set of all finite dimensional vector spaces over a field F. In this case, dimension is an isomorphism invariant; moreover, according to Theorem 2.13, the set with this function as its sole member is a complete set of invariants. There is a single member of each equivalence class that is determined by this invariant alone, *viz.*, $V_n(F)$. Because of the simplicity of this model for an n-dimensional space, we bestow upon it the title "canonical form under isomorphism," and upon the collection of spaces $V_n(F)$, $n = 1, 2, \ldots$, the title "a canonical set under isomorphism."

An alternative approach to the solution of the dominant problems for an equivalence relation is a direct attack upon the derivation of a canonical set, together with an algorithm for finding the canonical form of a given element. If this is solved, one has available the criterion that $a\mathcal{R}b$ if and only if a and b have the same canonical form. The material of Chap. 1 furnishes an excellent example of this. Indeed, starting with the set S of all consistent systems of linear equations in $x^1, x^2, \ldots, x^n$ with coefficients in a field F, and the definition that two systems are equivalent if and only if their solutions coincide, the principal topic there is the description of a canonical set (the echelon systems) and a procedure for obtaining the canonical form of (*i.e.*, the echelon system equivalent to) a given system. Once the canonical form of a system is known, the solutions of the system are available, so that for the problem at hand the canonical set of echelon systems is the simplest possible set of representatives.

PROBLEM

1. Decide which of the following relations are equivalence relations over the indicated set:

(*a*) $a \leqq |b|$ in the set of integers

(*b*) $|a - b| \leqq 1$ in the set of integers

(*c*) a is parallel to b in the set of all lines in the Euclidean plane

(*d*) $(a, b) = (c, d)$ if and only if $ad = bc$ in the set of all pairs (x, y) of integers for which $y \neq 0$

2.7. Direct Sums and Quotient Spaces. The operation of addition of subspaces has already been defined (Theorem 2.3), and we shall now discuss a special case which might arise if the question is asked as to when the dimension of the sum of two subspaces S and T is the sum of the respective dimensions: $d[S + T] = d[S] + d[T]$. According to Theorem 2.12, this is true if and only if $d[S \cap T] = 0$, which in turn holds if and only if $S \cap T = O$, the subspace consisting of the zero vector alone.

DEFINITION. *The sum of two subspaces S and T of a vector space is called a* direct sum, *in symbols $S \oplus T$, if $S \cap T = O$. Each summand is called the* complement *of the other in the sum.*

Theorem 2.16. If S and T are subspaces of V, the following statements are equivalent:

(i) The sum of S and T is direct.

(ii) Each vector in $S + T$ has a unique representation in the form $\sigma + \tau$ with σ in S and τ in T.

(iii) A basis for S together with a basis for T is a basis for $S + T$.

(iv) $d[S + T] = d[S] + d[T]$.

Proof. We shall prove the implications, (i) implies (ii) implies (iii) implies (iv) implies (i).

(i) *implies* (ii). If $S \cap T = O$, an equality $\sigma_1 + \tau_1 = \sigma_2 + \tau_2$ with σ_i in S and τ_i in T, $i = 1, 2$, when rewritten as $\sigma_1 - \sigma_2 = \tau_2 - \tau_1$ implies $\sigma_1 - \sigma_2 = \tau_2 - \tau_1 = 0$, whence the uniqueness follows.

(ii) *implies* (iii). A basis $\{\alpha_1, \alpha_2, \ldots, \alpha_m\}$ for S together with a basis $\{\beta_1, \beta_2, \ldots, \beta_n\}$ for T clearly spans $S + T$. Moreover, since the zero vector has the unique (by assumption) representation $0 + 0 = 0$, an equation of the form $\sum x^i\alpha_i + \sum y^i\beta_i = 0$ implies $\sum x^i\alpha_i = 0$ and $\sum y^i\beta_i = 0$. These in turn imply that $x^i = y^i = 0$, so that the α_i's together with the β_i's constitute an independent set.

(iii) *implies* (iv). Obvious.

(iv) *implies* (i). According to Theorem 2.12, (iv) implies that $d[S \cap T] = 0$, which in turn implies $S \cap T = O$.

The notion of direct sum is extended to more than two summands by the definition: A sum $S = S_1 + S_2 + \cdots + S_m$ of subspaces is a direct sum (in symbols, $S = S_1 \oplus S_2 \oplus \cdots \oplus S_m$) if the intersection of S_i with the sum of the remaining spaces is O, $i = 1, 2, \ldots, m$. This condition ensures the uniqueness of the representation of each vector ξ in S as a sum $\xi = \sum \xi_i$ with ξ_i in S_i. The preceding theorem can be easily extended to this case.

Now let S with basis $\{\alpha_1, \alpha_2, \ldots, \alpha_r\}$, $r < n$, be any subspace of a vector space V of dimension n. Extend this set to a basis for V by adjoining $\{\alpha_{r+1}, \alpha_{r+2}, \ldots, \alpha_n\}$ and set $T = [\alpha_{r+1}, \alpha_{r+2}, \ldots, \alpha_n]$. Then

$$V = S \oplus T$$

by virtue of (iii) of Theorem 2.16. This relation is also described by stating that V *is decomposed into the direct sum* of S and T. The choice of T is not unique for a fixed S since a basis for S can be extended to one for V in many ways.

EXAMPLE 2.9

$V_3(R^\#)$ is the direct sum of the two-dimensional subspace S, consisting of the vectors in a fixed plane through the origin, and the one-dimensional subspace T, generated by *any* one vector not in S.

The direct-sum decomposition of a vector space is an important tool in the study of a vector space V, since if $V = S_1 \oplus S_2 \oplus \cdots \oplus S_m$, properties of V can be deduced from those of the summands S_i.

Although a subspace S of a vector space V does not define a unique complement in V, all complements are isomorphic since they have the same dimension, namely, $d[V] - d[S]$.

Using S in a different way, it is possible to construct a unique vector space that is a model for these various complements. We begin by defining in V an equivalence relation in terms of S that is similar to the equivalence relation that we have called congruence modulo m in the set of integers. Because of this similarity the same notation will be used.

DEFINITION. *Let S denote a subspace of a vector space V. Two vectors ξ and η are* congruent modulo S, *in symbols $\xi \equiv \eta (\text{mod } S)$, if $\xi - \eta$ is in S.*

This is an equivalence relation over S. For $\xi \equiv \xi (\text{mod } S)$ is the statement that 0 is in S. That $\xi \equiv \eta$ implies $\eta \equiv \xi$ is the statement that, along with $\xi - \eta$, its negative $\eta - \xi$ is in S. Finally, $\xi \equiv \eta$ and $\eta \equiv \zeta$ imply that $\xi - \eta$ and $\eta - \zeta$, and hence their sum $\xi - \zeta$, is in S. This means that $\xi \equiv \zeta (\text{mod } S)$. As an equivalence relation, congruence modulo S determines a partition of the set V into so-called *residue classes* $\bar{\xi}$ modulo S. The residue class $\bar{\xi}$ containing ξ consists of all elements $\xi + \sigma$ with σ in S. This set is usually written $\xi + S$.

Next we shall define compositions such that the set V/S of residue classes modulo S, together with these compositions, is a vector space over F; the analogy with the definitions of compositions in the set J_p of Sec. 1.9 should be noted. Since $\xi \equiv \xi'$ implies $\xi + \zeta \equiv \xi' + \zeta$, it is possible to define addition of classes unambiguously by the rule

$$\bar{\xi} + \bar{\eta} = \overline{\xi + \eta}$$

To say that addition is unambiguously defined is to say that the definition is independent of the representative used for describing the classes. Let us verify this. If

$$\bar{\xi} = \overline{\xi'} \qquad \text{and} \qquad \bar{\eta} = \overline{\eta'}$$

so that

$$\xi \equiv \xi' \quad \text{and} \quad \eta \equiv \eta' (\text{mod } S)$$

then $\xi + \eta \equiv \xi' + \eta$ and $\eta + \xi' \equiv \eta' + \xi'$, upon adding η to the first and ξ' to the second of the above congruences. Since congruence is a transitive relation, it follows that $\xi + \eta \equiv \xi' + \eta'$ or $\overline{\xi + \eta} = \overline{\xi' + \eta'}$, which means that the sum $\bar{\xi}' + \bar{\eta}'$ determines the same class as $\bar{\xi} + \bar{\eta}$.

Since $\xi \equiv \xi' (\text{mod } S)$ implies that $c\xi \equiv c\xi' (\text{mod } S)$ for any scalar c, a scalar multiplication can be defined unambiguously for classes by the equation

$$c\bar{\xi} = \overline{c\xi} \qquad c \text{ in } F$$

It is a routine matter to prove that the set V/S together with the compositions

$$\bar{\xi} + \bar{\eta} = \overline{\xi + \eta} \qquad c\bar{\xi} = \overline{c\xi} \tag{1}$$

is a vector space over F. We call this space the *quotient space of V modulo S.*

Before presenting an example of this notion, we wish to point out an interesting relation between V and V/S. The make-up of an element of V/S suggests a "natural correspondence," φ let us say, on V onto V/S, namely

$$\varphi\xi = \bar{\xi} \qquad \text{for } \xi \text{ in } V$$

According to equation (1) above, φ is a linear function (see Sec. 2.5), but in general is not one-one (hence not an isomorphism), rather many-one. Such a correspondence between vector spaces is of sufficient importance to warrant a name.

DEFINITION. *A function φ on V onto V', both vector spaces over F, is called a* homomorphism *on V onto V' if it is preserved by the operations of addition and scalar multiplication, in other words, is linear. The space V' is called a* homomorphic image *of V; in symbols, $V \simeq V'$.*

The concept of homomorphism is weaker than that of isomorphism; every isomorphism is a homomorphism, but not conversely. We can now describe the quotient space V/S as a homomorphic image of V. The foregoing is summarized in the following theorem:

Theorem 2.17. If S is a subspace of V over F, the set V/S of residue classes $\bar{\xi}$ of V modulo S, together with the compositions

$$\bar{\xi} + \bar{\eta} = \overline{\xi + \eta} \qquad c\bar{\xi} = \overline{c\xi}$$

is a vector space over F. This quotient space is a homomorphic image of V under the correspondence $\xi \to \bar{\xi}$.

EXAMPLE 2.10

In $V_n(R^\#)$ let S be the subspace of all column vectors of the form $(0, 0, \ldots, 0, x^{r+1}, \ldots, x^n)^T$. Then $\alpha \equiv \beta(\text{mod } S)$ if and only if their first r components agree, and if $\alpha = (a^1, a^2, \ldots, a^r, \ldots, a^n)^T$, the equivalence class $\bar{\alpha}$ consists of all vectors whose first r components are $a^1, a^2, \ldots, a^r$, respectively. Thus we can effectively obtain V/S from S by ignoring the last $n - r$ components of vectors in $V_n(R^\#)$. If $n = 3$ and $r = 2$, so that the space is ordinary Euclidean space and S is the x^3 axis, the residue classes are the lines in $V_3(R^\#)$ parallel to the x^3 axis. If we take as the representative point in a line the point of intersection of the line with the x^1x^2 plane, then the formulas in Theorem 2.17 state that compositions for the vector space V/S of lines are the same as those for the vector space of points of the x^1x^2 plane. The correspondence $\xi \to \bar{\xi}$ of Theorem 2.17 is merely the mapping of all points of a line parallel to the x^3 axis on that line.

Now for the relation between V/S and a complement of S in V.

Theorem 2.18. Let S denote a subspace of V and T a complement of S in V. Then $V/S \cong T$; in particular $d[V/S] = d[V] - d[S]$.

Proof. If $\xi = \sigma_1 + \tau_1$ and $\eta = \sigma_2 + \tau_2$ with σ_i in S and τ_i in T, $i = 1, 2$, then $\xi \equiv \eta(\text{mod } S)$ if and only if $\tau_1 = \tau_2$ since $\xi - \eta = (\sigma_1 - \sigma_2) + (\tau_1 - \tau_2)$ in S implies $\tau_1 - \tau_2$ in S, which in turn implies $\tau_1 = \tau_2$ because $S \cap T = O$. Reversing these steps gives the converse. Thus there exists a one-one correspondence

$$\bar{\xi} \to \tau \qquad \text{if } \xi = \sigma + \tau$$

between V/S and T. This correspondence is an isomorphism since $\bar{\xi} \to \tau_1$ and $\bar{\eta} \to \tau_2$ imply $\overline{\xi + \eta} \to \tau_1 + \tau_2$ and $\overline{c\xi} \to c\tau_1$.

PROBLEM

1. In $V_4(R)$ let S be the subspace with basis $\{(2, 3, -1, 0), (1, -1, 0, 1)\}$. Describe V/S.

CHAPTER 3

BASIC OPERATIONS FOR MATRICES

3.1. The Matrix of a System of Linear Equations. In solving a system of linear equations by the elimination method of Chap. 1, it becomes evident that the unknowns serve merely as marks of position and that the computations are performed upon the array of coefficients. Once an order for the unknowns and the equations has been selected, it becomes a timesaving device to list the coefficients in an orderly array and operate upon the rows.

DEFINITION. *An* $m \times n$ matrix A *over a field* F *is a rectangular array of mn elements* a^i_j *in* F, *arranged in m rows and n columns as follows*:

$$(1) \qquad A = \begin{pmatrix} a^1_1 & a^1_2 & \cdots & a^1_n \\ a^2_1 & a^2_2 & \cdots & a^2_n \\ \cdot & \cdot & \cdots & \cdot \\ a^m_1 & a^m_2 & \cdots & a^m_n \end{pmatrix}$$

The line of coefficients

$$\rho^i = (a^i_1, a^i_2, \ldots, a^i_n)$$

is called the ith *row* of A, and the line

$$\gamma_j = \begin{pmatrix} a^1_j \\ a^2_j \\ \cdot \\ a^m_j \end{pmatrix}$$

is called the jth *column* of A. Thus an $m \times n$ matrix can be regarded as a collection of m row vectors ρ^i or, alternatively, as a collection of n column vectors γ_j. The entries of A are so indexed that the element a^i_j occurs at the intersection of the ith row and jth column. Without exception the upper index of a^i_j denotes the row, while the lower denotes the column in which it occurs. The whole matrix may be written compactly as

$$(a_j^i) \qquad i = 1, 2, \ldots, m; j = 1, 2, \ldots, n$$

A matrix with the same number of rows as columns is called *square,* and the common number is the *order* of the matrix. Since A in (1) is the array of coefficients of the unknowns in the typical system of linear equations (N) of Chap. 1, it is called the *coefficient matrix* of (N). If A is extended to $n + 1$ columns by adjoining the column vector $Y = (y^1, y^2, \ldots, y^m)^T$ of (N), the resulting matrix is called the *augmented matrix* of (N).

Sometimes we shall prefer to write the matrix A of (1) with double subscripts:

$$A = \begin{pmatrix} a_{11} & a_{12} & \cdots & a_{1n} \\ a_{21} & a_{22} & \cdots & a_{2n} \\ \cdot & \cdot & \cdots & \cdot \\ a_{m1} & a_{m2} & \cdots & a_{mn} \end{pmatrix}$$

When this is done, it will always be understood that a_{ij} is the element at the intersection of the ith row and jth column.

In this chapter the basic properties of, and operations for, matrices will be discussed in the light of their applications. In the first part of the chapter the applications to systems of linear equations provide the motivation; then a problem in connection with coordinate systems in vector spaces suggests additional notions for study.

3.2. Special Matrices. To expedite a systematic discussion of matrices, it is helpful to have available a certain amount of terminology, part of which we describe in this section.

Zero Matrix. An $m \times n$ matrix over F, each of whose entries is the zero element of F, is called a zero matrix and denoted by 0. No misunderstanding will result in using just this one symbol to denote every zero matrix.

Triangular Matrix. In the matrix A of (1), the line of elements a_1^1, a_2^2, . . . is called the *principal diagonal* of A. If A is square, these terms occur upon the geometric diagonal from the upper left corner to the lower right corner. A square matrix is called *upper triangular,* or *superdiagonal,* if all entries below the diagonal are zero; if all entries above the diagonal are zero, it is called *lower triangular,* or *subdiagonal.*

Diagonal Matrix. A square matrix $D = (d_j^i)$ is called diagonal if $d_j^i = 0$ for $i \neq j$, that is, if it is both upper and lower triangular. The following notation can be used for such a matrix:

$$D = \text{diag}\,(d_1^1, d_2^2, \ldots, d_n^n)$$

Scalar Matrix. A square matrix which is diagonal and in which the diagonal terms are equal to each other is called a scalar matrix.

Identity Matrix. A scalar matrix in which the common diagonal element is 1, the unit element of F, is called an identity matrix and will always be denoted by I. If necessary, a subscript will be added to indicate the number of rows and columns in I. For example,

$$I_3 = \begin{pmatrix} 1 & 0 & 0 \\ 0 & 1 & 0 \\ 0 & 0 & 1 \end{pmatrix}$$

In terms of the *Kronecker delta* δ_j^i (or δ_{ij}), which is defined by the equation

$$\delta_{ij} = \delta_j^i = \begin{cases} 1 & \text{if } i = j \\ 0 & \text{if } i \neq j \end{cases}$$

a diagonal matrix, scalar matrix, and identity matrix, respectively, can be written as

$$(\delta_j^i a_j^i) \qquad (\delta_j^i c) \qquad (\delta_j^i)$$

Submatrix. The subarray obtained from a matrix A by discarding one or more rows and/or one or more columns of A is called a submatrix of A. For example, if

$$A = \begin{pmatrix} a_1^1 & a_2^1 & a_3^1 & a_4^1 \\ a_1^2 & a_2^2 & a_3^2 & a_4^2 \\ a_1^3 & a_2^3 & a_3^3 & a_4^3 \end{pmatrix} \qquad B = \begin{pmatrix} a_1^1 & a_3^1 \\ a_1^3 & a_3^3 \end{pmatrix}$$

then B is a submatrix of A.

Transpose. The matrix obtained from a matrix A by interchanging the rows and columns of A is called the transpose of A and denoted by A^{T}. Thus, if $A = (a_j^i)$ and $A^{\mathrm{T}} = (b_j^i)$, then $b_j^i = a_i^j$. It is obvious that $(A^{\mathrm{T}})^{\mathrm{T}} = A$. Our agreement earlier, to write a column vector as a row vector with a superscript T, anticipated the definition of transpose.

Partition of a Matrix. Let A denote an $m \times n$ matrix (a_j^i), and let

$$m = m_1 + m_2 + \cdots + m_r \qquad n = n_1 + n_2 + \cdots + n_s$$

be any ordered partitions of m and n, respectively. Drawing horizontal lines between the rows numbered m_1 and $(m_1 + 1)$, $m_1 + m_2$ and $(m_1 + m_2 + 1)$, etc., as well as vertical lines between the columns numbered n_1 and $(n_1 + 1)$, $n_1 + n_2$ and $(n_1 + n_2 + 1)$, etc., partitions the given array into subarrays A_j^i such that A_j^i has m_i rows and n_j columns. We write

$$A = (A_j^i) \qquad i = 1, 2, \ldots, r; j = 1, 2, \ldots, s$$

and say that A has been partitioned into the submatrices A_j^i.

Below is indicated the partition of a 4×5 matrix accompanying the partitions $4 = 2 + 2$ and $5 = 1 + 2 + 2$.

$$A = \left(\begin{array}{c|cc|cc} a_1^1 & a_2^1 & a_3^1 & a_4^1 & a_5^1 \\ a_1^2 & a_2^2 & a_3^2 & a_4^2 & a_5^2 \\ \hline a_1^3 & a_2^3 & a_3^3 & a_4^3 & a_5^3 \\ a_1^4 & a_2^4 & a_3^4 & a_4^4 & a_5^4 \end{array}\right) = (A_j^i) \qquad \text{where } A_1^1 = \begin{pmatrix} a_1^1 \\ a_1^2 \end{pmatrix}, \text{ etc.}$$

This notion is useful as a notational device to indicate special features of a matrix. For example, the form of the matrices

$$\begin{pmatrix} a_{11} & a_{12} & \cdots & a_{1n} \\ \cdot & \cdot & \cdots & \cdot \\ a_{r1} & a_{r2} & \cdots & a_{rn} \\ 0 & 0 & \cdots & 0 \\ \cdot & \cdot & \cdots & \cdot \\ 0 & 0 & \cdots & 0 \end{pmatrix} \qquad \begin{pmatrix} 1 & 0 & 0 & 0 & \cdots & 0 \\ 0 & 1 & 0 & 0 & \cdots & 0 \\ 0 & 0 & 1 & 0 & \cdots & 0 \\ 0 & 0 & 0 & 0 & \cdots & 0 \\ \cdot & \cdot & \cdot & \cdot & \cdots & \cdot \\ 0 & 0 & 0 & 0 & \cdots & 0 \end{pmatrix}$$

can be indicated briefly by

$$\begin{pmatrix} A_1 \\ 0 \end{pmatrix} \qquad \begin{pmatrix} I_3 & 0 \\ 0 & 0 \end{pmatrix}$$

respectively, in this notation. Again, if in the above partition (A_j^i) of A, $r = s$, and each submatrix A_j^i is a zero matrix when $i \neq j$, we shall write

$$A = \operatorname{diag}(A_1^1, A_2^2, \ldots, A_r^r)$$

For instance,

$$\begin{pmatrix} I_3 & 0 \\ 0 & 0 \end{pmatrix} = \operatorname{diag}(I_3, 0)$$

As a final example, if A is the coefficient matrix of a linear system (N) and Y the column vector whose components are the right-hand members of the equations, the augmented matrix of the system can be written $(A\ Y)$.

Notice that if

$$A = (A_j^i) \qquad i = 1, 2, \ldots, r; j = 1, 2, \ldots, s$$

then

$$A^{\mathrm{T}} = (B_j^i),\ B_j^i = (A_i^j)^{\mathrm{T}} \qquad i = 1, 2, \ldots, s; j = 1, 2, \ldots, r$$

3.3. Row Equivalence of Matrices. It has been mentioned that performing the elementary operations of Chap. 1 upon the equations of a linear system amounts to operating upon the rows of the augmented matrix of the system. These operations (with those of type IV omitted since we prefer not to alter the dimensions of a matrix) are now described in terms of their effect upon a matrix and given the name of *elementary row transformations* of a matrix.

Type I. The interchange of two row vectors.

Type II. The multiplication of a row vector by a nonzero scalar.

Type III. The addition to any row vector of a scalar multiple of another row vector.

If ρ^i denotes the *i*th row of a matrix A, the following symbols will be used for the above transformations:

R_{ij} = the transformation which interchanges ρ^i and ρ^j

$R_i(c)$ = the transformation which replaces ρ^i by $c\rho^i$

$R_{ij}(c)$ = the transformation which replaces ρ^i by $\rho^i + c\rho^j$

The effect of each of these transformations can be annulled by one of the same type. Indeed, if R_{ij} is applied to A and then R_{ij} is applied to the new matrix, the result is just A. Similarly $R_i(c)$ followed by $R_i(c^{-1})$, as well as $R_{ij}(c)$ followed by $R_{ij}(-c)$, yields the original matrix.

DEFINITION. *Let A and B be $m \times n$ matrices over F such that B results from A by the application of a finite number of elementary row transformations to A. Then B is* row-equivalent *to A; in symbols, $A \sim B$.*

Theorem 3.1. Row equivalence is an equivalence relation over the set of all $m \times n$ matrices over F.

Proof. This is left as an exercise.

In view of the properties of an equivalence relation, we can use the phrase "A and B are row-equivalent" in place of "A is row-equivalent to B." Also, to show that $A \sim B$, it is sufficient to show that both are equivalent to a third matrix C.

Theorem 3.2. If A and B are $m \times n$ matrices having the form

$$A = \begin{pmatrix} A_1 & 0 \\ 0 & A_2 \end{pmatrix} \qquad B = \begin{pmatrix} B_1 & 0 \\ 0 & B_2 \end{pmatrix}$$

where $A_i \sim B_i$, $i = 1, 2$, then $A \sim B$.

Proof. This is left as an exercise.

Next we devise a canonical set for $m \times n$ matrices over a field under row equivalence. This may be done (following one of the procedures discussed in Sec. 2.6) using the notion of an echelon matrix, which is described below, and the theorem following which asserts that a class of row-equivalent matrices contains a single echelon matrix. Thus the echelon matrices constitute a canonical set for $m \times n$ matrices under row equivalence; an echelon matrix will be called a canonical form under row equivalence. To describe this representative, it is convenient to apply the definitions in Sec. 1.4 of length of a linear form and normalized form to row vectors; clearly this may be done.

DEFINITION. *An $m \times n$ matrix E is called an* echelon matrix *if the following conditions are satisfied:*

(*i*) *Each row vector of positive length is normalized.*

(*ii*) *If k_i is the length of the ith row, then $k_1 \leqq k_2 \leqq \cdots \leqq k_m$.*

(*iii*) *If $k_i > 0$, so that the (i, k_i)th entry in E is 1, the remaining entries in the k_ith column are zeros (and hence $k_{i-1} < k_i$).*

Of course this is a description of the coefficient matrix of an echelon system, with the exception that rows corresponding to zero forms are retained. For example, the echelon matrix row-equivalent to the coefficient matrix of Example 1.8 is

$$\begin{pmatrix} 0 & 0 & 0 & 0 & 0 & 0 & 0 \\ 0 & 0 & 0 & 0 & 0 & 0 & 0 \\ 1 & 2 & 1 & 0 & 0 & 0 & 0 \\ 0 & -1 & 0 & 1 & 0 & 0 & 0 \\ -3 & -5 & 0 & 0 & 1 & 1 & 0 \\ 2 & -3 & 0 & 0 & 0 & 0 & 1 \end{pmatrix}$$

Theorem 3.3. Each class of row-equivalent $m \times n$ matrices over F contains a unique echelon matrix.

Proof. The proof is like that of Theorem 1.3. However, we present it again to illustrate the use of row transformations. Let A denote a matrix in the equivalence class. If the last column contains a nonzero entry, a type I transformation will place it in the (m, n) position and a type II transformation will change it to 1. Using type III transformations, the remaining elements in the nth column can be made zero. Thus, in all cases A is row-equivalent to a matrix having either the form $(C \ 0)$ or $\begin{pmatrix} D & 0 \\ E & 1 \end{pmatrix}$.

In the first case the above step is repeated on C, while in the second case the above step is repeated on D. If D is replaced by a matrix with 1 in its last row and column, a multiple of the corresponding row in the full matrix can be added to the last row to obtain a zero in the $(m, n-1)$st position. Repetition of this procedure leads to the echelon form.

The proof of the uniqueness is essentially that of Theorem 1.6. The agreement to retain a zero row of a matrix compensates for the lack of type IV operations. Thus, if two echelon matrices are equivalent, it is possible to transform one into the other with row transformations. Using this fact, their identity can be established.

PROBLEM

1. For each of the following matrices compute the echelon matrix that is row-equivalent to it:

$$(a)\ \begin{pmatrix} 2 & 3 & -1 \\ 2 & 6 & -3 \\ 2 & 0 & 1 \end{pmatrix} \qquad (b)\ \begin{pmatrix} 2 & 3 & -1 & 0 \\ 1 & -1 & 0 & -1 \\ -3 & 4 & 6 & 2 \end{pmatrix}$$

$$(c)\ \begin{pmatrix} 2 & 3 & 0 & 4 \\ -1 & 1 & 3 & 2 \\ 4 & 0 & 2 & -3 \\ 2 & -3 & 1 & 6 \end{pmatrix}$$

3.4. Rank of a Matrix. It is desirable to introduce the notion of rank of a matrix A at this time. Since A is a collection of row vectors ρ^i, a definition can be phrased in terms of the maximum number of linearly independent row vectors or, what amounts to the same (Theorem 2.6), the dimension of the space $[\rho^1, \rho^2, \ldots, \rho^m]$. Then the rank would coincide with that of a system of linear equations having A as coefficient matrix. On the other hand, there is no reason to give preference to the interpretation of A in terms of row vectors over that in terms of column vectors γ_i.

DEFINITION. *The* $\begin{Bmatrix}\text{row}\\ \text{column}\end{Bmatrix}$ space *of an $m \times n$ matrix A over F is the subspace* $\begin{Bmatrix}[\rho^1, \rho^2, \ldots, \rho^m]\\ [\gamma_1, \gamma_2, \ldots, \gamma_n]\end{Bmatrix}$ *of* $\begin{Bmatrix}V_n(F)\\ V_m(F)\end{Bmatrix}$. *The* dimension *of the* $\begin{Bmatrix}\text{row}\\ \text{column}\end{Bmatrix}$ *space is called the* $\begin{Bmatrix}\text{row}\\ \text{column}\end{Bmatrix}$ rank *of A.*

Actually these two ranks agree. For just as the row rank of A is the rank of the homogeneous system

$$\text{(H)} \qquad \begin{aligned} a_1^1x^1 + a_2^1x^2 + \cdots + a_n^1x^n &= 0 \\ \cdots\cdots\cdots\cdots\cdots\cdots\cdots\cdots \\ a_1^mx^1 + a_2^mx^2 + \cdots + a_n^mx^n &= 0 \end{aligned}$$

so is the column rank of A the rank of the associated transposed system (H^T) with coefficient matrix A^T, and prior to Theorem 1.14 it was shown that (H) and (H^T) have the same rank. We shall state this result as a theorem and give a short proof which does not rely on Chap. 1.

Theorem 3.4. The row rank and column rank of a matrix A are one and the same number which is called the *rank* of A; in symbols, $r(A)$.

Proof. The theorem is trivial if the row rank r is equal to zero, since then $A = 0$; so assume that $r > 0$. Obviously, rearranging the rows of A does not disturb the row rank, nor does it disturb the column rank c. For any linear dependence $\sum_1^n \gamma_i x^i = 0$ among the columns γ_i of A determines a solution of (H), and rearranging rows of A simply rearranges these equations. So we may assume the first r rows of A are linearly independent and state that

$$A^* = \begin{pmatrix} a_1^1 & a_2^1 & \cdots & a_n^1 \\ \cdot & \cdot & \cdots & \cdot \\ a_1^r & a_2^r & \cdots & a_n^r \end{pmatrix}$$

has the same row rank as A. Now A^* has the same column rank as A since the system of equations

$$\text{(H*)} \qquad \begin{aligned} a_1^1x^1 + a_2^1x^2 + \cdots + a_n^1x^n &= 0 \\ \cdots\cdots\cdots\cdots\cdots\cdots\cdots\cdots \\ a_1^rx^1 + a_2^rx^2 + \cdots + a_n^rx^n &= 0 \end{aligned}$$

is equivalent to (H). To show this, observe first that every solution of (H) is a solution of (H*). But conversely, since $\{\rho^1, \rho^2, \ldots, \rho^r\}$ is a basis for the row space A, each remaining $\rho^j = \sum_1^r c_i \rho^i$, that is, each of the last $n - r$ equations of (H) is a linear combination of the first r. Hence any solution of (H*) is a solution of (H).

The column rank of A^* cannot exceed r (Theorem 2.7), and hence $c \leqq r$. Applying this result to A^T gives $r \leqq c$, and equality follows.

We state again for emphasis that the rank of a system of linear equations coincides with the rank of its coefficient matrix. Using the notion of a vector space together with the previous theorem, it is an easy matter to derive the necessary and sufficient conditions stated in Chap. 1 for the solvability of the system (N). In fact the equivalence of the following statements is quickly established:

(1) (N) is solvable.

(2) $Y = (y^1, y^2, \ldots, y^m)^T$ lies in the column space of A.

(3) The column rank of A equals the column rank of $(A\ Y)$.

(4) The row rank of A equals the row rank of $(A\ Y)$.

(5) The system is consistent.

Before announcing the next result concerning rank, we wish to introduce a new idea. In the case of matrices, it is convenient to have available, in addition to the definition of an invariant under an equivalence relation (see Sec. 2.6), that of an *invariant under a matrix transformation.* By a *transformation* τ *on a set* S *of* $m \times n$ *matrices,* we understand a function on S to S; for example, an elementary row transformation is a transformation on the set of all $m \times n$ matrices. Then a function f on S is said to be invariant under τ if $f(\tau A) = f(A)$ for all A in S. To illustrate this, consider the rank function on the set of $m \times n$ matrices, *i.e.*, the function r such that $r(A)$ is equal to the rank of A. Then the following theorem is a consequence of Lemma 1.2. However, we shall supply a shorter proof.

Theorem 3.5. Rank is an invariant function of matrices under elementary row transformations.

Proof. It is sufficient to show that if τ denotes a row transformation then the row space of τA coincides with that of A. This is obvious when τ is of type I. Also since every linear combination of either of the sets

$$\{\rho^1, \ldots, c\rho^i, \ldots, \rho^n\} \qquad \{\rho^1, \rho^2, \ldots, \rho^i + c\rho^k, \ldots, \rho^n\}$$

is a linear combination of the ρ^i's, the row space is not enlarged by a transformation of type II or III. But neither can it be diminished, for, if so, it would be possible to enlarge the row space of the transformed matrix by returning to the original matrix. This, however, we have just seen is impossible.

COROLLARY. The rank function of matrices is an invariant under row equivalence.

The proof of the following result can be supplied by the reader:

Theorem 3.6. If A is row equivalent to the echelon matrix E, then $r(A) = r(E)$. The latter number, in turn, is the number of nonzero rows of E.

This result, together with Theorem 3.5, provides a practical method for computing the rank of a matrix. Reduce the matrix to echelon form, and count the number of nonzero rows. As another application, we mention the problem of finding a basis for the subspace S spanned by a given set of row vectors. If the matrix A with these vectors as its rows is constructed and reduced to an echelon matrix E, the nonzero rows form a basis for S.

EXAMPLE 3.1

To find a basis for the subspace S of $V_5(R)$ generated by the set of row vectors $\{(-3, 3, 1, -3, 2), (0, 7, 1, -2, 3), (5, -11, 2, 4, -5), (2, -1, 4, -1, 0)\}$, we reduce the matrix

$$\begin{pmatrix} -3 & 3 & 1 & -3 & 2 \\ 0 & 7 & 1 & -2 & 3 \\ 5 & -11 & 2 & 4 & -5 \\ 2 & -1 & 4 & -1 & 0 \end{pmatrix}$$

to the echelon form

$$\begin{pmatrix} 0 & 0 & 0 & 0 & 0 \\ 1 & 0 & 1 & 0 & 0 \\ 2 & 1 & 0 & 1 & 0 \\ 1 & 3 & 0 & 0 & 1 \end{pmatrix}$$

The last three rows form a basis for S.

DEFINITION. *An* $n \times n$ *matrix* A *is called* nonsingular *if* $r(A) = n$; *otherwise,* A *is* singular.

According to Theorem 1.13, a system of linear equations with a nonsingular coefficient matrix has a unique solution. This is also a consequence of the following result:

Theorem 3.7. An $n \times n$ matrix A is nonsingular if and only if the echelon matrix row equivalent to A is I_n.

Proof. If $A \sim I_n$, the assertion follows from Theorem 3.5 and the independence of the unit vectors. Conversely, if A is nonsingular, the echelon matrix E equivalent to A has rank n. Consequently E contains no zero rows, and since the final nonzero entry in each of the n rows is 1, we conclude that $E = I_n$.

PROBLEM

1. Using Theorem 3.5, find a basis for the subspace S of $V_5(R)$ spanned by the row vectors $\{(3, 2, -1, 1, 0), (6, 2, 1, 1, 0), (2, -1, 1, 0, 2), (1, 5, -3, 2, -2), (0, 2, -3, 1, 0)\}$. Find another basis among the given set of generating elements.

3.5. Equivalent Matrices. The roles played by the transposed homogeneous system (H^{T}) in the discussion of the linear system (N) suggests the introduction of elementary column transformations for a matrix. We do this and observe that all results concerning row transformations have column analogues. It is of interest to determine just how far a matrix can be simplified using both row and column transformations, or, as we shall say, *elementary transformations.*

DEFINITION. *A matrix B obtainable from the matrix A by a finite number of elementary transformations is called* rationally equivalent *to A. In symbols,* $A \overset{E}{\sim} B$.

The next theorem settles all questions concerning this relation.

Theorem 3.8. Equivalence is an equivalence relation over the set of all $m \times n$ matrices A over F. If $r(A) = r$ then A is equivalent to a single matrix of the form J_r where

$$J_0 = 0 \qquad J_r = \begin{pmatrix} I_r & 0 \\ 0 & 0 \end{pmatrix} \qquad \text{if } r > 0$$

Here the row and/or column of zero submatrices in J_r may be absent. Moreover, two $m \times n$ matrices are equivalent if and only if they have the same rank. In the terminology of Sec. 2.6, the rank function constitutes a complete set of invariants under equivalence and the set of matrices $\{J_0, J_1, J_2, \ldots\}$ is a canonical set under equivalence.

Proof. The verification of the axioms for an equivalence relation is a routine matter, and we shall omit it. To prove the next statement, let A denote an $m \times n$ matrix, and reduce A to echelon form using row transformations. Then, with column transformations, all nonzero elements apart from the final 1 in a nonzero row, can be eliminated without disturbing other elements. The result is a matrix equivalent to A and which contains in each nonzero row and nonzero column exactly one nonzero term, *viz.*, a 1. By interchanging rows and columns if necessary, these 1's can be arranged as stated to obtain a matrix J_r equivalent to A. This former matrix is the only one of its kind in the class containing it since a different matrix of this form must have a different rank, and hence be inequivalent, since rank is preserved by elementary transformations.

Finally if $A \overset{E}{\sim} B$, evidently $r(A) = r(B)$. Conversely, if $r(A) = r(B) = r$, then both are equivalent to J_r and hence to each other.

In practice one need not adhere to the method of reduction to the canonical form outlined in the above proof. Rather, one will apply row and column transformations in any convenient order.

EXAMPLE 3.2

To reduce the matrix

$$A = \begin{pmatrix} 2 & 3 & -1 & 4 \\ 1 & 2 & 0 & 1 \\ 2 & -1 & 1 & 2 \\ 5 & 2 & 7 & 3 \end{pmatrix}$$

to canonical form under elementary transformations, we can first eliminate the 2, 3, 4 in the last column with row transformations of type III. We obtain

$$\begin{pmatrix} -2 & -5 & -1 & 0 \\ 1 & 2 & 0 & 1 \\ 0 & -5 & 1 & 0 \\ 2 & -4 & 7 & 0 \end{pmatrix}$$

Then the first 1 and 2 in the second row can be replaced with zeros using column transformations of type III. Next, the -1 and 7 in the third column can be eliminated with row transformations to obtain

$$\begin{pmatrix} -2 & -10 & 0 & 0 \\ 0 & 0 & 0 & 1 \\ 0 & -5 & 1 & 0 \\ 2 & 31 & 0 & 0 \end{pmatrix}$$

Since it is now clear that the four columns (or rows) are linearly independent, we conclude without further computation that $A \overset{E}{\sim} I_4$.

The reader with a knowledge of determinant theory may raise a question concerning the definition of rank of a matrix in that theory (as the order of a nonvanishing minor of maximal order) versus the one that we have adopted. That they are in agreement may be shown quickly, using the fact that rank, in either sense, is an invariant under elementary transformations. For then we may replace a matrix A by its canonical form J_r to conclude immediately that both definitions assign the same number, r, to A.

We also mention that our definition of nonsingularity for a matrix can be restated as $\det A \neq 0$ since both are equivalent to the condition $A \overset{E}{\sim} I_n$. Proofs for these remarks are to be found in Chap. 4.

After multiplication has been defined for matrices, we shall see that the foregoing results pertaining to equivalence of matrices can be put in alternative forms.

PROBLEM

1. For each of the matrices in the problem of Sec. 3.3, determine the matrix of the form J_r equivalent to it.

3.6. Matrix Multiplication. The occurrence of matrices in connection with problems originating in the study of vector spaces suggests methods for combining matrices in useful ways. We discuss one such problem now.

First a word about notation. In future discussions concerning vector spaces it will prove convenient to use the phrase *α-basis* to indicate a basis $\{\alpha_1, \alpha_2, \ldots, \alpha_n\}$ that has already been described. Also, the phrase *α-coordinates* will be used as an abbreviation for "coordinates relative to the α-basis."

Suppose that $\{\alpha_1, \alpha_2, \ldots, \alpha_n\}$ and $\{\beta_1, \beta_2, \ldots, \beta_n\}$ are two bases for a vector space V over F and that a vector ξ in V has α-coordinates $X = (x^1, x^2, \ldots, x^n)^T$ and β-coordinates $Y = (y^1, y^2, \ldots, y^n)^T$, so that

$$\xi = \sum x^i\alpha_i = \sum y^k\beta_k$$

We ask for the relationship between the α-coordinates and the β-coordinates. Since every vector in V can be represented as a linear combination of a basis for V, it is possible to find scalars b_j^k such that

$$
\begin{aligned}
\alpha_1 &= b_1^1\beta_1 + b_1^2\beta_2 + \cdots + b_1^n\beta_n \\
\alpha_2 &= b_2^1\beta_1 + b_2^2\beta_2 + \cdots + b_2^n\beta_n \\
&\cdots\cdots\cdots\cdots\cdots\cdots\cdots\cdots \\
\alpha_n &= b_n^1\beta_1 + b_n^2\beta_2 + \cdots + b_n^n\beta_n
\end{aligned}
\tag{2}
$$

or, in compact form,

$$\alpha_j = \sum_k b_j^k\beta_k \qquad j = 1, 2, \ldots, n$$

Then

$$\sum_k y^k\beta_k = \xi = \sum_j x^j\alpha_j = \sum_j x^j\left(\sum_k b_j^k\beta_k\right) = \sum_k\left(\sum_j b_j^k x^j\right)\beta_k$$

and from the uniqueness of the coordinates of a vector relative to a basis (Theorem 2.10), we conclude that

$$y^k = \sum_i b^k_i x^i \qquad k = 1, 2, \ldots, n$$

or, in expanded form,

$$\begin{aligned} y^1 &= b^1_1 x^1 + b^1_2 x^2 + \cdots + b^1_n x^n \\ y^2 &= b^2_1 x^1 + b^2_2 x^2 + \cdots + b^2_n x^n \\ &\cdots\cdots\cdots\cdots\cdots\cdots \\ y^n &= b^n_1 x^1 + b^n_2 x^2 + \cdots + b^n_n x^n \end{aligned} \tag{3}$$

Such a system of equations should be a familiar sight by now. However, it is clear from the context that we are assigning an entirely different interpretation to it than heretofore. If we think of transferring from the α-basis to the β-basis, this system gives the new coordinates Y in terms of the old coordinates X. Thus the systems (2) and (3) together imply that if the (old) α-basis is known in terms of the (new) β-basis, the (new) β-coordinates Y are known in terms of the (old) α-coordinates X. The rows of (2) are the columns of (3), *i.e.*, one coefficient matrix is the transpose of the other. Obviously, if the coefficient matrix $B = (b^k_i)$ of (3) is known, both systems are known [*e.g.*, in (2), b^k_i is the kth component of α_i with respect to the β-basis], so that a transfer from one basis to another is characterized by a matrix.

EXAMPLE 3.3

In the Euclidean plane let $(x^1, x^2)^T$ denote the coordinates of a point with respect to a pair of rectangular axes. If $(y^1, y^2)^T$ are the coordinates of the same point after a clockwise rotation of the axes through an angle θ, then, using equations (2) and (3), we deduce that

$$\begin{aligned} y^1 &= x^1 \cos\theta - x^2 \sin\theta \\ y^2 &= x^1 \sin\theta + x^2 \cos\theta \end{aligned}$$

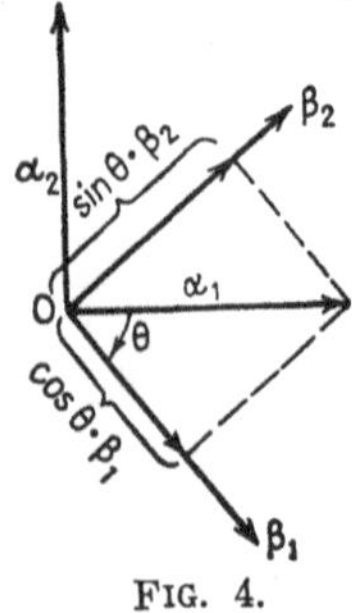

FIG. 4.

If $\{\alpha_1, \alpha_2\}$ and $\{\beta_1, \beta_2\}$ are the old and new bases, respectively, then, using equations (2) and (3), we deduce that

$$\begin{aligned} \alpha_1 &= \cos\theta\cdot\beta_1 + \sin\theta\cdot\beta_2 \\ \alpha_2 &= -\sin\theta\cdot\beta_1 + \cos\theta\cdot\beta_2 \end{aligned}$$

The vector diagram of Fig. 4 verifies the representation of α_1 in terms of β_1 and β_2.

Continuing the above problem let $\{\gamma_1, \gamma_2, \ldots, \gamma_n\}$ be a third basis for V and $\xi = \sum z^i\gamma_i$. If

$$\beta_k = \sum_i a_k^i\gamma_i \qquad k = 1, 2, \ldots, n$$

the foregoing results imply that the equations

$$z^i = \sum_k a_k^i y^k \qquad i = 1, 2, \ldots, n$$

express the γ-coordinates Z in terms of the β-coordinates Y. The matrix $A = (a_k^i)$ describes the transfer from the β-basis to the γ-basis. Consider now the transfer from the α-basis directly to the γ-basis. We know that the γ-coordinates Z can be expressed in terms of the α-coordinates X by a system of equations

$$z^i = \sum_j c_j^i x^j \qquad i = 1, 2, \ldots, n$$

where c_j^i is the ith component of α_j with respect to the γ-basis. Since

$$\alpha_j = \sum_k b_j^k\beta_k = \sum_k b_j^k\Big(\sum_i a_k^i\gamma_i\Big) = \sum_i\Big(\sum_k a_k^i b_j^k\Big)\gamma_i$$

we conclude that

$$c_j^i = \sum_k a_k^i b_j^k$$

The matrix $C = (c_j^i)$ which relates the α-coordinates X and the γ-coordinates Z we call the *product* of A by B; in symbols,

$$C = AB$$

Thus C is the matrix whose (i, j)th entry is the sum of the products of the entries of the ith row of A by the corresponding entries of the jth column of B.

We notice that the "row by column" rule for multiplication of square matrices can be extended to rectangular matrices for which the number of columns in the first agrees with the number of rows in the second. Thus if

$$A = (a_k^i) \text{ is } m \times n \qquad \text{and} \qquad B = (b_j^k) \text{ is } n \times p$$

we define AB to be the $m \times p$ matrix

$$C = (c_j^i) \qquad \text{where } c_j^i = \sum_{k=1}^{n} a_k^i b_j^k$$

Thus c_j^i is the sum of the products of the entries of the ith row of A by the corresponding entries of the jth column of B. The use of this rule is illustrated below.

If

$$A = \begin{pmatrix} 1 & 2 \\ 2 & 3 \\ 3 & 4 \end{pmatrix} \qquad B = \begin{pmatrix} 2 & 1 & 4 \\ 0 & 3 & 1 \end{pmatrix}$$

then

$$AB = \begin{pmatrix} 1 \cdot 2 + 2 \cdot 0 & 1 \cdot 1 + 2 \cdot 3 & 1 \cdot 4 + 2 \cdot 1 \\ 2 \cdot 2 + 3 \cdot 0 & 2 \cdot 1 + 3 \cdot 3 & 2 \cdot 4 + 3 \cdot 1 \\ 3 \cdot 2 + 4 \cdot 0 & 3 \cdot 1 + 4 \cdot 3 & 3 \cdot 4 + 4 \cdot 1 \end{pmatrix}$$

EXAMPLE 3.4

An illustration of a repeated change of bases (hence coordinates) occurs in the reduction of a quadric surface to standard form by repeated completion of squares. For instance, consider the quadric surface S defined by the equation

$$(x^1)^2 - (x^2)^2 + (x^3)^2 + 2x^1x^2 - 6x^1x^3 + 8x^2x^3 = 1$$

where $(x^1, x^2, x^3)^T = X$ are the coordinates of a point relative to a basis (the α-basis) for $V_3(R^{\#})$. Since the expression on the left can be written as

$$(x^1 + x^2 - 3x^3)^2 - 2(x^2)^2 + 14x^2x^3 - 8(x^3)^2$$

we introduce a new basis (the β-basis) in such a way that the new coordinates are given in terms of the old by the equations

$$\begin{aligned} y^1 &= x^1 + x^2 - 3x^3 \\ y^2 &= \phantom{x^1 + {}} x^2 \\ y^3 &= \phantom{x^1 + x^2 - 3{}} x^3 \end{aligned}$$

Relative to the β-basis, S is given by the equation

$$(y^1)^2 - 2(y^2)^2 + 14y^2y^3 - 8(y^3)^2$$
$$= (y^1)^2 - \tfrac{1}{2}(-2y^2 + 7y^3)^2 + \tfrac{33}{2}(y^3)^2 = 1$$

Again we introduce a new basis (the γ-basis), this time in such a way that the new coordinates are given in terms of the old by the equations

$$\begin{aligned} z^1 &= y^1 \\ z^2 &= -2y^2 + 7y^3 \\ z^3 &= \phantom{y^1 -2y^2 + 7{}} y^3 \end{aligned}$$

Relative to the γ-basis, S is given by the equation

$$(z^1)^2 - \tfrac{1}{2}(z^2)^2 + \tfrac{33}{2}(z^3)^2 = 1$$

The matrix which determines the final Z-coordinates in terms of the initial X-coordinates is the product of the two coefficient matrices in the above systems.

$$\begin{pmatrix} 1 & 0 & 0 \\ 0 & -2 & 7 \\ 0 & 0 & 1 \end{pmatrix} \begin{pmatrix} 1 & 1 & -3 \\ 0 & 1 & 0 \\ 0 & 0 & 1 \end{pmatrix} = \begin{pmatrix} 1 & 1 & -3 \\ 0 & -2 & 7 \\ 0 & 0 & 1 \end{pmatrix}$$

If we adopt the natural definition that two $m \times n$ matrices are *equal* if and only if they agree entry for entry, a great saving in space can be effected with a systematic use of matrix multiplication. For example, the familiar system (N) of Chap. 1 can be written

$$AX = Y \qquad \text{where } A = (a_j^i),\ X = \begin{pmatrix} x^1 \\ x^2 \\ \cdot \\ x^n \end{pmatrix},\ Y = \begin{pmatrix} y^1 \\ y^2 \\ \cdot \\ y^m \end{pmatrix}$$

Similarly, the system of equations (3) can be condensed to the single matrix equation $Y = BX$.

Moreover, the system of equations (2) can be written as a matrix equation. For this, however, a word of explanation is necessary. Although, in the above definition of the product of a matrix A by a matrix B, it was tacitly assumed that both A and B are matrices over a field, the definition has a meaning under more general circumstances. Indeed, the only requirement is that each expression of the type $\sum a_k^i b_j^k$ be defined. This is the case, for example, if A is an $m \times n$ matrix whose entries are vectors in a vector space V over F, and B is an $n \times p$ matrix over F; then the entries of AB are vectors. The one instance of this in which we are interested occurs when the first factor is a $1 \times n$ matrix $(\alpha_1, \alpha_2, \ldots, \alpha_n)$ whose entries form a basis for a vector space V. Then equation (2), which expresses the members of the α-basis in terms of the β-basis, can be written as

$$(\alpha_1, \alpha_2, \ldots, \alpha_n) = (\beta_1, \beta_2, \ldots, \beta_n)B$$

or, if we designate $(\alpha_1, \alpha_2, \ldots, \alpha_n)$ and $(\beta_1, \beta_2, \ldots, \beta_n)$ by simply α and β, respectively, we can write (2) and (3) in the cryptic form

$$\alpha = \beta B$$

$$Y = BX$$

Because of its efficiency, this notation will often be employed in future pages. With it, the successive change of bases which motivated our definition of matrix multiplication is described by

$$\begin{Bmatrix} \alpha = \beta B \\ Y = BX \end{Bmatrix} \text{ and } \begin{Bmatrix} \beta = \gamma A \\ Z = AY \end{Bmatrix} \text{ imply } \begin{Bmatrix} \alpha = \gamma(AB) \\ Z = (AB)X \end{Bmatrix}$$

The above relations among bases and the corresponding coordinates have also been summarized in the schematic diagram of Fig. 5 in the following sense: The basis (coordinate set) at the initial point of an arrow, when multiplied by the matrix at hand, yields the basis (coordinate set) at the final point of the arrow; for example, $\beta B = \alpha$ and $BX = Y$.

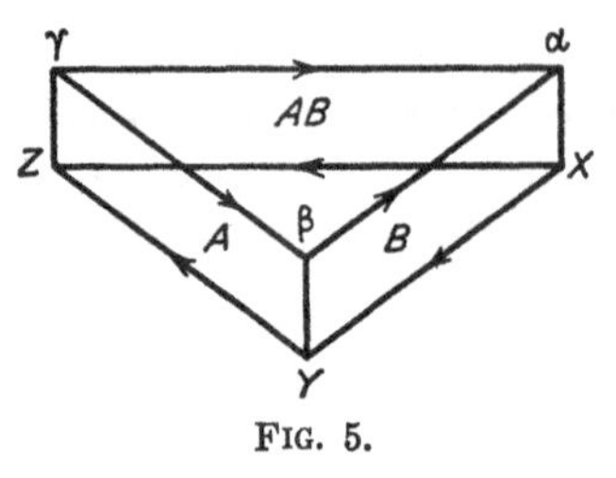

FIG. 5.

We conclude this section with a discussion of another matrix product of the type αB, where $\alpha = (\alpha_1, \alpha_2, \ldots, \alpha_n)$, that we shall use later. If $X = (x_1, x_2, \ldots, x_n)^T$, then αX is a 1×1 matrix whose sole entry is the vector $\sum \alpha_i x^i = \sum x^i \alpha_i$. Since a 1×1 matrix may be identified with its only entry, we may interpret αX as the vector with α-coordinates X. Some of the rules developed in the next section for multiplication of matrices with entries in a field are valid in a restricted sense for matrices of the type αB and αX; for example, the associative law $(\alpha B)C = \alpha(BC)$ is both meaningful and valid whenever BC is defined. The verification of this and similar rules, which will be suggested by those appearing in the next section, will be left to the reader.

PROBLEMS

1. Compute AB if

(a) $$A = \begin{pmatrix} 1 & 3 \\ 2 & -1 \\ 1 & 2 \end{pmatrix}, B = \begin{pmatrix} 1 & 2 & -3 & 1 \\ 4 & 3 & 2 & 0 \end{pmatrix}$$

(b) $$A = \begin{pmatrix} -2 & 4 & 3 & 2 \\ 3 & 5 & 6 & -1 \\ 4 & -2 & 0 & 3 \end{pmatrix}, B = \begin{pmatrix} 2 & 2 \\ 1 & -4 \\ 3 & 3 \\ 5 & 7 \end{pmatrix}$$

2. Compute AB, AC, and BC if

$$A = \begin{pmatrix} 2 & 3 & -1 \\ 2 & 7 & 3 \end{pmatrix}, \quad B = \begin{pmatrix} 4 & -2 & 1 \\ 3 & 2 & 5 \\ 4 & 0 & -1 \end{pmatrix}, \quad C = \begin{pmatrix} 3 & 7 & 2 \\ 1 & 0 & 1 \\ 2 & 4 & 5 \end{pmatrix}$$

3. If $-(x^1)^2 + 5(x^2)^2 - (x^3)^2 + 2x^1x^2 + 10x^1x^3 + 2x^2x^3 = 1$ is the equation of a locus S in $V_3(R^*)$ relative to the α-basis, use the method of Example 3.4 to determine another basis (the γ-basis) relative to which S is described by an equation containing only squares.

3.7. Properties of Matrix Multiplication. In this section we shall collect the basic properties and computational rules for matrix multiplication of matrices over a field F.

Rule 1. If A is $m \times n$,

$$I_m A = A I_n = A \qquad 0A = 0 \qquad A0 = 0$$

Rule 2. If A is $m \times n$, then AB and BA are defined if and only if B is $n \times m$. If this condition is met, $AB \neq BA$ in general, even if $m = n$. For example

$$\begin{pmatrix} 1 & 0 \\ 0 & 0 \end{pmatrix} \begin{pmatrix} 0 & 1 \\ 0 & 0 \end{pmatrix} \neq \begin{pmatrix} 0 & 1 \\ 0 & 0 \end{pmatrix} \begin{pmatrix} 1 & 0 \\ 0 & 0 \end{pmatrix}$$

Rule 3. $(AB)^T = B^T A^T$

To verify this, set $C = AB$ and $C^T = (\bar{c}^i_j)$. Then

$$\bar{c}^i_j = c^j_i = \sum_k a^j_k b^k_i = \sum_k b^k_i a^j_k = \sum_k \bar{b}^i_k \bar{a}^k_j$$

if $A^T = (\bar{a}^k_j)$ and $B^T = (\bar{b}^i_k)$.

Rule 4. The associative law holds for multiplication:

$$A(BC) = (AB)C$$

For

$$(AB)C = ((a^i_j)(b^j_k))(c^k_l) = (\sum_k(\sum_j a^i_j b^j_k)c^k_l)$$

and

$$A(BC) = (a^i_j)((b^j_k)(c^k_l)) = (\sum_j a^i_j(\sum_k b^j_k c^k_l))$$

and both of these final terms are equal to $(\sum_{j,k} a^i_j b^j_k c^k_l)$ since the associative and distributive laws hold in F.

Rule 5. If A is an $m \times n$ matrix and $B = \text{diag}\,(b_1, b_2, \ldots, b_m)$ is an $m \times m$ diagonal matrix, then BA is obtained from A by multiplying the ith row of A by b_i, $i = 1, 2, \ldots, m$. If B is $n \times n$, AB is obtained from A by multiplying the jth column of A by b_j, $j = 1, 2, \ldots, n$.

Rule 6. Any n-rowed square scalar matrix $C = \text{diag}(a, a, \ldots, a)$ commutes with every n-rowed square matrix A, that is, $AC = CA$. Conversely, the scalar matrices are the only ones which commute with all matrices.

The first statement follows from Rule 5. For the converse, if a matrix B commutes with every A, then in particular it must commute with the matrix E_j^i, which has a 1 as its (i, j)th entry and zeros elsewhere. This assumption for all i, j forces A to be a scalar matrix, as the reader may verify.

The remaining rules relate multiplication of matrices with other compositions for matrices. The following definitions of these compositions have their origin in the trivial observation that an $m \times n$ matrix defines an mn-tuple in a variety of ways.

DEFINITION. *The* sum, $A + B$, *of the* $m \times n$ *matrices* $A = (a_j^i)$ *and* $B = (b_j^i)$ *over* F, *is the matrix* $(a_j^i + b_j^i)$. *The* scalar product cA *of* A *by the scalar* c *is the matrix* (ca_j^i).

It is immediately seen that the set of all $m \times n$ matrices over F, together with addition and scalar multiplication, is a vector space of dimension mn over F. In particular, addition of matrices has the following properties because addition in F satisfies corresponding properties:

$$A + B = B + A \qquad A + (B + C) = (A + B) + C \qquad A + 0 = A$$

Using scalar multiplication, we may write a scalar matrix diag $(a, a, \ldots, a)$ as

$$\text{diag}(a, a, \ldots, a) = aI$$

which makes our next rule seem plausible.

Rule 7. The scalar matrices aI of a given order with a in F, together with the compositions of matrix addition and multiplication, constitute a field isomorphic to F.

That the system in question is a field is easily verified. To demonstrate the isomorphism, consider the correspondence $a \rightarrow aI$ of the scalar a to the scalar matrix aI. Clearly it is one-one. Moreover, if $b \rightarrow bI$, then

$$a + b \rightarrow (a + b)I = aI + bI \qquad \text{and} \qquad ab \rightarrow abI = aI \cdot bI$$

which completes the proof.

Rule 8. If the product AB is defined and both A and B are partitioned in any way, subject to the restriction that the ordered partition of the columns of A agrees with that of the rows of B, the product AB can be computed by the usual "row by column" rule using the submatrices of A and B defined by the partition as the elements.

For example, suppose that the 5×5 matrix A and the 5×3 matrix B

are partitioned so that

$$A = \begin{array}{c} \begin{array}{ccc} 1 & 2 & 2 \end{array} \\ \begin{pmatrix} A_1^1 & A_2^1 & A_3^1 \\ A_1^2 & A_2^2 & A_3^2 \end{pmatrix} \begin{array}{c} 2 \\ 3 \end{array} \end{array} \qquad B = \begin{array}{c} \begin{array}{cc} 2 & 1 \end{array} \\ \begin{pmatrix} B_1^1 & B_2^1 \\ B_1^2 & B_2^2 \\ B_1^3 & B_2^3 \end{pmatrix} \begin{array}{c} 1 \\ 2 \\ 2 \end{array} \end{array}$$

where the bordering integers indicate the dimensions of the various submatrices. Since the number of columns in A is equal to the number of rows in B, the product AB is defined. Moreover, since the column partition of A agrees with the row partition of B, the product of (A_k^i) and (B_j^k) is defined. In addition, any product $A_k^i B_j^k$ that occurs, regarded as a matrix product, has a sense, as does any sum of such products. The rule states that, if these sums and products are computed as matrix sums and products, the resulting matrix of submatrices is AB.

We shall prove this rule for only the following important case:

$$A = \begin{pmatrix} A_1^1 & A_2^1 \\ A_1^2 & A_2^2 \end{pmatrix} \qquad B = \begin{pmatrix} B_1^1 & B_2^1 \\ B_1^2 & B_2^2 \end{pmatrix} \tag{4}$$

This result can be deduced from the following two special cases of it which we shall prove:

(a) If $A = \overset{r \quad n-r}{(A_1 \quad A_2)}$ and $B = \begin{pmatrix} B^1 \\ B^2 \end{pmatrix} \begin{array}{c} r \\ n-r \end{array}$, then $AB = (A_1B^1 + A_2B^2)$

(b) If $A = \begin{pmatrix} A^1 \\ A^2 \end{pmatrix}$ and $B = (B_1 \; B_2)$, then $AB = \begin{pmatrix} A^1B_1 & A^1B_2 \\ A^2B_1 & A^2B_2 \end{pmatrix}$

To establish (a), write

$$AB = \left(\sum_{k=1}^{n} a_k^i b_j^k \right) \quad \text{as} \quad \left(\sum_{k=1}^{r} a_k^i b_j^k + \sum_{k=r+1}^{n} a_k^i b_j^k \right)$$

The proof of (b) is similar.

If now in the partition of A and B in (4) we set

$$A_1 = \begin{pmatrix} A_1^1 \\ A_1^2 \end{pmatrix} \qquad A_2 = \begin{pmatrix} A_2^1 \\ A_2^2 \end{pmatrix} \qquad B^1 = (B_1^1 \; B_2^1) \qquad B^2 = (B_1^2 \; B_2^2)$$

so that A and B have the form required in (a), we conclude that

$$AB = (A_1B^1 + A_2B^2)$$

But now

$$A_1B^1 = \begin{pmatrix} A_1^1 \\ A_1^2 \end{pmatrix} (B_1^1\ B_2^1)$$

and likewise A_2B^2 have the form required in (b), and if these are expanded according to that rule, the desired result follows.

The usefulness of this multiplication rule will become apparent in the future in dealing with special products.

PROBLEMS

1. If $A = \begin{pmatrix} 2 & 3 \\ 1 & 4 \end{pmatrix}$, verify that $A^2 - 6A + 5I_2 = 0$.

2. If A is a square matrix of the form $\begin{pmatrix} B & 0 \\ C & I \end{pmatrix}$ where I is an identity matrix, derive a formula for A^n where n is a positive integer.

3. If the product

$$\begin{pmatrix} A_1 & 0 \\ A_3 & A_4 \end{pmatrix} \begin{pmatrix} B_1 & 0 \\ 0 & B_2 \end{pmatrix}$$

is defined for the indicated partitioning and is equal to an identity matrix, show that A_1B_1 (assuming that it is a square matrix) and A_4B_2 are identity matrices.

4. In rule 6 for multiplication of matrices, E_j^i is defined as the $n \times n$ matrix with 1 as its (i, j)th entry and 0's elsewhere. Show that these matrices, $i, j = 1, 2, \ldots, n$, form a basis for the vector space of all $n \times n$ matrices over a field F. Also verify the rule $E_j^i E_l^k = \delta_j^k E_l^i$.

5. Show that I, A, B, AB form a basis for the vector space of all 2×2 matrices over R if

$$A = \begin{pmatrix} 1 & 0 \\ 2 & 1 \end{pmatrix} \qquad B = \begin{pmatrix} 2 & 1 \\ 0 & 1 \end{pmatrix}$$

6. Let $U = E_2^1 + E_3^2 + \cdots + E_n^{n-1}$, where E_j^i is the $n \times n$ matrix of Prob. 4. Show that $U^n = 0$.

7. If

$$A = \begin{pmatrix} 2 & 3 & -1 & 4 \\ 1 & 2 & 0 & 1 \\ 2 & 3 & 1 & 2 \\ -1 & 2 & 1 & 0 \end{pmatrix} \qquad B = \begin{pmatrix} 2 & 3 & 1 \\ 1 & 5 & 2 \\ 2 & 1 & 3 \\ 0 & 2 & 4 \end{pmatrix}$$

compute AB directly as well as when partitioned in some suitable fashion.

8. Find all matrices B such that $AB = BA$ if

$$(a) \quad A = \begin{pmatrix} 1 & 0 & 0 \\ 0 & 0 & 1 \\ 0 & 1 & 0 \end{pmatrix} \qquad (b) \quad A = \begin{pmatrix} 2 & 1 & 0 \\ 3 & -1 & 2 \\ 5 & 0 & 2 \end{pmatrix}$$

9. Prove that any two matrices of the form $\begin{pmatrix} a & b \\ -b & a \end{pmatrix}$, where a and b are real numbers, are commutative. Show that the correspondence $\begin{pmatrix} a & b \\ -b & a \end{pmatrix} \rightarrow a + bi$ between these matrices and the complex numbers is an isomorphism.

3.8. Elementary Matrices. The concept of equivalence of $m \times n$ matrices can be developed in terms of matrix multiplication. Before doing this we mention some further properties of matrix multiplication.

Theorem 3.9. If the product AB is defined, $r(AB)$ does not exceed the rank of either factor.

Proof. We shall use rule 8 of the preceding section to compute AB, after partitioning B into its row vectors ρ^i. The equation

$$\begin{pmatrix} a_1^1 & a_2^1 & \cdots & a_n^1 \\ a_1^2 & a_2^2 & \cdots & a_n^2 \\ \cdot & \cdot & \cdots & \cdot \\ a_1^m & a_2^m & \cdots & a_n^m \end{pmatrix} \begin{pmatrix} \rho^1 \\ \rho^2 \\ \cdot \\ \rho^n \end{pmatrix} = \begin{pmatrix} a_1^1\rho^1 + a_2^1\rho^2 + \cdots + a_n^1\rho^n \\ \cdot \\ \cdot \\ a_1^m\rho^1 + a_2^m\rho^2 + \cdots + a_n^m\rho^n \end{pmatrix}$$

exhibits the rows of AB as linear combinations of the rows ρ^i of B. Hence the row space of AB is included in the row space of B, that is, $r(AB) \leq r(B)$. Writing AB as

$$(\gamma_1 \quad \gamma_2 \quad \cdots \quad \gamma_n) \begin{pmatrix} b_1^1 & b_2^1 & \cdots & b_p^1 \\ \cdot & \cdot & \cdots & \cdot \\ b_1^n & b_2^n & \cdots & b_p^n \end{pmatrix}$$

exhibits the columns of AB as linear combinations of the columns γ_i of A, which establishes the inequality $r(AB) \leq r(A)$.

Theorem 3.10. If τ denotes any elementary row transformation, then $\tau(AB) = (\tau A)B$, where τA is the matrix obtained from A by applying τ.

Proof. We examine $\tau(a_k^i)(b_j^k)$ for each type of transformation.

Type I. The superscript of a_k^i is the row index in both A and AB; so an interchange of rows in A produces the same interchange in the order of rows in AB.

Type II. If the ith row of A is multiplied by c, the (i, j)th element of AB becomes $\sum ca_k^i b_j^k$, that is, the ith row of AB is multiplied by c.

Type III. Let the elements of the ith row of A be replaced by

$$a_1^i + ca_1^l\,,\ a_2^i + ca_2^l\,,\ \ldots,\ a_n^i + ca_n^l$$

Then the members of the ith row of AB become

$$\sum_k (a_k^i + ca_k^l)b_1^k\,,\ \ldots,\ \sum_k (a_k^i + ca_k^l)b_n^k$$

or

$$\sum_k a_k^i b_1^k + c\sum_k a_k^l b_1^k\,,\ \ldots,\ \sum_k a_k^i b_n^k + c\sum_k a_k^l b_n^k$$

This is the same result as would have been obtained from the application of the same transformation to AB.

DEFINITION. *A matrix obtained from an identity matrix by the application of any one elementary row transformation is called an* elementary matrix.

For example, each of the following is an elementary matrix, having been obtained from I_4 by the transformations R_{24}, $R_2(3)$, and $R_{32}(4)$, respectively (see Sec. 3.3):

$$\begin{pmatrix} 1 & 0 & 0 & 0 \\ 0 & 0 & 0 & 1 \\ 0 & 0 & 1 & 0 \\ 0 & 1 & 0 & 0 \end{pmatrix} \qquad \begin{pmatrix} 1 & 0 & 0 & 0 \\ 0 & 3 & 0 & 0 \\ 0 & 0 & 1 & 0 \\ 0 & 0 & 0 & 1 \end{pmatrix} \qquad \begin{pmatrix} 1 & 0 & 0 & 0 \\ 0 & 1 & 0 & 0 \\ 0 & 4 & 1 & 0 \\ 0 & 0 & 0 & 1 \end{pmatrix}$$

The usefulness of such matrices is indicated in the next result.

Theorem 3.11. Every elementary transformation τ on the rows of a matrix can be performed by multiplying A on the left by the corresponding elementary matrix.

Proof. Using Theorem 3.10, $\tau A = \tau(IA) = (\tau I)A$.

There can be no misunderstanding if we identify an elementary matrix by the symbol that designates the elementary row transformation used to obtain the matrix from the identity matrix. According to Sec. 3.3 the following equations hold for these matrices:

$$R_{ij}R_{ij} = I$$

$$R_i(c)R_i(c^{-1}) = R_i(c^{-1})R_i(c) = I$$

$$R_{ij}(c)R_{ij}(-c) = R_{ij}(-c)R_{ij}(c) = I$$

DEFINITION. *A matrix B is called an* inverse *of a matrix A if $AB = BA = I$.*

The above equations may now be summarized in the following way:

Theorem 3.12. Each elementary matrix has an inverse which is an elementary matrix of the same type.

It is clear that only a square matrix can have an inverse. Moreover, if an inverse of A exists, it is unique since if, along with B, the matrix C is an inverse, multiplying the equation $I = AB$ on the left by C gives

$$C = C(AB) = (CA)B = IB = B$$

Thus A^{-1} is a suitable notation for the inverse of A if it exists. Notice that if A and B have inverses and are of the same order, then AB has an inverse. It is $B^{-1}A^{-1}$, since $AB(B^{-1}A^{-1}) = A(BB^{-1})A^{-1} = AIA^{-1} = I$ and, likewise, $(B^{-1}A^{-1})AB = I$. This establishes the rule: *The inverse of a product of two matrices is equal to the product of the inverses of the factors in reverse order* (assuming the existence of these quantities):

$$(AB)^{-1} = B^{-1}A^{-1}$$

Also, if A has an inverse, so does A^T. It is $(A^{-1})^T$. For by rule 3, Sec. 3.7,

$$A^T(A^{-1})^T = (A^{-1}A)^T = I^T = I$$

and

$$(A^{-1})^T A^T = (AA^{-1})^T = I^T = I$$

An interest in matrices having inverses could be motivated by our next theorem, which uses the notation introduced at the end of Sec. 3.6.

Theorem 3.13. If $\{\alpha_1, \alpha_2, \ldots, \alpha_n\}$ is a basis for a vector space V, then the components of the row matrix αA, where $\alpha = (\alpha_1, \alpha_2, \ldots, \alpha_n)$, form a basis for V if and only if A has an inverse.

Proof. Let $\beta = (\beta_1, \beta_2, \ldots, \beta_n)$ denote the matrix product αA. If A has an inverse, the equation $\beta = \alpha A$ implies $\alpha = \beta A^{-1}$; hence V is spanned by $\{\beta_1, \beta_2, \ldots, \beta_n\}$, and since the number of vectors in the set is n, it is a basis by Theorem 2.9. Conversely, if α and $\beta = \alpha A$ are bases, the α_i's are expressible in terms of the β_j's, for example, $\alpha = \beta B$. Hence $\beta = \beta BA$, or $BA = I$. Also $\alpha = \beta B = \alpha AB$ so that $AB = I$ and A has an inverse.

Those matrices for which there exists an inverse are described next.

Theorem 3.14. The following statements about a square matrix A are equivalent to each other:

(i) A is nonsingular (or $\det A \neq 0$).

(ii) A is a product of elementary matrices.

(iii) A has an inverse.

Proof. (i) *implies* (ii). If A is nonsingular, it is row-equivalent to I_n(Theorem 3.7); hence by Theorem 3.11 there exist elementary matrices such that

$$E_k E_{k-1} \cdots E_1 A = I$$

Multiplying this equation on the left by $E_k^{-1}, E_{k-1}^{-1}, \ldots, E_1^{-1}$ in turn gives

$$A = E_1^{-1} E_2^{-1} \cdots E_k^{-1}$$

which, in view of Theorem 3.12, gives (ii).

(ii) *implies* (iii). Let $A = F_1 F_2 \cdots F_l$ be a product of elementary matrices F_i. Then $F_l^{-1} F_{l-1}^{-1} \cdots F_1^{-1} = (F_1 F_2 \cdots F_l)^{-1}$ is an inverse for A, as a direct computation shows.

(iii) *implies* (i). If A^{-1} exists, the equation $A^{-1}A = I_n$ implies (Theorem 3.9) that $r(A) = n$, or A is nonsingular by definition.

The method of proof of this theorem supplies the following practical method for computing the inverse of a nonsingular matrix: *To find the inverse of a nonsingular matrix A, apply to I in the same order those row transformations which reduce A to I.* For if

$$E_k E_{k-1} \cdots E_1 A = I$$

then

$$A = E_1^{-1} E_2^{-1} \cdots E_k^{-1}$$

and

$$A^{-1} = E_k E_{k-1} \cdots E_1 = (E_k E_{k-1} \cdots E_1) I$$

EXAMPLE 3.5

The matrix

$$A = \begin{pmatrix} 1 & 1 & 1 \\ 2 & -3 & 2 \\ -1 & -3 & -2 \end{pmatrix}$$

is nonsingular. Below we have indicated a sequence of row transformations which reduce A to I_3; next the inverse is obtained by applying the same sequence to I_3.

$$A \to \begin{pmatrix} 1 & 1 & 1 \\ 0 & -5 & 0 \\ 0 & -2 & -1 \end{pmatrix} \to \begin{pmatrix} 1 & 1 & 1 \\ 0 & 1 & 3 \\ 0 & -2 & -1 \end{pmatrix} \to \begin{pmatrix} 1 & 0 & -2 \\ 0 & 1 & 3 \\ 0 & 0 & 5 \end{pmatrix}$$

$$\to \begin{pmatrix} 1 & 0 & 0 \\ 0 & 1 & 0 \\ 0 & 0 & 1 \end{pmatrix} = I_3$$

$$I_3 \to \begin{pmatrix} 1 & 0 & 0 \\ -2 & 1 & 0 \\ 1 & 0 & 1 \end{pmatrix} \to \begin{pmatrix} 1 & 0 & 0 \\ -5 & 1 & -3 \\ 1 & 0 & 1 \end{pmatrix} \to \begin{pmatrix} 6 & -1 & 3 \\ -5 & 1 & -3 \\ -9 & 2 & -5 \end{pmatrix}$$

$$\to \begin{pmatrix} \frac{12}{5} & -\frac{1}{5} & 1 \\ \frac{2}{5} & -\frac{1}{5} & 0 \\ -\frac{9}{5} & \frac{2}{5} & -1 \end{pmatrix} = A^{-1}$$

The proof of the following result is left as an exercise:

Theorem 3.15. Two $m \times n$ matrices A and B are row-equivalent if and only if $B = PA$ for a nonsingular matrix P. P is the matrix obtained from I by application, in the same order, of those row transformations which reduce A to B.

Although the following result is merely another version of Theorem 3.15, we prefer to list it separately:

Theorem 3.16. If P is nonsingular, then $r(PA) = r(A)$.
Proof. According to Theorem 3.15, PA is row-equivalent to A, hence has the same rank as A by Theorem 3.5.

As for effecting column transformations of a matrix A by matrix multiplication, we point out that column transformations of A are row transformations of A^{T}, and conversely. Consequently a matrix B can be obtained from A by a sequence of column transformations if and only if B^{T} can be obtained from A^{T} by a sequence of row transformations. Thus if B is column-equivalent to A, $B^{\mathrm{T}} = PA^{\mathrm{T}}$ according to Theorem 3.15, or $B = AP^{\mathrm{T}} = AQ$, let us say, where $Q = P^{\mathrm{T}}$ is nonsingular. Conversely, if $B = AQ$ for a nonsingular Q, B is column-equivalent to A. Hence, application of column transformations amounts to postmultiplication by nonsingular factors, *i.e.*, products of elementary matrices. The effect of the various elementary matrices as postmultipliers can be obtained as follows:

A, after interchanging ith and jth columns becomes $(R_{ij}A^T)^T = AR_{ij}^T = AR_{ij}$.

A, after multiplying ith column by c becomes $(R_i(c)A^T)^T = AR_i^T(c) = AR_i(c)$.

A, after adding c times jth column to ith becomes $(R_{ij}(c)A^T)^T = AR_{ij}^T(c) = AR_{ji}(c)$.

We conclude that the elementary matrices produce the three types of column transformations.

It is now possible to state an alternative criterion for the equivalence of two $m \times n$ matrices.

Theorem 3.17. Two $m \times n$ matrices A and B over F are rationally equivalent if and only if $B = PAQ$ for suitable nonsingular matrices P and Q. Here $P(Q)$ is obtained by applying in the same order to I those row (column) transformations used to reduce A to B. In particular, if $r(A) = r$, there exist matrices P, Q such that (see Theorem 3.8)

$$PAQ = J_r$$

EXAMPLE 3.6

Determine P and Q such that $PAQ = J_r$ if

$$A = \begin{pmatrix} 3 & -7 & 3 & -4 \\ 2 & -3 & 1 & 0 \\ 3 & -2 & 0 & 4 \end{pmatrix}$$

Below are indicated in turn (i) several stages in the reduction of A to J_2, (ii) the development of P from I_3 using the row transformations applied to A, and (iii) the development of Q from I_4 using the column transformations applied to A. To assist the reader we remark that the row transformations used to obtain the first indicated stage in the reduction of A are $R_{31}(-1)$, $R_{12}(-1)$, $R_{21}(-2)$, and $R_{32}(-1)$ in order of application.

$$A \overset{\text{row}}{\rightarrow} \begin{pmatrix} 1 & -4 & 2 & -4 \\ 0 & 5 & -3 & 8 \\ 0 & 0 & 0 & 0 \end{pmatrix} \overset{\text{col}}{\rightarrow} \begin{pmatrix} 1 & 0 & 0 & 0 \\ 0 & 5 & -3 & 8 \\ 0 & 0 & 0 & 0 \end{pmatrix}$$

$$\overset{\text{col}}{\rightarrow} \begin{pmatrix} 1 & 0 & 0 & 0 \\ 0 & 1 & -3 & 8 \\ 0 & 0 & 0 & 0 \end{pmatrix} \overset{\text{col}}{\rightarrow} \begin{pmatrix} 1 & 0 & 0 & 0 \\ 0 & 1 & 0 & 0 \\ 0 & 0 & 0 & 0 \end{pmatrix}$$

$$I_3 \to \begin{pmatrix} 1 & -1 & 0 \\ -2 & 3 & 0 \\ 1 & -3 & 1 \end{pmatrix} = P$$

$$I_4 \to \begin{pmatrix} 1 & 4 & -2 & 4 \\ 0 & 1 & 0 & 0 \\ 0 & 0 & 1 & 0 \\ 0 & 0 & 0 & 1 \end{pmatrix} \to \begin{pmatrix} 1 & 0 & -2 & 4 \\ 0 & -1 & 0 & 0 \\ 0 & -2 & 1 & 0 \\ 0 & 0 & 0 & 1 \end{pmatrix} \to \begin{pmatrix} 1 & 0 & -2 & 4 \\ 0 & -1 & -3 & 8 \\ 0 & -2 & -5 & 16 \\ 0 & 0 & 0 & 1 \end{pmatrix} = Q$$

Instead of a display of the type used above for determining matrices P and Q such that $PAQ = J_r$, it is less demanding on one's memory to start with what might be called an "L-array," consisting of the three matrices A, I_m, and I_n (assuming that A is $m \times n$), arranged as shown below:

$$\begin{matrix} I_n & \\ A & I_m \end{matrix}$$

If the row transformations intended for A are applied to the matrix $(A\ I_m)$ and the column transformations intended for A are applied to $(I_n\ A)^{\mathrm{T}}$, then, when A is reduced to J_r, the array assumes the form

$$\begin{matrix} Q & \\ J_r & P \end{matrix}$$

thereby displaying the desired P and Q.

It was remarked, following Theorem 3.8, that in a reduction of the above type the order of application of row and column transformations is immaterial. This fact, since we now know that such transformations can be effected by pre- and postmultiplication with elementary matrices, is seen to be a consequence of the associative law for multiplication of matrices. For if E and F denote elementary matrices, the equation $(EA)F = E(AF)$ proves our assertion.

Earlier we saw that a nonsingular matrix A can be reduced to the equivalent identity matrix I using row transformations alone, and that simultaneously A^{-1} can be constructed. In general, A can be reduced to I more quickly if both row and column transformations are used. Simultaneously, according to Theorem 3.17, one may construct matrices P and Q such that $PAQ = I$, which yields the formula $A^{-1} = QP$. This method of computing A^{-1} is often faster in practice than the earlier one, particularly when used in conjunction with an L-array.

EXAMPLE 3.7

The matrix

$$A = \begin{pmatrix} 1 & 1 & 1 & -1 \\ 2 & -3 & -2 & 2 \\ -1 & -3 & -2 & 2 \\ 3 & 1 & -1 & 0 \end{pmatrix}$$

is found to be nonsingular upon reduction to canonical form. The row and column transformations used in the reduction produce the following P and Q:

$$P = \begin{pmatrix} 1 & 0 & 0 & 0 \\ -5 & 1 & -3 & 0 \\ 3 & -\frac{2}{3} & \frac{5}{3} & 0 \\ -5 & 2 & -4 & -1 \end{pmatrix} \qquad Q = \begin{pmatrix} 1 & -1 & -2 & 0 \\ 0 & 1 & 1 & 0 \\ 0 & 0 & 1 & 1 \\ 0 & 0 & 0 & 1 \end{pmatrix}$$

Thus

$$A^{-1} = QP = \begin{pmatrix} 0 & \frac{1}{3} & -\frac{1}{3} & 0 \\ -2 & \frac{1}{3} & -\frac{4}{3} & 0 \\ -2 & \frac{4}{3} & -\frac{7}{3} & -1 \\ -5 & 2 & -4 & -1 \end{pmatrix}$$

PROBLEMS

1. Compute the inverse of each of the following matrices making use of the method of Example 3.5:

$$(a) \quad \begin{pmatrix} 0 & 0 & 1 \\ 1 & 2 & 6 \\ 2 & 3 & 4 \end{pmatrix} \qquad (b) \quad \begin{pmatrix} 2 & -1 & 1 & 6 \\ 1 & 0 & 2 & 3 \\ -1 & 1 & 1 & -2 \\ 0 & 1 & 2 & 1 \end{pmatrix}$$

2. Show that the following variation of the method of Example 3.5 will produce the inverse of a nonsingular matrix A. Apply to the $n \times 2n$ matrix $(I_n\ A)$ those row transformations which reduce A to I_n; the resulting matrix is then $(A^{-1}\ I_n)$.

3. Let A be a matrix of the form $\begin{pmatrix} A_1 & A_2 \\ A_3 & A_4 \end{pmatrix}$, where A_1 is nonsingular. Show that a nonsingular matrix P can be found such that $PA = \begin{pmatrix} A_1 & A_2 \\ 0 & A_5 \end{pmatrix}$.

4. Determine values for P and Q such that $PAQ = J_r$, if

$$(a) \quad A = \begin{pmatrix} 3 & 5 & 10 & 4 \\ 0 & 1 & 2 & 1 \\ 1 & -1 & 2 & 0 \end{pmatrix} \qquad (b) \quad A = \begin{pmatrix} -1 & 2 & -2 & 0 \\ 0 & -1 & 1 & 1 \\ -3 & 1 & -4 & 3 \\ 0 & 2 & -1 & -1 \end{pmatrix}$$

5. Suppose that A and B are 3×3 matrices such that $r(A) = r(B) = 2$. Show that $AB \neq 0$. HINT: There exist nonsingular matrices P, Q such that $PAQ = J_2$, so that the assumption $AB = 0$ implies that $(PAQ)(Q^{-1}B) = 0$. Show that this leads to a contradiction.

6. Defining the *nullity* $n(A)$ of an $n \times n$ matrix A as $n - r(A)$, *Sylvester's law of nullity* reads

$$\max\{n(A), n(B)\} \leqq n(AB) \leqq n(A) + n(B)$$

In terms of rank, this becomes

$$r(A) + r(B) - n \leqq r(AB) \leqq \min\{r(A), r(B)\}$$

As such, Theorem 3.9 proves the right inequality. Prove the left one using the following suggestions: First show that if, in an $n \times n$ matrix of rank r, s rows are selected, the rank of the matrix so obtained is greater than or equal to $r + s - n$. Then prove the inequality for the case $A = J_s$. Extend this to the general case using a P, Q such that $PAQ = J_s$.

7. Show that if a square matrix A is known to have a unique left inverse B (that is, $BA = I$), then B is also a right inverse. HINT: Consider $A^* = AB - I + B$.

8. Using the final statement in Theorem 3.17, show that every $n \times n$ matrix can be written as a product of elementary matrices and matrices of the form $R_i(0)$. Notice that this result generalizes (ii) of Theorem 3.14.

9. In Theorem 3.14 deduce that (i) implies (iii) without using elementary matrices.

CHAPTER 4

DETERMINANTS

4.1. Introduction. This paragraph is addressed to those readers who have encountered determinants previously. Undoubtedly these functions were introduced as a device for expressing the solution of a system of n linear equations in n unknowns, as well as investigating the consistency of any linear system. An understanding of Chap. 1 should dispel the illusion of the necessity of determinants in connection with any aspect of a system of linear equations. However, the notion is useful and consequently deserving of our attention. The following treatment may be different from the one previously encountered by the student. If so, it is hoped that he will find it less tiresome and less involved in uninteresting details. In this connection we ask only that the discussion be approached with an open mind.

The concept of a determinant might arise in the following way: If the rank of a system of n linear equations in n unknowns

$$AX = Y \tag{1}$$

is n, we know that a unique solution X_0 exists. After a moment's reflection upon the procedure used to obtain the solved form of (1) according to the method of Chap. 1, it becomes clear that each x_0^i is expressible as a linear combination of the y^i's where the coefficients are obtained from the a_j^i's by rational operations. Hence, if the a_j^i's are regarded as variables, the coefficient of each y^i is a function of these variables. Whereas the method of Chap. 1 gives the value of each of these functions for a specified set of values for the a_j^i's, it does not explicitly determine the form of the functions themselves. The goal of determinant theory is to obtain explicit representations for the functions that enter into the formula expressing x_0^i in terms of the a_j^i's and y^i's.†

Next we describe a problem that appears to be less difficult than the

†The state of affairs is comparable with the following: Using the method of completing the square, any quadratic equation can be solved. However, it is not until the quadratic formula is developed that we have an explicit representation for the roots of a general quadratic in terms of its coefficients.

foregoing. Given a system of n homogeneous linear equations in n unknowns, with coefficients in F,

$$AX = 0 \tag{2}$$

to devise a function on the set of square matrices over F to F that vanishes at A if and only if the system has a nontrivial solution—in other words, to devise an eliminant for the system.† Of course, we may determine whether a specified system has a nontrivial solution by applying the methods of Chap. 1.

It is this second problem that we shall investigate. This investigation leads to the so-called "determinant function" of nth-order matrices. With this function at our disposal it is then possible to solve the first problem. That is, the coefficients of the y^i's become values of the determinant function.

4.2. Postulates for a Determinant Function. Our second problem calls for the construction of a certain type of function on $n \times n$ matrices over F whose values are in F. At the moment, any such function will be called a determinant and denoted by det. The function value at A will be written det$A$ and called the determinant of the matrix A. If it is desired to indicate the elements of A, the field element det$A$ will be written $\det(a_j^i)$ or

$$\det\begin{pmatrix} a_1^1 & a_2^1 & \cdots & a_n^1 \\ a_1^2 & a_2^2 & \cdots & a_n^2 \\ \cdot & \cdot & \cdots & \cdot \\ a_1^n & a_2^n & \cdots & a_n^n \end{pmatrix}$$

A determinant function of nth-order matrices is said to be of nth order.

Since an nth-order matrix is an ordered set of n row vectors, an nth-order determinant may be regarded as a function of n variables $\alpha^1, \alpha^2, \ldots, \alpha^n$ in $V_n(F)$. With this viewpoint function values will be denoted by

$$\det\,(\alpha^1, \alpha^2, \ldots, \alpha^n)^T$$

If we fix all row vectors except the ith, a matrix A with a variable ith row is defined; a determinant function when restricted to such a collection of matrices will be designated by $\det_i^A$ and the function value at α, a specific choice for the ith row of A, by $\det_i^A \alpha$.

†In view of Theorem 1.13, together with the definition of a singular matrix, the problem can be rephrased thus: Construct a function on the set of square matrices over F to F that vanishes at A if and only if A is singular.

We now enumerate the minimum requirements for a determinant function together with our justification for the choice. It will develop that there is exactly one function meeting these requirements, so that eventually the word "determinant" will acquire a precise meaning.

DEFINITION. *A function det of the row vectors α^i of nth-order matrices A is a* determinant *if it satisfies the following postulates:*

D_1. *Viewed as a function of any one row it is homogeneous:*

$$\det_i^A(c\alpha^i) = c \cdot \det_i^A \alpha^i \qquad \text{all } c \text{ in } F$$

D_2. *With the same viewpoint*

$$\det_i^A(\alpha^i + \alpha^k) = \det_i^A \alpha^i \qquad i \neq k$$

D_3. *Its value is* 1 *if* $\alpha^i = \epsilon^i$, *the ith unit vector,* $i = 1, 2, \ldots, n$. *In other words,*

$$\det I = 1$$

These postulates have been chosen with the second problem of the previous section in mind. Thus, if $c \neq 0$, D_1 states that a system derived from (2) by a type II row operation is equivalent to (2). If $c = 0$, the ith equation of (2) vanishes; hence the resulting system has a nontrivial solution, and we demand that the determinant of the corresponding matrix vanish. Postulate D_2, together with D_1, indicates that a type III row operation on (2) yields an equivalent system. Finally, in D_3 appears an instance of our demand that the coefficient matrix of a system with only the trivial solution have a nonzero determinant. The value 1 has been chosen because not only is this a nonzero element in every field but in addition a multiplicative property of determinant functions (see D_{11}) assumes its simplest form with this choice.

The question as to whether or not determinant functions exist will not be settled until later. Now we shall derive several consequences of the postulates; in this way we shall obtain further information about such functions *if they exist.*

D_4. *If one row of A is the zero vector, then* $\det A = 0$. *That is,*

$$\det_i^A 0 = 0$$

This follows from D_1 upon setting $c = 0$.

D_5. *The addition to one row of a multiple of another row does not alter the value of a determinant. That is,*

$$\det_i^A(\alpha^i + c\alpha^k) = \det_i^A \alpha^i \qquad i \neq k$$

We prove this for $c \neq 0$, which is the only case of any interest. Application of D_1 and D_2 gives

$$\det(\ldots, \alpha^i, \ldots, \alpha^k, \ldots)^T = \frac{1}{c}\det(\ldots, \alpha^i, \ldots, c\alpha^k, \ldots)^T$$

$$= \frac{1}{c}\det(\ldots, \alpha^i + c\alpha^k, \ldots, c\alpha^k, \ldots)^T$$

$$= \det(\ldots, \alpha^i + c\alpha^k, \ldots, \alpha^k, \ldots)^T$$

where we have indicated only those arguments which enter into the computation.

D_6. *If the set of row vectors* $\{\alpha^1, \alpha^2, \ldots, \alpha^n\}$ *is linearly dependent, then* $\det(\alpha^1, \alpha^2, \ldots, \alpha^n)^T = 0$.

By assumption some one row vector, for example α^i, is a linear combination of the remaining rows. Repeated use of D_5 gives $\det A = \det_i^A 0$, which is 0 by D_4.

Observe in particular that D_6 implies that a matrix with two equal rows has zero determinant.

D_7. *A determinant function* $\det_i^A$ *is additive. That is,*

$$\det_i^A(\beta + \gamma) = \det_i^A\beta + \det_i^A\gamma$$

There is no loss of generality if we prove this for $i = 1$. Since this case has the advantage that the notation is less cumbersome, we shall so restrict the proof. First there is the trivial case where the set $\{\alpha^2, \alpha^3, \ldots, \alpha^n\}$ is linearly dependent. Then each of the sets $\{\beta + \gamma, \alpha^2, \ldots, \alpha^n\}$ and $\{\beta, \alpha^2, \ldots, \alpha^n\}$ and $\{\gamma, \alpha^2, \ldots, \alpha^n\}$ is dependent so that each term in D_7 is zero by D_6. Next assume that $\{\alpha^2, \alpha^3, \ldots, \alpha^n\}$ is a linearly independent set, and extend it to a basis for $V_n(F)$ by adjoining a vector α^*. Then β and γ can be written as linear combinations of this basis:

$$\beta = c_1\alpha^* + c_2\alpha^2 + \cdots + c_n\alpha^n$$

$$\gamma = d_1\alpha^* + d_2\alpha^2 + \cdots + d_n\alpha^n$$

If we now replace

$$\beta \text{ by } \beta - \sum_{k=2}^{n} c_k\alpha^k = c_1\alpha^* \qquad \text{in } \det_1^A\beta$$

$$\gamma \text{ by } \gamma - \sum_{k=2}^{n} d_k\alpha^k = d_1\alpha^* \qquad \text{in } \det_1^A\gamma$$

$$\beta + \gamma \text{ by } \beta + \gamma - \sum_{k=2}^{n} (c_k + d_k)\alpha^k = (c_1 + d_1)\alpha^* \qquad \text{in } \det_1^A(\beta + \gamma)$$

the respective determinant values are unchanged according to D_5 used repeatedly. Hence, according to D_1,

$$\det_1^A \beta = c_1 \cdot \det_1^A \alpha^*$$

$$\det_1^A \gamma = d_1 \cdot \det_1^A \alpha^*$$

$$\det_1^A (\beta + \gamma) = (c_1 + d_1) \cdot \det_1^A \alpha^*$$

From these the desired result follows upon adding the first two equations and comparing the result with the third.

D_8. *An interchange of two rows in* $\det(\alpha^1, \alpha^2, \ldots, \alpha^n)^T$ *changes the sign.*

This is seen from the following computation where only those arguments, *viz.*, the ith and jth, that enter into the computation have been indicated:

$$\begin{aligned} \det(\ldots, \alpha^i, \ldots, \alpha^j, \ldots)^T &= \det(\ldots, \alpha^i, \ldots, \alpha^i + \alpha^j, \ldots)^T \\ &= \det(\ldots, -\alpha^j, \ldots, \alpha^i + \alpha^j, \ldots)^T \\ &= -\det(\ldots, \alpha^j, \ldots, \alpha^i, \ldots)^T \end{aligned}$$

D_9. *Let* $i_1, i_2, \ldots, i_n$ *denote an arrangement of the superscripts* $1, 2, \ldots, n$. *Then*

$$\det(\alpha^{i_1}, \alpha^{i_2}, \ldots, \alpha^{i_n})^T = \pm\det(\alpha^1, \alpha^2, \ldots, \alpha^n)^T$$

where the prefixed algebraic sign is uniquely determined by the given arrangement of subscripts if $\det(\alpha^1, \alpha^2, \ldots, \alpha^n)^T \neq 0$.

It is clear that any arrangement $i_1, i_2, \ldots, i_n$ of $1, 2, \ldots, n$ can be reverted to the latter order by a finite number of interchanges of pairs. This can be done in an unlimited number of ways; we mention only one systematic method. If $i_1 \neq 1$, interchange 1 and i_1 to obtain an arrangement $1, j_2, j_3, \ldots, j_n$. If $j_2 \neq 2$, interchange 2 and j_2, etc.

Hence if a set of r interchanges of pairs rearranges $i_1, i_2, \ldots, i_n$ in the order $1, 2, \ldots, n$, the set of r corresponding interchanges of pairs of rows in $\det(\alpha^{i_1}, \alpha^{i_2}, \ldots, \alpha^{i_n})^T$ changes this value to $(-1)^r\det(\alpha^1, \alpha^2, \ldots, \alpha^n)^T$ by D_8. If a second set of s interchanges also reduces $i_1, i_2, \ldots, i_n$ to $1, 2, \ldots, n$ so that

$$\begin{aligned} \det(\alpha^{i_1}, \alpha^{i_2}, \ldots, \alpha^{i_n})^T &= (-1)^s\det(\alpha^1, \alpha^2, \ldots, \alpha^n)^T \\ &= (-1)^r\det(\alpha^1, \alpha^2, \ldots, \alpha^n)^T \end{aligned}$$

it follows that $(-1)^r = (-1)^s$ if $\det(\alpha^1, \alpha^2, \ldots, \alpha^n)^T \neq 0$.

The significance of D_9 can be stated as follows, in terms of the equation $(-1)^r = (-1)^s$: If $\det(\alpha^1, \alpha^2, \ldots, \alpha^n) \neq 0$, then the algebraic sign in D_9 is plus or minus according as an even or an odd number of interchanges of pairs will revert the arrangement $i_1, i_2, \ldots, i_n$ to $1, 2, \ldots, n$.

PROBLEMS

1. Show that $n!$ arrangements $i_1, i_2, \ldots, i_n$ of $1, 2, \ldots, n$, $n > 1$, are possible and that for one half of these, $\det(\alpha^{i_1}, \alpha^{i_2}, \ldots, \alpha^{i_n})^{\mathrm{T}} = \det(\alpha^1, \alpha^2, \ldots, \alpha^n)^{\mathrm{T}}$, while for the other half, a minus sign occurs.

2. In the Euclidean plane let α_j be the position vector of a point P_j with rectangular coordinates $(a_j^1, a_j^2)^{\mathrm{T}}$, $j = 1, 2$. If $V(\alpha_1, \alpha_2)$ is the area of the parallelogram with α_1 and α_2 as adjacent edges, show that if the algebraic sign of V is disregarded, it satisfies axioms D_1, D_2, and D_3, restated for columns. (This example is continued at the end of Sec. 4.4.)

3. (An alternate approach to determinant functions.) In view of the footnote on page 87 and the fact that a product of two nth-order matrices is singular if and only if at least one factor is singular, one might approach the second problem of Sec. 4.1 as follows: To find a function d of nth-order matrices A such that $d(AB) = d(A)d(B)$. If we further postulate that d is to be nonconstant and the simplest polynomial in the entries of a matrix, it is possible to show that $d(A) = \det A$. For this, recall the results in Prob. 8, Sec. 3.8. Then prove that

(*a*) The condition $d(AB) = d(A)d(B)$ for all A and all B is implied by the condition $d(AB) = d(A)d(B)$ for A arbitrary and B one of I, R_{ij}, $R_i(c)$, $R_{ij}(c)$, $R_i(0)$.

(*b*) $d(0) = 0$, $d(I) = 1$, $d(R_{ij}(c)) = 1$, $d(R_{ij}) = -1$, $d(R_i(c)) = c^k$, where $k > 0$.

(*c*) The function d satisfies D_1, D_2, D_3 with $k = 1$ in (*b*).

4.3. Further Properties of Determinants. In order to obtain further properties of a det function, we introduce a second nth-order matrix $B = (b_j^i)$ into the discussion. Let $\{\beta^1, \beta^2, \ldots, \beta^n\}$ denote its row vectors and form

$$C = AB$$

with A regarded as an $n \times n$ matrix and B as an $n \times 1$ matrix to obtain C as a column matrix of its row vectors

$$\gamma^i = a_1^i\beta^1 + a_2^i\beta^2 + \cdots + a_n^i\beta^n \qquad i = 1, 2, \ldots, n \tag{3}$$

We compute $\det(\gamma^1, \gamma^2, \ldots, \gamma^n)^{\mathrm{T}}$ by applying D_7 (which obviously can be extended to the case of n summands) to γ^1 to decompose this value into n summands; then in each of these we do the same to γ^2, etc. This gives

$$\begin{aligned} \det AB &= \det(\gamma^1, \gamma^2, \ldots, \gamma^n)^{\mathrm{T}} \\ &= \sum_{j_1, \ldots, j_n} \det(a_{j_1}^1\beta^{j_1}, a_{j_2}^2\beta^{j_2}, \ldots, a_{j_n}^n\beta^{j_n})^{\mathrm{T}} \\ &= \sum_{j_1, \ldots, j_n} a_{j_1}^1 a_{j_2}^2 \cdots a_{j_n}^n \det(\beta^{j_1}, \beta^{j_2}, \ldots, \beta^{j_n})^{\mathrm{T}} \end{aligned} \tag{4}$$

where each index j_k runs independently from 1 to n. Since for any set of index values in which two are equal, $\det(\beta^{j_1}, \beta^{j_2}, \ldots, \beta^{j_n})^{\mathrm{T}} = 0$, we retain only those terms in (4) for which $j_1, j_2, \ldots, j_n$ is an arrangement of 1, 2, . . . , n. Using D_9, this gives

$$\det AB = \Big(\sum_{j_1,\ldots,j_n} \pm a^1_{j_1} a^2_{j_2} \cdots a^n_{j_n}\Big) \det(\beta^1, \beta^2, \ldots, \beta^n)^T \tag{5}$$

where $j_1, j_2, \ldots, j_n$ vary over all $n!$ arrangements of 1, 2, . . . , n and where $\pm$ stands for the sign associated with that arrangement (see the remark at the end of the preceding section).

Many conclusions can be derived from (5). First, it is important to observe that this formula is a consequence of the postulates D_1 and D_2 alone. Thus, if in (5) we specialize B to the identity matrix, we obtain

$$\det A = c \cdot \sum_{j_1,\ldots,j_n} \pm a^1_{j_1} \cdots a^n_{j_n} \qquad c = \det I \tag{6}$$

as the explicit formula for functions satisfying D_1 and D_2, provided they exist at all. We now introduce D_3 into the discussion, substitute $\det I = 1$ in (6), and obtain our next result.

D_{10}. *If determinants exist, then the value of a determinant function at a matrix A is uniquely determined by postulates D_1, D_2, and D_3 and an explicit formula is*

$$\det A = \sum_{j_1,\ldots,j_n} \pm a^1_{j_1} a^2_{j_2} \cdots a^n_{j_n} \tag{7}$$

where the summation extends over the $n!$ arrangements $j_1, j_2, \ldots, j_n$ of 1, 2, . . . , n and the prefixed sign is plus or minus according as an even or an odd number of interchanges of pairs reverts $j_1, j_2, \ldots, j_n$ to 1, 2, . . . , n.

A straightforward calculation will verify that the function defined by (7) meets all requirements for a determinant. However, we prefer to present later an existence proof by induction that has the advantage of providing a practical method for evaluating determinants.

We return now to formula (5), and substitute (7) in it to obtain an important multiplicative property of determinants.

D_{11}. *The determinant of the product of two matrices is the product of the determinants of the factors:*

$$\det AB = \det A \cdot \det B \tag{8}$$

Although up to this point we have discussed the det function exclusively as a function of the row vectors of a matrix, every conclusion has its column analogue. This will be deduced from our next conclusion.

D_{12}. *The determinant of a matrix is equal to the determinant of its transpose:*

$$\det A^T = \det A$$

For proof we define a function f of nth-order matrices by the equation

$$f(A) = \det A^{\mathrm{T}}$$

and verify that f satisfies D_1, D_2, and D_3 to conclude that $f(A) = \det A$ and hence that $\det A = \det A^{\mathrm{T}}$. To obtain these properties of f, we use elementary matrices and a consequence of D_{11}, namely, $\det AB = \det BA$. The following two chains of equalities show that f satisfies D_1 and D_2, respectively:

$$f(R_i(c)A) = \det A^{\mathrm{T}}R_i(c) = \det R_i(c)A^{\mathrm{T}} = c \cdot \det A^{\mathrm{T}} = c \cdot f(A)$$

$$f(R_{ik}(1)A) = \det A^{\mathrm{T}}R_{ki}(1) = \det R_{ki}(1)A^{\mathrm{T}} = \det A^{\mathrm{T}} = f(A)$$

Finally, the equations

$$f(I) = f(I^{\mathrm{T}}) = \det I = 1$$

demonstrate that f obeys D_3. This completes the proof.

D_{13}. *All properties of a determinant, when regarded as a function of row vectors, are valid when it is regarded as a function of column vectors.*

PROBLEMS

1. If D_3 is weakened to read: Its value is $c \neq 0$ if α^i is the unit vector ϵ^i, $i = 1, 2, \ldots, n$, what product formula for a determinant function can be derived?

2. Prove D_{12} by verifying that the determinant function with values given by formula (7) satisfies D_1, D_2, and D_3 stated for columns instead of rows.

3. Show that if the "row by column" rule for multiplying matrices is replaced by either the "row by row" or "column by column" rule, D_{11} is valid.

4. Use formula (7) to write out the six terms with proper signs which occur in the determinant of a third-order matrix.

5. Check D_{11} for the case of second-order matrices directly.

6. Prove that if A is a skew-symmetric matrix ($A = -A^{\mathrm{T}}$) of odd order, $\det A = 0$.

7. Prove that if each entry of the nth-order matrix $A = (a_j^i)$ is a differentiable function of x, $a_j^i = a_j^i(x)$, then $\det A$ is a differentiable function of x and its derivative is the sum of n determinants obtained from A by successively replacing the first, second, etc., row by the derivatives of the entries of that row.

4.4. A Proof of the Existence of Determinants. For a one-rowed matrix (a_1^1) the element a_1^1 is the determinant. We assume the existence of $(n-1)$st-order determinants and prove the existence of nth-order determinants. We can associate with an nth-order matrix $A = (a_j^i)$ the determinants of certain $(n-1)$-rowed submatrices in the following way: Let a_j^i denote a fixed element in A and A_i^j denote $(-1)^{i+j}$ times the determinant of the submatrix obtained by deleting the row and column containing a_j^i. The element A_i^j is called the *cofactor of* a_j^i. Consider now the function d of nth-order matrices whose value at A is

$$d(A) = a_j^1 A_1^j + a_j^2 A_2^j + \cdots + a_j^n A_n^j \qquad 1 \leq j \leq n \tag{9}$$

This is the sum of the products of the elements of the jth column of A and their respective cofactors.

We shall prove that d satisfies D_1, D_2, and D_3, from which it will follow that $d = \det$. The symbols d_i^A and $d_i^A(\alpha)$ will have their usual meaning. For D_1 and D_2 the straightforward computation of $d_i^A(c\alpha^i)$ and $d_i^A(\alpha^i + \alpha^k)$ establishes the desired equalities. Moreover, D_3 holds. For if we set $\alpha^i = \epsilon^i$, $i = 1, 2, \ldots, n$, then $a_j^i = 0$ for all $j \neq i$ while $a_i^i = A_i^i = 1$. Substitution in (9) gives $d(I) = 1$.

This completes the proof of the existence of a determinant function. Owing to the uniqueness property D_{10} it follows that $\det A$ is given by (9), the so-called *development of a determinant according to its jth column.* The advantage of this existence proof is that it simultaneously gives a systematic procedure for expressing the determinant of an nth-order matrix in terms of n determinants of $(n - 1)$st-order matrices. That this is worth while is realized as soon as one notices how rapidly $n!$, the number of terms in the expanded form (7), increases with n. Later in the chapter the systematic use of (9) in computations will be discussed. First we wish to generalize (9) to the following property of determinants:

D_{14}. *The sum of the products of the elements of a row (column) in A by the corresponding cofactors of a row (column) in A is equal to* $\det A$ *or* 0, *according as the cofactors selected belong to the row (column) used, or a different row (column). That is,*

$$a_j^1 A_1^i + a_j^2 A_2^i + \cdots + a_j^n A_n^i = \begin{cases} \det A & \text{if } i = j \\ 0 & \text{if } i \neq j \end{cases} \tag{10}$$

and

$$a_1^i A_j^1 + a_2^i A_j^2 + \cdots + a_n^i A_j^n = \begin{cases} \det A & \text{if } j = i \\ 0 & \text{if } j \neq i \end{cases} \tag{11}$$

These formulas can be written more compactly using the *Kronecker delta* δ_j^i, as follows:

$$\sum_k a_j^k A_k^i = \delta_j^i \det A \tag{10'}$$

$$\sum_k a_k^i A_j^k = \delta_j^i \det A \tag{11'}$$

In the way of proofs for (10) and (11), notice that the first part of (10) is formula (9). To establish the second part, replace the ith column in A by the jth column, and develop according to this new column. Since for $i \neq j$ the determinant of this new matrix is 0, the result follows. Formula (11) follows upon an interchange of the rows and the columns in A together with property D_{12}.

PROBLEMS

1. In view of Prob. 2, Sec. 4.2, one might tentatively adopt the following requirements for a function V of n column vectors $\alpha_1, \alpha_2, \ldots, \alpha_n$ such that $V(\alpha_1, \alpha_2, \ldots, \alpha_n)$ is the volume of a parallelotope (the n-dimensional analogue of a parallelogram) with $\alpha_1, \alpha_2, \ldots, \alpha_n$ as adjacent edges: V should satisfy postulates D_1, D_2, and D_3 stated for columns with D_1 modified to read $V_i^A(c\alpha_i) = |c| V_i^A(\alpha_i)$, and finally $V(\alpha_1, \alpha_2, \ldots, \alpha_n) \geqq 0$. Show that there exists a unique function, namely, $|\det A|$ with A the matrix whose columns are $\alpha_1, \alpha_2, \ldots, \alpha_n$, meeting these requirements. Thus this is the volume formula to adopt.

2. Using the previous problem, set up a formula for the volume of an n-dimensional simplex (the generalization of a triangle in two dimensions and a tetrahedron in three dimensions). For the three-dimensional case check your result as follows: Show that the equation of the plane containing the point P_j with rectangular coordinates $(a_j^1, a_j^2, a_j^3)^T$, $j = 1, 2, 3$, may be written in the form

$$f(x^1, x^2, x^3) = \det \begin{pmatrix} a_1^1 & a_2^1 & a_3^1 & x^1 \\ a_1^2 & a_2^2 & a_3^2 & x^2 \\ a_1^3 & a_2^3 & a_3^3 & x^3 \\ 1 & 1 & 1 & 1 \end{pmatrix} = 0 .$$

Next show that the cofactor of x^i is the area of the projection of the triangle $P_1P_2P_3$ upon the x^jx^k plane, where i, j, k is an arrangement of 1, 2, 3, and that the square root of the sum of the squares of these cofactors equals the area of this triangle. Finally, prove that the volume of the tetrahedron $P_0P_1P_2P_3$, where P_0 has the rectangular coordinates (x_0^1, x_0^2, x_0^3), is $\frac{1}{6}f(x_0^1, x_0^2, x_0^3)$.

3. If

$$f(n) = \det \begin{pmatrix} a_1 & 1 & 0 & 0 & \cdots & 0 \\ -1 & a_2 & 1 & 0 & \cdots & 0 \\ 0 & -1 & a_3 & 1 & \cdots & 0 \\ \cdot & \cdot & \cdot & \cdot & \cdots & \cdot \\ 0 & 0 & 0 & 0 & \cdots & a_n \end{pmatrix}$$

show that $f(n) = a_n f(n-1) + f(n-2)$.

4. Show that if $\det A$ is regarded as a function of the variables a_j^i, then $\partial \det A / \partial a_j^i = A_i^j$.

4.5. The Adjoint. The combinations of terms that occur in (10)′ and (11)′ resemble those which appear in the product of two matrices. Experimentation leads us to introduce the following matrix:

DEFINITION. *The* adjoint *of a square matrix A, in symbols adj$A$, is the matrix (A_j^i) with the cofactor A_j^i of a_i^j in the (i, j)th position.*

Now (10)′ and (11)′ can be written in the compact form

$$\text{adj}A \cdot A = A \cdot \text{adj}A = \det A \cdot I$$

where the final matrix is the scalar matrix with $\det A$ as the common principal diagonal element. It follows that if $\det A \neq 0$, A has an inverse:

$$A^{-1} = \frac{1}{\det A}\,\text{adj}A \tag{12}$$

Conversely, if A has an inverse, the multiplicative property of determinants when applied to the equation $AA^{-1} = I$ gives $\det A \cdot \det A^{-1} = 1$ so that $\det A \neq 0$. This result coupled with Theorem 3.14 establishes the equivalence of the following properties of a square matrix A: $\det A \neq 0$, A^{-1} exists, A is nonsingular.

It is now possible to verify that determinants provide answers to the questions raised at the beginning of the chapter. The first question was that of determining explicit formulas for the unique solution of the system $AX = Y$, where A is nonsingular, and in particular a representation for the function whose values enter into the formula for x_0^i as a linear combination of the y^i's. Starting with $AX_0 = Y$, where A is nonsingular, gives

$$X_0 = A^{-1}Y = \frac{1}{\det A}\,\text{adj}A \cdot Y$$

using (12). Equating corresponding entries of the initial and final matrix gives the desired result, *viz.*,

$$x_0^i = \frac{A_1^i}{\det A}\,y^1 + \frac{A_2^i}{\det A}\,y^2 + \cdots + \frac{A_n^i}{\det A}\,y^n \qquad i = 1, 2, \ldots, n \tag{13}$$

Incidentally, this manner of obtaining the solution can be interpreted as an instance of the elimination method of Chap. 1 for any system having a nonsingular coefficient matrix. Recalling the corollary to Theorem 1.2, we choose for the multiplier c_k the value A_k^i, $k = 1, 2, \ldots, n$, to obtain as the new ith equation

$$\det A \cdot x^i = A_1^i y^1 + A_2^i y^2 + \cdots + A_n^i y^n$$

from which follows the existence of a unique solution, given by (13).

The above result is usually summed up in the following statement, known as *Cramer's rule*:

Theorem 4.1. Let $AX = Y$ be a system of nonhomogeneous equations with $\det A \neq 0$. Let Y_k be the matrix derived from A by replacing its kth column by Y. Then the system has the unique solution

$$\left(\frac{\det Y_1}{\det A}, \frac{\det Y_2}{\det A}, \ldots, \frac{\det Y_n}{\det A}\right)^T$$

Proof. Develop the numerator $\det Y_i$ according to its ith column to obtain the right member of (13).

EXAMPLE 4.1

Consider a set of n functions $\{f_1, f_2, \ldots, f_n\}$ each of $n + m$ variables $y_1, \ldots, y_n, x_1, \ldots, x_m$. Under well-known conditions the system of n equations

$$(14) \qquad f_k(Y, X) = 0 \qquad k = 1, 2, \ldots, n; \qquad Y = (y_1, y_2, \ldots, y_n)^T; X = (x_1, x_2, \ldots, x_m)^T$$

implicitly determines a set of n functions $F_1, F_2, \ldots, F_n$ of X having first partial derivatives and such that

$$Y = F(X) \qquad F(X) = (F_1(X), F_2(X), \ldots, F_n(X))^T$$

throughout a region which includes a point $P(Y^0, X^0)$, where every f_k vanishes. One of these conditions is that the functional determinant

$$\det \begin{pmatrix} \frac{\partial f_1}{\partial y_1} & \frac{\partial f_1}{\partial y_2} & \cdots & \frac{\partial f_1}{\partial y_n} \\ \frac{\partial f_2}{\partial y_1} & \frac{\partial f_2}{\partial y_2} & \cdots & \frac{\partial f_2}{\partial y_n} \\ \cdot & \cdot & \cdots & \cdot \\ \frac{\partial f_n}{\partial y_1} & \frac{\partial f_n}{\partial y_2} & \cdots & \frac{\partial f_n}{\partial y_n} \end{pmatrix}$$

not vanish at P. This determinant is called the *Jacobian* of the n functions f_k with respect to the n variables y_l and is usually abbreviated to

$$J \quad \text{or} \quad \frac{\partial(f_1, f_2, \ldots, f_n)}{\partial(y_1, y_2, \ldots, y_n)}$$

Now it is precisely the nonvanishing of J that permits one to compute the first partial derivatives $\partial y_l / \partial x_j$ using Cramer's rule. Indeed, partial differentiation of (14) with respect to x_j gives the system of linear equations

$$(15) \qquad \sum_{l=1}^{n} \frac{\partial f_k}{\partial y_l} \cdot \frac{\partial y_l}{\partial x_j} = -\frac{\partial f_k}{\partial x_j} \qquad k = 1, 2, \ldots, n$$

whose coefficient matrix is precisely J. Thus when $J \neq 0$, the solution of (15) for a particular derivative $\partial y_l / \partial x_j$ may be written

$$\frac{\partial y_l}{\partial x_j} = -\frac{\dfrac{\partial(f_1, f_2, \ldots, f_l, \ldots, f_n)}{\partial(y_1, y_2, \ldots, x_j, \ldots, y_n)}}{J}$$

PROBLEMS

1. Devise a proof of the equivalence of the properties $\det A \neq 0$ and that A is nonsingular, which does not involve the mention of A^{-1}.

2. Show that if A is an nth-order matrix of rank $n - 1 > 0$, $\operatorname{adj} A$ has rank 1.

3. Show that $\det(\operatorname{adj} A) = (\det A)^{n-1}$ if A is of order n.

4. Show that $\operatorname{adj}(\operatorname{adj} A) = (\det A)^{n-2} A$ if $n > 2$ and $\operatorname{adj}(\operatorname{adj} A) = A$ if $n = 2$.

5. Solve Probs. 3 and 5 in Sec. 1.5 by Cramer's rule.

6. Let

$$\sum_{j=1}^{n} a_j^i x^j = 0 \qquad i = 1, 2, \ldots, n - 1$$

be a system of $n - 1$ homogeneous equations in n unknowns. Let A_j equal $(-1)^{j+1}$ times the determinant of the matrix obtained by suppressing the jth column of the coefficient matrix. Show that $(A_1, A_2, \ldots, A_n)^T$ is a solution. A system of n equations in n unknowns which has rank n can be solved by this method. Rewrite

$$\sum_{j=1}^{n} a_j^i x^j = y^i \qquad \text{as} \qquad \sum_{j=1}^{n} a_j^i x^j - y^i x^{n+1} = 0 \qquad i = 1, 2, \ldots, n$$

and determine solutions for which $x^{n+1} = 1$. Prove Cramer's rule in this way.

4.6. Matrices with Determinant Zero. Recalling the second problem proposed at the beginning of the chapter, *viz.*, to devise a function of nth-order matrices which vanishes at A if and only if the homogeneous system $AX = 0$ has a nontrivial solution, we claim that the determinant function is a solution. This is an immediate consequence of the equivalence of the nonvanishing of $\det A$ and the nonsingularity of A mentioned in the previous section. For clearly the latter property of A is synonymous with $AX = 0$ having only the trivial solution. A final rephrasing of this same result is stated next.

Theorem 4.2. $\det A = 0$ if and only if the row (or column) vectors of A constitute a linearly dependent set.

One might inquire next as to whether or not this theorem could be given a quantitative phrasing in the event $\det A = 0$, that is, if a "degree of vanishing" could be defined which would enable one to predict the "degree of dependence" among the row vectors of A. We shall answer the complementary question which relates nonvanishing with independence, indeed not only for square matrices but, more generally, for rectangular matrices. The new notion that is needed in this connection is described below:

DEFINITION. *Let $A = (a_j^i)$ be an $m \times n$ matrix with entries in a field F.* *A* minor *of A is the determinant of any square submatrix of A. The* order *of a minor is the order of that submatrix used to compute the minor.*

For example, if A is square, the cofactor A_i^j of a_j^i is a signed minor of A. The result we have in mind now follows:

Theorem 4.3. Let r' denote the order of a nonzero minor of maximum order of the $m \times n$ matrix $A \neq 0$. If $A = 0$, set $r' = 0$. Then, in every case $r' = r(A)$.

Proof. We may restrict our attention to the case where $A \neq 0$, in which case $r' > 0$. Then we may assume the notation to be chosen so that

$$B = \begin{vmatrix} a_1^1 & a_2^1 & \cdots & a_{r'}^1 \\ a_1^2 & a_2^2 & \cdots & a_{r'}^2 \\ \cdot & \cdot & \cdots & \cdot \\ a_1^{r'} & a_2^{r'} & \cdots & a_{r'}^{r'} \end{vmatrix}$$

is a nonzero minor of order r'. Then the set of row vectors $\{\alpha^1, \alpha^2, \ldots, \alpha^{r'}\}$ of A is linearly independent. For a linear dependence among them would imply a linear relation among the rows of B and would therefore imply $\det B = 0$. Thus $r \geqq r'$ since r may be interpreted as the dimension of the row space of A.

If $r' = m$, then $r \leqq r'$ and consequently $r = r'$. If $r' < m$, then every matrix

$$B_t^s = \begin{pmatrix} a_1^1 & \cdots & a_{r'}^1 & a_t^1 \\ \cdot & \cdots & \cdot & \cdot \\ a_1^{r'} & \cdots & a_{r'}^{r'} & a_t^{r'} \\ a_1^s & \cdots & a_{r'}^s & a_t^s \end{pmatrix} \qquad \begin{matrix} s = r' + 1, \ldots, m \\ t = 1, 2, \ldots, n \end{matrix}$$

has vanishing determinant. For if $t \leqq r'$, B_t^s has two equal columns, while if $t > r'$, B_t^s is a submatrix of order $r' + 1$ in A and hence has determinant 0 by assumption. The cofactor of the element a_t^i in the last column of B_t^s is independent of t, and we may denote it by A_i^s. Developing $\det B_t^s$ according to its last column gives

$$A_1^s a_t^1 + A_2^s a_t^2 + \cdots + A_{r'}^s a_t^{r'} + A_s^s a_t^s = 0$$

Since this holds for $t = 1, 2, \ldots, n$

$$A_1^s \alpha^1 + A_2^s \alpha^2 + \cdots + A_{r'}^s \alpha^{r'} + A_s^s \alpha^s = 0$$

By assumption $A_s^s = \det B \neq 0$, so that every vector α^s, $s = r' + 1, \ldots, m$, is linearly dependent upon the vectors $\alpha^1, \alpha^2, \ldots, \alpha^{r'}$. This again gives $r \leqq r'$ and consequently $r = r'$.

PROBLEMS

1. Let A be an nth-order matrix with entries in a field F and λ an indeterminate over F, that is, a symbol which we assume can be combined with elements of F and which commutes with all elements of $F(a\lambda = \lambda a$ for a in $F)$. Then $\det(\lambda I_n - A)$ is a polynomial in λ called the characteristic polynomial of A. Prove that the coefficient of $(-1)^k\lambda^{n-k}$ in the characteristic polynomial of A is the sum of all principal minors of order k of A. A *principal minor* of A is the determinant of a submatrix whose principal diagonal is along the principal diagonal of A. For example, the diagonal elements a_i^i are principal minors of order 1, and their cofactors A_i^i are principal minors of order $n - 1$.

2. Regarding the rank of a matrix as the order of a nonzero minor of maximum order, show that the rank of a matrix is unaltered by the application of any elementary transformation upon the matrix.

3. Let $A = (a_j^i)$ be an nth-order matrix of real numbers having the properties (i) $a_i^i > 0$; (ii) $a_j^i < 0$ for $i \neq j$; (iii) $\sum_{j=1}^n a_j^i > 0$. Give an indirect proof that $\det A \neq 0$, using the fact that if $\det A = 0$, the system $AX = 0$ has a nontrivial solution.

4.7. Evaluation of Determinants. The rapidity with which $n!$, the number of terms in formula (7) for $\det A$, increases with n has already been mentioned. Thus any suggestions for reducing the labor in evaluating a determinant function should be welcome. In this connection it is obvious that zero entries at strategic locations in a matrix may often simplify the computation of its determinant. For example, the determinant of the nth-order matrix

$$\begin{pmatrix} a_1^1 & a_2^1 & \cdots & a_n^1 \\ 0 & a_2^2 & \cdots & a_n^2 \\ \cdot & \cdot & \cdots & \cdot \\ 0 & a_2^n & \cdots & a_n^n \end{pmatrix}$$

is seen to equal a_1^1 times the cofactor A_1^1 upon development according to the first column. Such an example suggests the investigation of methods for altering entries in a matrix in such ways that the alterations in its determinant can be predicted. The elementary transformations serve us well in this respect, since on the one hand drastic simplifications in a matrix are possible with them, and on the other hand properties D_8, D_1, and D_5 describe the change in a determinant value produced by an elementary row transformation of types I, II, and III, respectively, upon the matrix. Moreover, property D_{13} states that column transformations affect the determinant value as do their row analogues. Thus if a matrix $A \neq 0$ is reduced to its canonical form under equivalence, its determinant can be modified at each step to obtain

$$\det A = a \cdot \det\begin{pmatrix} I_r & 0 \\ 0 & 0 \end{pmatrix}$$

where a is a nonzero field element. If $r < n$, the order of A, then $\det A = 0$, while if $r = n$, $\det A = a$.

Frequently one can simplify the evaluation of a determinant by means of the following result, which, although it is a special case of a theorem due to Laplace, is sufficient in most applications:

Theorem 4.4. If A and B are square matrices, then

$$\det \begin{pmatrix} A & 0 \\ C & B \end{pmatrix} = \det A \cdot \det B$$

Proof. We consider the determinant function of order $n+m$ and throughout the proof restrict it to matrices of the form $\begin{pmatrix} A & 0 \\ C & B \end{pmatrix}$, where n, m are the orders of A, B, respectively. Regarded as a function of its first n rows alone, it satisfies axioms D_1 and D_2 so that its value at $\begin{pmatrix} A & 0 \\ C & B \end{pmatrix}$ is, according to (6), $c_1 \cdot \det A$, where c_1 is the determinant of $\begin{pmatrix} A & 0 \\ C & B \end{pmatrix}$ upon replacement of A by the unit matrix:

$$\det \begin{pmatrix} A & 0 \\ C & B \end{pmatrix} = c_1 \cdot \det A \qquad c_1 = \det \begin{pmatrix} I_n & 0 \\ C & B \end{pmatrix}$$

Regarded as a function of its last m columns, our determinant function again satisfies D_1 and D_2 so that

$$c_1 = \det B \cdot \det \begin{pmatrix} I_n & 0 \\ C & I_m \end{pmatrix}$$

Subtracting multiples of the rows of I_n from C, we can replace C by 0 to find that the last determinant is 1. Hence $c_1 = \det B$, and the proof is completed.

By induction and transposition we quickly obtain the following generalization:

COROLLARY. If a square matrix A can be partitioned into submatrices A_j^i in such a way that each A_i^i is square and $A = (A_j^i)$ is upper or lower triangular, then

$$\det A = \prod \det A_i^i$$

One situation where the foregoing remarks concerning evaluation of determinants may be of little value is that of a determinant whose entries are functions of several variables. To evaluate such a determinant, one must frequently rely upon his ingenuity, often using induction and the determination of factors as the principal tools. We discuss two well-known examples of this type

EXAMPLE 4.2

Show that the Vandermonde determinant

$$V(x_1, x_2, \ldots, x_n) \equiv \det \begin{pmatrix} 1 & 1 & \cdots & 1 \\ x_1 & x_2 & \cdots & x_n \\ x_1^2 & x_2^2 & \cdots & x_n^2 \\ \cdot & \cdot & \cdots & \cdot \\ x_1^{n-1} & x_2^{n-1} & \cdots & x_n^{n-1} \end{pmatrix}$$

is the product of all differences $x_i - x_j$ for which $i > j$, independently of whatever values are assigned to the x_i's; that is,

$$V \equiv \prod_{1 \le j < i \le n} (x_i - x_j)$$

The identity is clear for the case $n = 2$; assume it holds for the case $n - 1$. Operate on the above matrix by subtracting x_1 times the $(n - 1)$st row from the nth row, then subtracting x_1 times the $(n - 2)$nd row from the $(n - 1)$st row, etc., to obtain

$$V \equiv \det \begin{pmatrix} 1 & 1 & \cdots & 1 \\ 0 & x_2 - x_1 & \cdots & x_n - x_1 \\ 0 & x_2(x_2 - x_1) & \cdots & x_n(x_n - x_1) \\ \cdot & \cdot & \cdots & \cdot \\ 0 & x_2^{n-2}(x_2 - x_1) & \cdots & x_n^{n-2}(x_n - x_1) \end{pmatrix}$$

$$\equiv (x_2 - x_1)(x_3 - x_1) \cdots (x_n - x_1) \det \begin{pmatrix} 1 & 1 & \cdots & 1 \\ x_2 & x_3 & \cdots & x_n \\ x_2^2 & x_3^2 & \cdots & x_n^2 \\ \cdot & \cdot & \cdots & \cdot \\ x_n^{n-2} & x_3^{n-2} & \cdots & x_n^{n-2} \end{pmatrix}$$

Since by assumption the desired result holds for the last determinant, we conclude it holds for V.

EXAMPLE 4.3

Verify that

$$\det\begin{pmatrix} x_1 & x_2 & x_3 & \cdots & x_n \\ x_n & x_1 & x_2 & \cdots & x_{n-1} \\ x_{n-1} & x_n & x_1 & \cdots & x_{n-2} \\ \cdot & \cdot & \cdot & \cdots & \cdot \\ x_2 & x_3 & x_4 & \cdots & x_1 \end{pmatrix} = \prod_{j=1}^{n} C_j$$

where

$$C_j = x_1 + \omega_j x_2 + \omega_j^2 x_3 + \cdots + \omega_j^{n-1} x_n$$

and $\omega_1, \omega_2, \ldots, \omega_n$ are the n nth roots of unity. If the above matrix, call it C, is multiplied on the right by

$$W = \begin{pmatrix} 1 & \omega_2^{n-1} & \cdots & \omega_n \\ \omega_1 & 1 & \cdots & \omega_n^2 \\ \omega_1^2 & \omega_2 & \cdots & \omega_n^3 \\ \cdot & \cdot & \cdots & \cdot \\ \omega_1^{n-1} & \omega_2^{n-2} & \cdots & 1 \end{pmatrix}$$

C_j appears as a common factor of the elements in the jth column of CW. Moreover, if this factor is removed, in every case the resulting matrix is W; thus

$$\det CW = \det C \cdot \det W = \prod_j C_j \det W$$

Finally, $\det W \neq 0$, since upon factoring out ω_2^{n-1} from the second column, ω_3^{n-2} from the third, etc., we obtain

$$\det W = \omega_2^{n-1}\omega_3^{n-2} \cdots \omega_n V(\omega_1, \omega_2, \ldots, \omega_n) \neq 0$$

PROBLEMS

1. Show that

$$\det\begin{pmatrix} 1 & 1 & 1 & \cdots & 1 \\ 1 & 2 & 2^2 & \cdots & 2^{n-1} \\ 1 & 3 & 3^2 & \cdots & 3^{n-1} \\ \cdot & \cdot & \cdot & \cdots & \cdot \\ 1 & n & n^2 & \cdots & n^{n-1} \end{pmatrix} = 1!\,2!\,3! \cdots (n-1)!$$

2. Show that

$$\det \begin{pmatrix} \binom{0}{0} & \binom{1}{0} & \binom{2}{0} & \cdots & \binom{n-1}{0} \\ \binom{1}{1} & \binom{2}{1} & \binom{3}{1} & \cdots & \binom{n}{1} \\ \binom{2}{2} & \binom{3}{2} & \binom{4}{2} & \cdots & \binom{n+1}{2} \\ \cdot & \cdot & \cdot & \cdots & \cdot \\ \binom{n-1}{n-1} & \binom{n}{n-1} & \binom{n+1}{n-1} & \cdots & \binom{2n-2}{n-1} \end{pmatrix} = 1$$

where

$$\binom{r}{s} = \frac{r!}{(r-s)!s!} \qquad \text{if } r \geqq s$$

CHAPTER 5

BILINEAR AND QUADRATIC FUNCTIONS AND FORMS

5.1. Linear Functions. The classes of functions indicated in the chapter title have widespread utility, as the examples and problems will show. Usually these functions are studied without the aid of vector spaces. However, a vector space is the proper setting for their study, since the procedures followed are entirely natural and easily explained from this viewpoint. For the moment, we shall say merely that bilinear and quadratic functions are examples of functions with arguments in a vector space V over a field F and values in F, in brief, *functions on V to F*.

In order that the reader may become acquainted with our point of view, we shall begin by discussing an example of a linear function† on a vector space to its coefficient field. Let $V = P_3(C)$, the space of all polynomials ξ of degree less than or equal to 2 in a real variable t with complex coefficients. Define the function f on V to C as follows: $f(\xi)$ is equal to the value of ξ at $t = 1$. Clearly f is linear. Relative to the basis $\{\alpha_1 = 1, \alpha_2 = t, \alpha_3 = t^2\}$ for V, we have

$$f(\xi) = f_\alpha(X) = x^1 + x^2 + x^3 \qquad \text{if } \xi = \sum x^i\alpha_i \text{ and } X = (x^1, x^2, x^3)^T$$

Here f_α, a function on $V_3(C)$ to C, which is defined by the equation in which it appears, may be used to compute the values of the original function on V to C. Using the terminology introduced below, we call f_α the representation of f relative to the α-basis: its value at X is $f(\xi)$ if ξ has α-coordinates X.

If, instead of the α-basis, we choose the basis $\{\beta_1 = 1, \beta_2 = t - 1, \beta_3 = (t - 1)^2\}$ for V, then

$$f(\eta) = f_\beta(Y) = y_1 \qquad \text{if } \eta = \sum y^i\beta_i \text{ and } Y = (y^1, y^2, y^3)^T$$

It is to be observed that the representation f_β of f relative to the β-basis for V is (i) different from the previous representation and (ii) simpler in form than f_α.

Consider next the case of an arbitrary linear function g on an n-dimensional vector space V over F to the scalar field F. As in the foregoing

†The reader should review Sec. 2.5 for the definitions of a linear function, isomorphism, etc.

example, it is easy to construct a function on $V_n(F)$ to F that assumes the same values as g. Indeed, if a basis $\{\alpha_1, \alpha_2, \ldots, \alpha_n\}$ is chosen for V, define g_α on $V_n(F)$ to F as follows:

$$g_\alpha(X) = g(\xi) \qquad \text{if } \xi = \sum x^i\alpha_i \tag{1}$$

Concerning the structure of g_α, we observe that

$$g_\alpha(X) = g(\sum x^i\alpha_i) = \sum g(x^i\alpha_i) = \sum g(\alpha_i)x^i$$

Since $g(\alpha_i)$, $i = 1, 2, \ldots, n$, are fixed field elements, g_α is a *linear form in* $x^1, x^2, \ldots, x^n$ in view of the following well-established definition:

DEFINITION. *A function whose value at $x^1, x^2, \ldots, x^n$ is the sum of a finite number of terms of the type $a(x^1)^{k_1}(x^2)^{k_2} \cdots (x^n)^{k_n}$, where k_i is 0 or a positive integer, is called a* form of degree k in $x^1, x^2, \ldots, x^n$ *if all terms have the same total degree $k = \sum k_i$.*

Usually a form of degree 1 is called linear; of degree 2, quadratic; etc.

In order to describe precisely the relationship between g and g_α, we introduce the isomorphism φ between V and $V_n(F)$ defined by the α-basis. Thus

$$\varphi(\xi) = X \qquad \text{if } \xi \text{ has } \alpha\text{-coordinates } X$$

Substituting this in (1) gives

$$g_\alpha(\varphi(\xi)) = g(\xi) \qquad \text{all } \xi \text{ in } V$$

so that $g_\alpha\varphi = g$. The schematic diagram in Fig. 6 [where it is indicated how to go from V to F via $V_n(F)$] should prove helpful in visualizing this relationship.

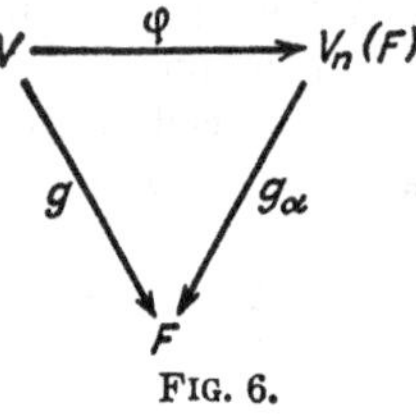

FIG. 6.

We call the linear form g_α *the representation of g relative to the α-basis. Its value at X is $g(\xi)$ if ξ has α-coordinates X.* If in the foregoing we interpret g as any function on V to F, we may still conclude that $g = g_\alpha\varphi$ where g_α is a function on $V_n(F)$ to F, such that $g_\alpha(X) = g(\xi)$ if ξ has α-coordinates X. The function g_α is called a *representation of g*. We shall say that *g is represented by* or *reduces to g_α*. The intention of a representation is to simplify the computation of values of a given function, and possibly to place in evidence nonapparent properties of the function. Of course the initial function may, itself, be defined on $V_n(F)$.

Above, with g linear, observe that the values of the representation g_α (and hence the values of g) are known everywhere, as soon as g is known at n linearly independent vectors. If a different basis $\{\beta_1, \beta_2, \ldots, \beta_n\}$ is chosen for V, a different representation g_β of g is obtained in general, since the coefficients in $g_\beta(X)$ will be $g(\beta_1), g(\beta_2), \ldots, g(\beta_n)$. Because of the simplicity of the example, there is little reason to choose one representation (*i.e.*, choose one basis for V) in preference to another. However, in the following section we shall see that for the case of a more complicated function, one representation may have decided preference over another.

PROBLEM

1. Show that if f is any linear function on V over F to F, there exists a representation f_α on $V_n(F)$ to F such that $f_\alpha(X)$ is equal to the first component of X for all X.

5.2. Bilinear Functions and Forms. A more interesting class of functions which utilize the linearity property is now described.

DEFINITION. *A function f on $V \times V$ to F, where V is a vector space over the field F, is called* bilinear *if it is linear in each argument. That is,*

$$f(\xi + \zeta, \eta) = f(\xi, \eta) + f(\zeta, \eta)$$
$$f(\xi, \eta + \zeta) = f(\xi, \eta) + f(\xi, \zeta)$$
$$f(c\xi, \eta) = f(\xi, c\eta) = cf(\xi, \eta)$$

for all ζ in V and all c in F.

In order to obtain representations for such functions, we imitate the procedure followed in the previous section.

Theorem 5.1. Let f be a bilinear function on $V \times V$ to F, where V is a vector space over F. Let $\{\alpha_1, \alpha_2, \ldots, \alpha_n\}$ and $\{\beta_1, \beta_2, \ldots, \beta_n\}$ be any bases for V. Then relative to these bases there exists a *representation* $f_{\alpha\beta}$ *of* f, that is, a function on $V_n(F) \times V_n(F)$ to F such that $f_{\alpha\beta}(X, Y) = f(\xi, \eta)$ where ξ has α-coordinates X and η has β-coordinates Y. Moreover,

$$f(\xi, \eta) = f_{\alpha\beta}(X, Y) = \sum_{i,j=1}^{n} a_{ij}x^iy^j = X^TAY$$

where

$$a_{ij} = f(\alpha_i, \beta_j) \qquad A = (a_{ij})$$

Such a form $f_{\alpha\beta}$ of degree 2 in X and Y is called a *bilinear form.*

Proof. Let $\varphi(\psi)$ be the isomorphism of V to $V_n(F)$ determined by the $\alpha(\beta)$-basis. Define $f_{\alpha\beta}$ as follows: If

$$X = \varphi(\xi) \qquad Y = \psi(\eta)$$

then

$$f_{\alpha\beta}(X, Y) = f(\xi, \eta)$$

Thus

$$f(\xi, \eta) = f_{\alpha\beta}(X, Y) = f\Big(\sum_i x^i\alpha_i, \sum_j y^j\beta_j\Big) = \sum_{i,j} x^if(\alpha_i, \beta_j)y^j = \sum a_{ij}x^iy^j$$

A direct computation shows that this value of $f_{\alpha\beta}$ can be written as the matrix product $X^T A Y$. This completes the proof.

This result implies that once bases are chosen, a bilinear function is completely determined by its values at n^2 "points" (α_i, β_j), $i, j = 1, 2, \ldots, n$. In other words, f is completely determined by the matrix A. We call A the *matrix of the form* $f_{\alpha\beta}$ and its rank the *rank of the form.* If A is nonsingular or singular, the form is labeled similarly.

If now we choose other bases in V, we expect f to be represented by a different form since we shall evaluate it at different points.

DEFINITION. *Two bilinear forms are* equivalent *if and only if they are representations of one and the same bilinear function.*

Obviously this is an equivalence relation over the set of bilinear forms in $2n$ variables over F. Interestingly enough, with the matrix theory developed so far, this relation can be restated in more familiar terms.

Theorem 5.2. Two bilinear forms are equivalent if and only if their respective matrices A and B are equivalent ($A \overset{E}{\sim} B$), or alternatively if and only if the forms have the same rank.

Proof. Suppose that $f_{\alpha\beta}$ is the representation of f relative to the α, β-bases, while $f_{\gamma\delta}$ is the representation relative to the γ, δ-bases. According to Theorem 3.13 and the results in Sec. 3.6, there exist nonsingular matrices P and Q such that

$$\left.\begin{aligned} \gamma &= \alpha P^T \\ X &= P^T U \end{aligned}\right\} \quad \text{and} \quad \left\{\begin{aligned} \delta &= \beta Q \\ Y &= QV \end{aligned}\right.$$

so that

$$f(\xi, \eta) = f_{\alpha\beta}(X, Y) = X^T A Y = U^T(PAQ)V = f_{\gamma\delta}(U, V)$$

We conclude that $B = PAQ$ is equivalent to A by Theorem 3.17.

Conversely, suppose that f and g are two bilinear forms with

$$f(X, Y) = X^T A Y \qquad g(X, Y) = X^T B Y \qquad B = PAQ$$

where P and Q are nonsingular. Then f may be regarded as a representation $f_{\epsilon\epsilon}$ of itself relative to the ϵ, ϵ-bases and g regarded as a representation of f relative to the ϵP^T, ϵQ-bases. Thus f and g are equivalent since both are representations of the bilinear function f. The final assertion of the theorem is simply another criterion for the equivalence of two matrices according to Theorem 3.8.

The above theorem has several important consequences. For example, it makes possible the definition of the *rank* of a bilinear function as the rank of any representation and the conclusion that two bilinear functions are indistinguishable if and only if they have the same rank. Also, it suggests a canonical set of bilinear forms under equivalence, or in other words, a canonical set of representations of bilinear functions. These representations are described next.

Theorem 5.3. Corresponding to a bilinear function f of rank r on V, there are bases α, β of V relative to which $f_{\alpha\beta}(X, Y) = \sum_1^r x^i y^i$. This representation $f_{\alpha\beta}$ we regard as the *canonical representation* of f.

Proof. The matrix of $f_{\alpha\beta}$, namely, $\text{diag}(I_r, 0)$, has rank r, and consequently $f_{\alpha\beta}$ is a representation of f.

The two preceding theorems can be formulated for bilinear forms without any explicit reference to vector spaces. For this, only a change in point of view is necessary, *viz.*, we emphasize coordinates rather than bases. We have learned (Sec. 3.6) that accompanying a change of basis in a vector space over F there is a change in the coordinates of a vector which is described by a system of linear equations with a nonsingular matrix B, expressing one set of coordinates X in terms of the other set U:

$$X = BU \qquad B \text{ nonsingular}$$

The replacement of the variables $X = (x^1, x^2, \ldots, x^n)^T$ in a function f by new variables $U = (u^1, u^2, \ldots, u^n)^T$ related to the x^i's as above is described as a *nonsingular substitution on X*. Thus, the transition from a given bilinear form f to an equivalent form amounts to a nonsingular substitution on the two sets of variables. Hence Theorem 5.3 can be restated as follows:

Theorem 5.4. With a suitable nonsingular substitution

$$X = P^T U \qquad Y = QV$$

a bilinear form of rank r reduces to the form whose value at U, V is $\sum_1^r u^i v^i$. The coefficients in the substitution can be restricted to any field that contains the coefficients of the matrix of the original form.

Remark. In order to shorten statements of results in the future, we adopt the following convention: The phrase "the form f whose value at X, Y is $f(X, Y) = \sum a_{ij} x^i y^j$" will be written "the form $f = \sum a_{ij} x^i y^j$."

EXAMPLE 5.1

Reduce the form

$$f = 2x^1y^1 - 3x^1y^2 + x^1y^3 - x^2y^1 + 5x^2y^3 - 6x^3y^1 + 3x^3y^2 + 19x^3y^3$$

to the form $\sum u^i v^i$ by a substitution having rational coefficients.

The matrix of f is

$$A = \begin{pmatrix} 2 & -3 & 1 \\ -1 & 0 & 5 \\ -6 & 3 & 19 \end{pmatrix}$$

Using the proof of Theorem 5.2 as our guide, let us find nonsingular matrices P, Q such that

$$PAQ = \operatorname{diag}(I_r, 0)$$

since the substitution $X = P^{\mathrm{T}}U$, $Y = QV$ will reduce $X^{\mathrm{T}}AY$ to $\sum_1^r u^iv^i$. To find a P and Q we proceed as in Example 3.6. The computations are listed below:

$$A \rightarrow \begin{pmatrix} -1 & 0 & 5 \\ 2 & -3 & 1 \\ -6 & 3 & 19 \end{pmatrix} \rightarrow \begin{pmatrix} -1 & 0 & 5 \\ 0 & -3 & 11 \\ 0 & 3 & -11 \end{pmatrix} \rightarrow \begin{pmatrix} 1 & 0 & -5 \\ 0 & -3 & 11 \\ 0 & 0 & 0 \end{pmatrix}$$

$$\rightarrow \begin{pmatrix} 1 & 0 & -5 \\ 0 & 1 & -\frac{11}{3} \\ 0 & 0 & 0 \end{pmatrix} \rightarrow \begin{pmatrix} 1 & 0 & 0 \\ 0 & 1 & 0 \\ 0 & 0 & 0 \end{pmatrix}$$

$$P: I \rightarrow \begin{pmatrix} 0 & 1 & 0 \\ 1 & 0 & 0 \\ 0 & 0 & 1 \end{pmatrix} \rightarrow \begin{pmatrix} 0 & 1 & 0 \\ 1 & 2 & 0 \\ 0 & -6 & 1 \end{pmatrix} \rightarrow \begin{pmatrix} 0 & -1 & 0 \\ 1 & 2 & 0 \\ 1 & -4 & 1 \end{pmatrix}$$

$$\rightarrow \begin{pmatrix} 0 & -1 & 0 \\ -\frac{1}{3} & -\frac{2}{3} & 0 \\ 1 & -4 & 1 \end{pmatrix} = P$$

$$Q: I \rightarrow \begin{pmatrix} 1 & 0 & 5 \\ 0 & 1 & \frac{11}{3} \\ 0 & 0 & 1 \end{pmatrix} = Q$$

Then the substitution $X = P^{\mathrm{T}}U$, $Y = QV$ reduces $X^{\mathrm{T}}AY$ to $u^1v^1 + u^2v^2$.

PROBLEMS

1. Reduce the bilinear form

$$x^1y^1 - x^1y^2 + x^1y^3 + 4x^2y^1 - 3x^2y^2 + 2x^2y^3 - x^3y^1 + 2x^3y^2 - 3x^3y^3$$

to canonical form by a nonsingular substitution with rational coefficients on the two sets of variables.

2. The same reduction is required for the bilinear form

$$2x^1y^1 - x^1y^2 + 2x^1y^3 + x^1y^4 - 2x^2y^1 + 3x^2y^2 - 3x^2y^3 + 5x^2y^4 + x^3y^1 - x^3y^3 + 2x^3y^4 + 4x^4y^1 - x^4y^2 + x^4y^3$$

5.3. Quadratic Functions and Forms. If in a bilinear function defined on a vector space V we equate the two arguments, the resulting function of one variable is called a *quadratic function.* Specializing Theorem 5.1, we obtain the result: Let X be the column vector of coordinates of a vector in V relative to the α-basis. Corresponding to this basis there is a representation q_α of a quadratic function q on V to F whose value at X is

$$q_\alpha(X) = X^{\mathrm{T}}AX = \sum_{i,j=1}^{n} a_{ij}\, x^i x^j$$

The function q_α is a *quadratic form.*

DEFINITION. *A square matrix is called* symmetric (skew-symmetric) *if* $A^{\mathrm{T}} = A$ $(A^{\mathrm{T}} = -A)$.

The matrix of a quadratic form can always be chosen as a symmetric matrix. For example, the term $6x^1x^2$ involving mixed letters in the form

$$(x^1, x^2)\begin{pmatrix} 1 & 2 \\ 4 & 2 \end{pmatrix}\begin{pmatrix} x^1 \\ x^2 \end{pmatrix} = (x^1)^2 + 6x^1x^2 + 2(x^2)^2$$

can be broken up into the equal parts $3x^1x^2$ and $3x^1x^2$ and the form written as

$$(x^1, x^2)\begin{pmatrix} 1 & 3 \\ 3 & 2 \end{pmatrix}\begin{pmatrix} x^1 \\ x^2 \end{pmatrix}$$

A proof for the general case can be based upon this simple idea or upon the following result, which is of some interest:

Theorem 5.5. A square matrix A is the sum of a symmetric matrix and a skew-symmetric matrix. This representation is unique.

Proof. Write

$$A = \tfrac{1}{2}(A + A^{\mathrm{T}}) + \tfrac{1}{2}(A - A^{\mathrm{T}}) = B + C$$

let us say, where $B = \frac{1}{2}(A + A^{\mathrm{T}})$ and $C = \frac{1}{2}(A - A^{\mathrm{T}})$. Since $(A \pm A^{\mathrm{T}})^{\mathrm{T}} = A^{\mathrm{T}} \pm A$, it is clear that B is symmetric and C is skew-symmetric. If $A = B_1 + C_1$ is another such decomposition, we have $A^{\mathrm{T}} = B_1^{\mathrm{T}} + C_1^{\mathrm{T}} = B_1 - C_1$, and it follows that $A + A^{\mathrm{T}} = 2B_1$ and $A - A^{\mathrm{T}} = 2C_1$, or $B = B_1$ and $C = C_1$. This completes the proof.

Now if A is the matrix of a quadratic form q, we may write $A = B + C$

as above to get

$$q(X) = X^TAX = X^T(B + C)X = X^TBX + X^TCX$$

Since X^TCX is equal to zero, the desired result follows.

In the next theorem we continue with the specialization of previous results.

Theorem 5.6. Two quadratic forms are equivalent (*i.e.*, representations of one and the same quadratic function) if and only if their matrices A and B are related by an equation of the form $B = PAP^T$, where P is nonsingular. Two matrices so related are called *congruent* (in symbols, $A \overset{C}{\sim} B$). Congruence is an equivalence relation over the set of all $n \times n$ matrices, for which the accompanying partition is finer than that based on equivalence of matrices.

Proof. The first statement is a specialization of Theorem 5.2. Referring to the proof of that theorem, if in the form X^TAX we substitute $X = P^TU$, the new form has the matrix PAP^T. The verification of the equivalence relation postulates for the congruence relation is straightforward, and we omit it.

The matrices

$$\begin{pmatrix} 1 & 0 \\ 0 & 0 \end{pmatrix} \qquad \begin{pmatrix} 0 & 0 \\ 1 & 0 \end{pmatrix}$$

which are equivalent are easily seen to be incongruent. Since, on the other hand, congruent matrices are equivalent, it follows that an equivalence class of $n \times n$ matrices is composed of one or more congruence classes; hence the latter constitutes a *finer* partition. This completes the proof.

One observes the difference in form between the definition of congruence and that of equivalence. The former is analogous to the characterization of equivalence obtained in Theorem 3.17. A characterization of congruence analogous to the definition of equivalence can be formulated. It reads as follows: $A \overset{C}{\sim} B$ if and only if B can be obtained from A by a sequence of transformations, each of which is an elementary row transformation followed by the corresponding column transformation.† To verify this, we recall that a nonsingular matrix P is equal to a product of elementary matrices $E_1, E_2, \ldots, E_k$ and conversely, and consider the matrix product

$$PAP^T = E_k(\cdots(E_2(E_1AE_1^T)E_2^T)\cdots)E_k^T$$

†It is obvious from this formulation of congruence, just as from the definition, that congruence is a restricted type of equivalence.

Our assertion follows upon observing that if

$E_p = R_{ij}$, then E_p^T as a postmultiplier interchanges the ith and jth columns,

$E_p = R_i(c)$, then E_p^T as a postmultiplier multiplies the ith column by c,

$E_p = R_{ij}(c)$, then E_p^T as a postmultiplier adds c times the jth column to the ith.

We now propose to discuss for congruence the inevitable topic that arises in the study of an equivalence relation—the determination of a canonical set and a practical formulation of the relation (*e.g.*, in the form of a complete set of invariants). Results in this direction are incomplete for congruence; the notable results are valid for only the class of symmetric matrices, to which we shall confine our discussion after several preliminary remarks.

A study of the matrices congruent to a given matrix A is a study of A under *congruence transformations*, *i.e.*, functions of the form $A \to PAP^T$ where P is nonsingular. According to the above characterization of congruence, any congruence transformation is equal to a product of congruence transformations of the form, an elementary row transformation followed by the corresponding column transformation. Two types of such "elementary congruence transformations" which we shall use systematically in the study of symmetric matrices are described next.

DEFINITION. *A C_I transformation is an elementary row transformation of type I followed by the corresponding column transformation. A C_{III} transformation is a type III transformation $R_{ij}(c)$ with $i > j$ followed by the corresponding column transformation.*

In order to describe the effect of each type of C transformation on a matrix, two definitions are needed.

DEFINITION. *A* principal submatrix of order k *of an nth-order matrix $A = (a_{ij})$ is a submatrix obtained by deleting $n - k$ rows and the corresponding columns of A. A* principal minor *of A is the determinant of a principal submatrix of A. The principal minor of order k of A computed from the first k rows and columns of A will be denoted by $D_k(A)$, or simply D_k :*

$$D_k(A) = \det\begin{pmatrix} a_{11} & a_{12} & \cdots & a_{1k} \\ a_{21} & a_{22} & \cdots & a_{2k} \\ \cdot & \cdot & \cdots & \cdot \\ a_{k1} & a_{k2} & \cdots & a_{kk} \end{pmatrix} \qquad k = 1, 2, \ldots, n$$

Lemma 5.1. Let A denote a square matrix and B a matrix obtained from A by a C_I transformation. Then the set of principal minors of B is the same as that of A. If B is obtained from A by a C_{III} transformation, then $D_k(B) = D_k(A)$ for all k.

Proof. The verification is left as an exercise.

Before beginning a study of congruence in the set of $n \times n$ symmetric matrices over F, an important remark is in order. Since a matrix congruent to a symmetric matrix is symmetric (indeed, if $A = A^T$, then $(PAP^T)^T = PA^TP^T = PAP^T$), the set of $n \times n$ symmetric matrices is a sum of certain congruence classes in the set of *all* $n \times n$ matrices. Consequently, the definition of the equivalence relation induced in the set of $n \times n$ symmetric matrices by congruence (which is defined over the set of all $n \times n$ matrices) reads exactly as the general definition, *viz.*, $A \overset{C}{\sim} B$ if and only if there exists a nonsingular P such that $B = PAP^T$.

Without regard to the features indicated in Lemma 5.1 of C transformations, it is easily seen that they are adequate to reduce a symmetric matrix of positive rank to an essentially diagonal form which, although not unique, will play the role of a preferred form in our exposition.

Lemma 5.2. A symmetric matrix A of positive rank can be reduced with C transformations to a congruent matrix

$$A^* = \operatorname{diag}(A_1, A_2, \ldots, A_s, 0) \tag{2}$$

where each A_i has the form

$$(a) \text{ or } A(a) = \begin{pmatrix} 0 & a \\ a & 0 \end{pmatrix} \qquad a \neq 0$$

Moreover this reduction may be effected with a sequence of C_I transformations followed by a sequence of C_{III} transformations.

Proof. In order to establish an algorithm for this reduction, two remarks are necessary. First, if A contains a diagonal element $a \neq 0$, it can be brought to the (1, 1) position by a C_I transformation and then the first row and column cleared by C_{III} transformations to obtain the matrix

$$\begin{pmatrix} a & 0 \\ 0 & B \end{pmatrix} \overset{C}{\sim} A$$

Second, if every diagonal element of A is 0 but the (i, j)th entry $(i \neq j)$ is $a \neq 0$, A can be reduced to the congruent matrix having $A(a)$ in its upper left-hand corner. Then the first two rows and columns can be cleared to obtain a matrix

$$\begin{pmatrix} A(a) & 0 \\ \mathbf{0} & C \end{pmatrix}$$

congruent to A. The application of one or more such steps will reduce A to (2).

To prove the final assertion, we assume that the reduction of A is carried out as outlined above. Let $\tau_1, \tau_2, \ldots, \tau_t$ denote the transformations used, numbered in order of application, so that A^* in (2) may be described by (in the notation of Theorem 3.10)

$$\tau_t \tau_{t-1} \cdots \tau_1 A \tag{3}$$

A C_{I} transformation R_{kl}† in this sequence, if preceded by a C_{III} transformation $R_{ij}(c)$, can be interchanged with it if $i \neq k, l$. If $i = k$ (respectively l) the interchange can still be made, provided i is replaced by l (respectively k) in $R_{ij}(c)$. With our prescribed order of reducing A, j is always less than both k and l, hence is never altered. In this way each C_{I} transformation applied to A can be shifted to the right of every C_{III} transformation in (3). This completes the proof.

Our only result concerning the existence of a diagonal matrix congruent to a symmetric matrix, that is independent of the field concerned, follows next. The diagonal matrix described is not uniquely determined, so that we do not obtain a canonical set under congruence.

Theorem 5.7. Each symmetric matrix of rank r is congruent to a diagonal matrix having r nonzero diagonal elements.

Proof. We may assume that $r > 0$. Then, according to Lemma 5.2, it is only necessary to show that the matrix $A(a)$ has this property. The row transformation $R_{12}(1)$ and its column analogue, followed by $R_{21}(-\frac{1}{2})$ and its column analogue, establishes this result.

COROLLARY. A quadratic form $X^{\text{T}}AX$ of rank r can be reduced to the form $\sum_1^r d_i(u^i)^2$, $d_i \neq 0$, with a suitable nonsingular substitution, where the coefficients in the substitution can be restricted to any field containing the coefficients of the form. In particular, a quadratic function of rank r can be represented by such a form.

Proof. Since A may be assumed symmetric, there exists a nonsingular matrix P, such that $PAP^{\text{T}} = \text{diag}(d_1, d_2, \ldots, d_r, 0, \ldots, 0)$. Then the substitution $X = P^{\text{T}}U$ effects the desired reduction in $X^{\text{T}}AX$.

The substitution $X = P^{\text{T}}U$ used above is known as soon as P is known. In turn, P can be obtained by applying to I those row transformations used in the reduction of A to I. As an alternative method for the reduction

†This notation is adequate to describe the interchange of the kth and lth rows followed by the interchange of the corresponding columns.

of a quadratic form, one may complete squares repeatedly as in Example 3.4. One then obtains the new variables in terms of the old. Before presenting an example which illustrates both methods of procedure, it is appropriate to mention that the substitution accompanying a C_{I} transformation is merely a renumbering of the variables, while that accompanying a C_{III} transformation may be called a triangular substitution, since it is of the form

$$x^k = u^k + \text{a linear form in } u^{k+1}, \ldots, u^n \qquad k = 1, 2, \ldots, n$$

EXAMPLE 5.2

Find a nonsingular substitution over R which reduces the quadratic form $X^{\text{T}}AX$ where

$$A = \begin{pmatrix} -1 & 1 & 2 & 0 \\ 1 & -1 & -2 & 2 \\ 2 & -2 & -3 & 1 \\ 0 & 2 & 1 & 0 \end{pmatrix}$$

to the diagonal form $\sum d_i(u^i)^2$. The row transformations

$$R_{43}(2), R_{34}, R_{42}(-1), R_{23}, R_{31}(2), R_{21}(1) \tag{4}$$

reading from right to left, when applied together with their column analogues to A, yield the matrix

$$\operatorname{diag}(-1, 1, -1, 4)$$

These row transformations when applied to I give

$$P = \begin{pmatrix} 1 & 0 & 0 & 0 \\ 2 & 0 & 1 & 0 \\ -2 & 0 & -1 & 1 \\ -3 & 1 & -2 & 2 \end{pmatrix}$$

Thus $PAP^{\text{T}} = \operatorname{diag}(-1, 1, -1, 4)$, and the substitution $X = P^{\text{T}}U$ reduces $X^{\text{T}}AX$ to

$$-(u^1)^2 + (u^2)^2 - (u^3)^2 + 4(u^4)^2$$

To carry out the same reduction on the form itself, the first step is to make the substitution

$$u^1 = x^1 - x^2 - 2x^3$$

which reduces the form to $-(u^1)^2 + (x^3)^2 + 4x^2x^4 + 2x^3x^4$. Next we successively substitute

$$\begin{aligned} u^2 &= \quad\quad x^3 + x^4 \\ u^3 &= -2x^2 \quad + x^4 \\ u^4 &= \quad x^2 \end{aligned}$$

to obtain the same result. Finally, to illustrate the last assertion in Lemma 5.2, we mention that the sequence of transformations (4) can be rearranged as

$$R_{43}(2),\ R_{32}(-1),\ R_{21}(2),\ R_{41}(1),\ R_{34}\ ,\ R_{23}$$

where now all the C_{I} transformations are applied first.

It is natural to ask whether a further substitution will reduce the form of the above example to $(v^1)^2 + (v^2)^2 + (v^3)^2 + (v^4)^2$. The answer to this question depends upon the field, F, of scalars, which has remained in the background of our discussion. If F is the real field, the answer is no, since then $\sum(v^i)^2$ is never negative, whereas $-(u^1)^2 + (u^2)^2 - (u^3)^2 + 4(u^4)^2$ is sometimes negative. However, if F is the complex field, it is clear that the substitution

$$\begin{aligned} \sqrt{-1}\,u^1 &= v^1 \\ u^2 &= v^2 \\ \sqrt{-1}\,u^3 &= v^3 \\ 2u^4 &= v^4 \end{aligned}$$

reduces one to the other.

In general, the form $\sum_1^r d_i(u^i)^2$ of Theorem 5.7 can be reduced with a substitution over F to a sum of squares if every d_i has a square root in F. The substitution

$$\begin{aligned} \sqrt{d_i}u^i &= v^i \qquad i = 1, 2, \ldots, r \\ u^i &= v^i \qquad i = r + 1, \ldots, n \end{aligned}$$

will effect the reduction. This proves our next theorem.

Theorem 5.8. Let F denote a field with the property that every element of F has a square root in F. Then a quadratic form of rank r over F can be reduced to a sum of squares $\sum_1^r (v^i)^2$ by a nonsingular substitution over F. Stated for matrices, a symmetric matrix of rank r over F is congruent to $\mathrm{diag}(I_r\ ,\ 0)$. Two symmetric matrices over F are congruent if and only if they have the same rank.

Thus, for the set of $n \times n$ symmetric matrices over a field of the type in Theorem 5.8, rank is a complete set of invariants under congruence and the matrices $\text{diag}(I_r, 0)$ constitute a canonical set.

PROBLEMS

1. Verify the assertions made prior to Example 5.2 concerning the substitutions accompanying a C_{I} and a C_{III} transformation respectively.

2. Reduce the quadratic form

$$9(x^1)^2 - 6x^1x^2 + 6x^1x^3 + (x^2)^2 - 2x^2x^3 + (x^3)^2$$

to a form $\sum d_i(u^i)^2$ by a nonsingular substitution with rational coefficients.

3. Reduce the quadratic form

$$\bar{3}(x^1)^2 + \bar{4}x^1x^2 + \bar{2}x^1x^3 + \bar{2}x^1x^4 + \bar{2}(x^2)^2 + \bar{2}x^2x^3 + \bar{2}x^2x^4 + (x^3)^2$$

with coefficients in J_5 to a form $\sum d_i(u^i)^2$.

4. Verify the statements made below equation (3) in the proof of Lemma 5.2.

5. State and prove the analogue of Lemma 5.2 for skew-symmetric matrices.

6. Prove that the rank of a skew-symmetric matrix is an even integer $2t$ and that the matrices

$$\text{diag}(E_1, E_2, \ldots, E_t, 0), \quad E_i = \begin{pmatrix} 0 & 1 \\ -1 & 0 \end{pmatrix}$$

form a canonical set under congruence. Finally, show that two skew-symmetric matrices are congruent if and only if they are equivalent.

5.4. Further Properties of Symmetric Matrices. The usefulness of the following theorems is due to the relation between symmetric matrices and quadratic forms. They are proved with the aid of Lemmas 5.1 and 5.2.

Theorem 5.9. A symmetric matrix A of rank $r > 0$ has at least one nonzero principal minor of order r.

Proof. According to Lemma 5.2, A may be reduced to A^* in (2) using a sequence of C_{I} transformations followed by C_{III} transformations. Hence A^* may be reduced to A using a sequence of C_{III} transformations followed by C_{I} transformations. Since $D_r(A^*) \neq 0$, the assertion follows from Lemma 5.1.

Theorem 5.10. If a symmetric matrix A has a principal submatrix P of order r with nonzero determinant and every principal submatrix of order $r + 1$, $r + 2$, respectively, which contains P has zero determinant, then $r(A) = r$.

Proof. By the first hypothesis together with Lemma 5.2, there exists a sequence of C_{I} transformations which reduces A to a matrix B such that $D_r(B) \neq 0$ and only C_{III} transformations are necessary to reduce B to A^* in (2). Then $r(A^*)$, hence that of A, is r. For $D_r(A^*) \neq 0$, and if $r(A^*) >$

r, at least one of $D_{r+1}(A^*)$, $D_{r+2}(A^*)$ is nonzero, which contradicts the second hypothesis.

Theorem 5.11. Using C_{I} transformations, a symmetric matrix A of rank $r > 0$ may be reduced to a matrix S such that no two consecutive numbers in the sequence

$$1, D_1(S), D_2(S), \ldots, D_r(S)$$

are zero. We call S a *regular symmetric matrix.*

Proof. Apply to A those C_{I} transformations which when followed by appropriate C_{III} transformations will reduce A to A^* in (2).

COROLLARY. The (symmetric) matrix of a quadratic form can be made regular upon a suitable renumbering of the variables.

Further development of this manner of reasoning depends upon the specialization of the field to the real field and the relation between symmetric matrices and quadratic forms. This appears in Sec. 5.5. Finally, we mention that an analogous development is possible for Hermitian matrices; this is discussed in Sec. 5.7.

PROBLEMS

1. Prove by induction the following theorem, which provides an alternative proof of Theorem 5.9: If in a symmetric matrix every minor of order $r + 1$ and every principal minor of order r vanishes, then every minor of order r vanishes.

2. Show that in a symmetric matrix of rank $r > 0$ the sum of the principal minors of order r is not zero.

3. Extend Theorems 5.9 and 5.10 to skew-symmetric matrices.

5.5. Quadratic Functions and Forms over the Real Field. These functions deserve special attention because of their occurrence in geometry, as well as in the formulation of many physical concepts. Examples are postponed until the next section in order that all of the main results of the chapter can be presented without interruption. We shall first prove the fundamental theorem for real quadratic functions.

Theorem 5.12. Let V denote a vector space over $R^\#$ and q a quadratic function of rank r on V to $R^\#$. Then q can be represented by the form

$$(v^1)^2 + (v^2)^2 + \cdots + (v^p)^2 - (v^{p+1})^2 - \cdots - (v^r)^2 \tag{5}$$

where the number p of positive squares is uniquely determined by q (*Sylvester's law of inertia*). We call (5) the *canonical representation* of a real quadratic function. Stated for quadratic forms, any quadratic form over $R^\#$ can be reduced by a nonsingular substitution over $R^\#$ to a uniquely determined form of the type (5). Finally, stated for symmetric matrices, a real symmetric matrix A of rank r is congruent to a uniquely determined diagonal matrix of the type

$$\text{diag}(I_p\,,\ -I_{r-p}\,,\ 0) \tag{6}$$

Proof. According to Theorem 5.7, q is represented by a form $U^{\text{T}}DU$, where $D = \text{diag}(d_1\,,\ d_2\,,\ \ldots\,,\ d_r\,,\ 0,\ \ldots\,,\ 0)$ relative to a suitable basis for V. With a change in the order of the basis vectors (if necessary) we may assume $d_1\,,\ d_2\,,\ \ldots\,,\ d_p > 0$ and $d_{p+1}\,,\ \ldots\,,\ d_r < 0$. Upon a change in basis defined by the matrix

$$\begin{aligned} P &= P^{\text{T}} \\ &= \text{diag}\left(\frac{1}{\sqrt{d_1}}\,,\ \ldots\,,\ \frac{1}{\sqrt{d_p}}\,,\ \frac{1}{\sqrt{-d_{p+1}}}\,,\ \ldots\,,\ \frac{1}{\sqrt{-d_r}}\,,\ 1,\ \ldots\,,\ 1\right) \end{aligned}$$

q is represented by (5).

To prove the uniqueness, suppose that, besides (5), q is also represented by

$$(w^1)^2 + (w^2)^2 + \cdots + (w^{p'})^2 - (w^{p'+1})^2 - \cdots - (w^r)^2 \tag{7}$$

hence that (5) and (7) are equivalent forms. Assume that $p' < p$. Now $q(\xi) \geqq 0$ for those ξ's corresponding to n-tuples $(v^1, v^2, \ldots, v^n)^{\text{T}}$ such that $v^{p+1} = \cdots = v^r = 0$ in accordance with (5). These ξ's constitute an $n - (r - p)$-dimensional subspace S_1 of V. Similarly, from (7), $q(\xi) < 0$ for each $\xi \neq 0$ corresponding to an n-tuple $(w^1, w^2, \ldots, w^n)^{\text{T}}$ such that $w^1 = \cdots = w^{p'} = w^{r+1} = \cdots = w^n = 0$. These conditions determine an $(r - p')$-dimensional subspace S_2 of V. Now $d[S_1] + d[S_2] = n + (p - p') > n$, which means that (Theorem 2.12) $S_1 \cap S_2 \neq O$. Hence S_1 and S_2 have a nonzero vector ξ in common. For this common vector $q(\xi) \geqq 0$ by (5) but is less than 0 by (7). The assumption $p' > p$ would lead to a similar contradiction, and so $p = p'$. This completes the proof.

In the form (5) and the matrix (6), the number of negative terms is uniquely determined by q and A, respectively; thus, the same is true of the number $s = p - (r - p) = 2p - r$, which is called the *signature* of q and A, respectively. Since, conversely, r and s determine p, it follows that a quadratic function on an n-dimensional space over R^* is characterized by two numbers, its rank r and signature s. Alternatively, we can say that for the set of real quadratic forms in $x^1, x^2, \ldots, x^n$, rank and signature constitute a complete set of invariants under equivalence;

the forms of the type (5) constitute a canonical set determined by these invariants. Finally, interpreted for matrices, our results state that for the set of $n \times n$ symmetric matrices over R^*, rank and signature constitute a complete set of invariants under congruence and define a canonical set consisting of the matrices (6); in particular, two symmetric matrices over R^* are congruent if and only if they have the same rank and signature.

If the matrix of a real quadratic form is regular, the signature can be computed using the following result:

Theorem 5.13. Let S denote a regular symmetric matrix of rank r; hence in the sequence

$$1,\ D_1(S),\ D_2(S),\ \ldots,\ D_r(S)$$

no two consecutive numbers vanish. If $D_i = 0$, then $D_{i-1}D_{i+1} < 0$. Moreover, the signature of S is equal to the number of permanences minus the number of variations of sign in this sequence, where either sign may be affixed to a D_i which vanishes.

Proof. It is easily seen that, using C_{III} transformations alone, S can be reduced to a congruent matrix

$$S^* = \operatorname{diag}(S_1\ ,\ S_2\ ,\ \ldots\ ,\ S_k\ ,\ 0)$$

where each S_i has the form

$$(b) \text{ or } \begin{pmatrix} 0 & b \\ b & c \end{pmatrix} \qquad b \neq 0$$

In turn S^* can be reduced to a diagonal matrix

$$D = \operatorname{diag}(d_1\ ,\ d_2\ ,\ \ldots\ ,\ d_r\ ,\ 0,\ \ldots\ ,\ 0)$$

by reducing each 2×2 block in S^* to diagonal form using C_{I} and C_{III} transformations. Now on the one hand the numbers $D_i(S)$ can be computed from S^*; on the other hand the signature can be computed from D, so it is S^* and D that we shall compare.

First it is clear from an examination of S^* that if $D_i = 0$, $D_{i+1} = -b^2D_{i-1}$, which proves the first assertion. Next we observe that for any $D_k \neq 0$, $D_k = d_1d_2 \cdots d_k$. Consequently if neither member of the pair $\{D_{i-1}\ ,\ D_i\}$ vanishes, $d_i < 0$ if and only if D_{i-1} and $D_i = D_{i-1}d_i$ have opposite signs. If $D_i = 0$, the triplet $\{D_{i-1}\ ,\ D_i\ ,\ D_{i+1}\}$ exhibits one permanence and one variation of sign if D_i is counted as positive or negative. On the other hand, one of $d_i\ ,\ d_{i+1}$ is positive and the other negative. The second assertion then follows.

It is convenient to have available a classification of real quadratic forms and functions in terms of the algebraic signs of their values. In this connection the following definitions are introduced:

DEFINITION. *If, apart from* $q(0)$, *which is always* 0, *the values of a real quadratic function or form* q

$$\left.\begin{cases}\textit{always have the same sign}\\ \textit{include } 0, \textit{ otherwise all have the same sign}\\ \textit{include both positive and negative numbers}\end{cases}\right\}\ \textit{then } q \textit{ is called } \left\{\begin{matrix}\text{definite}\\ \text{semidefinite}\\ \text{indefinite}\end{matrix}\right\}$$

If q *is definite or semidefinite and the nonzero values are positive (negative), then* q *is further classified as* positive (negative).

The same classification is frequently applied to symmetric matrices. The reader can easily supply a proof of the following result:

Theorem 5.14. A real quadratic function or form is positive definite, positive semidefinite, or indefinite, respectively, if and only if its canonical form is

$$\sum_1^n (v^i)^2 \qquad \sum_1^r (v^i)^2, \quad r < n \qquad \sum_1^p (v^i)^2 - \sum_{p+1}^r (v^i)^2, \quad p < r \leqq n$$

The following theorems are useful in classifying quadratic forms:

Theorem 5.15. A real quadratic form q with matrix A is positive definite if and only if the members of the sequence of principal minors

$$1,\ D_1(A),\ D_2(A),\ \ldots,\ D_n(A)$$

are all positive. The form is negative definite if and only if this sequence exhibits n variations in sign; in other words, $D_1(A) < 0$, and the succeeding terms alternate in sign.

Proof. To prove the necessity of the first statement, observe that $q(1, 0, 0, \ldots, 0)^T = a_{11} = D_1(A) > 0$ if q is positive definite. With $a_{11} \neq 0$ the first row and column of A can be cleared with C_{III} transformations to obtain the matrix

$$\begin{pmatrix} a_{11} & 0 \\ 0 & B \end{pmatrix} \qquad B = (b_{ij})$$

of a form q^* equivalent to q. Since q^* is positive definite, $b_{11} > 0$; moreover $D_2(A) = D_1(A)b_{11} > 0$. Continuing in this way, we obtain the desired result.

The converse follows from Theorem 5.13, which is applicable since A is obviously regular. The proof of the second part of the theorem is similar to the foregoing.

The proof of the next result is left to the reader.

Theorem 5.16. A real quadratic form of rank $r < n$ in $x^1, x^2, \ldots, x^n$ is positive semidefinite if and only if the variables can be renumbered so that the principal minors

$$1, D_1, D_2, \ldots, D_r$$

of the corresponding matrix are all positive. The form is negative semidefinite if and only if the variables can be renumbered so that the signs in the above sequence alternate.

EXAMPLE 5.3

Along with the rank and signature of a real quadratic function q, the number $r - p$ of negative signs in its canonical form (5) is also of importance. The number of negative signs in (5) is called the *index* of q. The interest in the index is due to its characterization as the dimension of the subspace of largest dimension on which q is negative definite, since this version remains useful in the infinite dimensional analogue consisting of quadratic functionals on an infinite dimensional space.

To establish this characterization of the index of a quadratic function q on an n-dimensional space V to $R^\#$, suppose that the canonical form of q is f, where

$$f(X) = \sum_1^p (x^i)^2 - \sum_{p+1}^r (x^i)^2$$

Let S denote a subspace of dimension s of V on which q is negative definite; then f is negative definite on S', the s-dimensional image of S in $V_n(R^\#)$ under an isomorphism between V and $V_n(R^\#)$. If T is the $(n + p - r)$-dimensional subspace of $V_n(R^\#)$ defined by

$$x^{p+1} = \cdots = x^r = 0$$

then $d[S' \cap T] \geqq s + p - r$. If $r < s + p$, $S' \cap T$ contains a nonzero vector X, and at X, f is both negative and nonnegative. From this contradiction we conclude that $r \geqq s + p$ or $s \leqq r - p$, the index of q. On the other hand, f is negative definite on the $(r - p)$-dimensional space defined by

$$x^1 = \cdots = x^p = x^{p+1} = \cdots = x^n = 0$$

Hence the upper bound, $r - p$, for the dimensions of subspaces on which

q is negative is attained by at least one space. Thus the desired characterization of the index $r - p$ is established.

PROBLEMS

1. Prove directly that the real quadratic form $ax^2 + 2bxy + cy^2$ is positive definite if and only if $a > 0$ and $ac - b^2 > 0$.

2. Verify the following statements: The matrix A of a positive semidefinite form of rank r can be expressed in the form $B^T B$, where B is a real matrix of rank r. Conversely, if B is any real $m \times n$ matrix of rank r, $B^T B$ is the matrix of a positive semidefinite form of rank r.

3. Show that the quadratic form $5(x^1)^2 - 4x^1x^2 + 3(x^2)^2 - 2x^2x^3 + (x^3)^2$ is positive definite.

4. Show that the quadratic form with matrix

$$\begin{pmatrix} -24 & -3 & 3 & 2 \\ -3 & -15 & 2 & -3 \\ 3 & 2 & -15 & 3 \\ 2 & -3 & 3 & -24 \end{pmatrix}$$

is negative definite.

5. Supply a proof for Theorem 5.16.

5.6. Examples. In this section we discuss several examples which indicate the widespread occurrence of real quadratic functions and forms.

EXAMPLE 5.4

A quadratic function q is frequently used in geometry to define a *figure* or *locus*, found by setting the function equal to a constant c. The figure consists of all solutions ξ of the equation $q(\xi) = c$. For instance, in the Euclidean plane, let $(x^1, x^2)^T$ denote the rectangular coordinates of a point. A quadratic function over this space is represented by a quadratic form $a_{11}(x^1)^2 + 2a_{12}x^1x^2 + a_{22}(x^2)^2$, and the locus of the equation

$$a_{11}(x^1)^2 + 2a_{12}x^1x^2 + a_{22}(x^2)^2 = c$$

is a conic section. In higher dimensional Euclidean spaces, the locus of the equation $X^TAX = c$ is a type of so-called "quadric surface." Since we shall discuss only those quadric surfaces with an equation of the type $X^TAX = c$, "quadric surface" will always refer to the locus of such an equation. The *central* quadric surfaces are those for which A is nonsingular. For example, in three dimensions there are three types of central quadrics: the ellipsoid, the cone, and the hyperboloid (see Sec. 8.9). The presence of mixed terms in the equation of such a surface indicates that no set of principal axes of the surface (the generalization of the major and minor axes of an ellipse) is coincident with the axes of the rectangular reference coordinate system.

Using the theory developed thus far, we can analyze quadric surfaces as follows: Matters are simplified with the observation that since an equation $q(\xi) = c$ and a nonzero scalar multiple $aq(\xi) = ac$ have identical loci, it is sufficient to consider the values 0 and 1 for c. But then our study of quadratic forms shows that, *relative to a suitably chosen set of axes,* a quadric surface has an equation of the form

$$(x^1)^2 + \cdots + (x^p)^2 - (x^{p+1})^2 - \cdots - (x^r)^2 = 0 \text{ or } 1 \qquad p \leqq r \leqq n$$

For example, in the plane case, this gives the following possibilities:

$$\begin{array}{cc} r = 2 & r = 1 \\ (x^1)^2 + (x^2)^2 = 0,\ 1 & (x^1)^2 = 0,\ 1 \\ (x^1)^2 - (x^2)^2 = 0,\ 1 & -(x^1)^2 = 0,\ 1 \\ -(x^1)^2 - (x^2)^2 = 0,\ 1 & \end{array}$$

Thus the possible loci are a circle, a hyperbola, two intersecting lines, one point, a pair of parallel lines, and, finally, no locus.

In the above example it should be clearly understood that the simplicity of our analysis is possible provided only that we are prepared to make certain sacrifices. We can elaborate this point now if the reader will permit us to rely on his knowledge of analytic geometry to the extent of the notions of the angle between two vectors and the length of a vector. In dealing with a function, quadratic or otherwise, over some Euclidean space it is customary to think in terms of a rectangular-coordinate system, which means referring vectors to a basis consisting of mutually perpendicular vectors of unit length (a so-called *orthonormal basis*). We know that the reduction of a function by a nonsingular substitution implies the selection of a new basis; it is the orthonormal basis that may be sacrificed to get a simple representation. The following example illustrates the state of affairs:

EXAMPLE 5.5

Consider the conic section

$$\begin{pmatrix} x^1 \\ x^2 \end{pmatrix}^{\mathrm{T}} \begin{pmatrix} 5 & 2 \\ 2 & 8 \end{pmatrix} \begin{pmatrix} x^1 \\ x^2 \end{pmatrix} = 9$$

whose graph is the ellipse shown in Fig. 7, where $(x^1, x^2)^T$ are the rectangular coordinates of a point in the plane (hence the coordinates of a vector relative to the unit vectors ϵ_1 and ϵ_2).

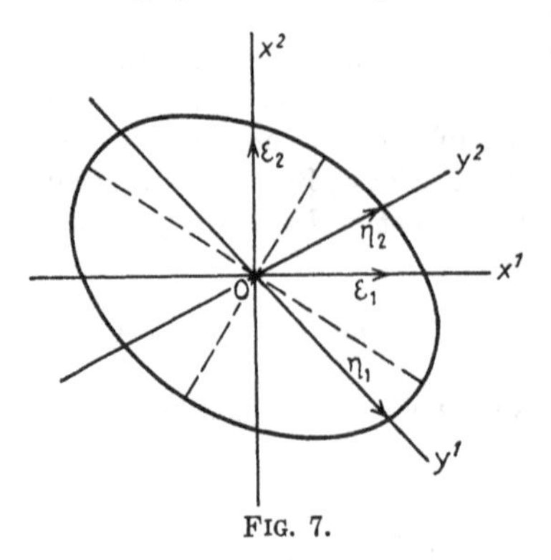

FIG. 7.

The substitution

$$\begin{pmatrix} x^1 \\ x^2 \end{pmatrix} = \begin{pmatrix} 1 & 1 \\ -1 & \frac{1}{2} \end{pmatrix} \begin{pmatrix} y^1 \\ y^2 \end{pmatrix}$$

reduces the above equation to

$$(y^1)^2 + (y^2)^2 = 1$$

The corresponding oblique axes have been drawn on the figure. It is clear that the new basis vectors η_1, η_2, which are related to ϵ_1, ϵ_2 by the equations

$$(\eta_1, \eta_2) = (\epsilon_1, \epsilon_2)\begin{pmatrix} 1 & 1 \\ -1 & \frac{1}{2} \end{pmatrix}$$

are neither perpendicular nor of unit length.

However, there is a reduction of the original equation to a sum of squares which corresponds to a transfer to another orthonormal basis. For example, the substitution

$$\begin{pmatrix} x^1 \\ x^2 \end{pmatrix} = \frac{1}{\sqrt{5}} \begin{pmatrix} 1 & -2 \\ 2 & 1 \end{pmatrix} \begin{pmatrix} z^1 \\ z^2 \end{pmatrix}$$

which is a rotation of the original basis into coincidence with the principal axes of the ellipse, reduces the equation to $9(z^1)^2 + 4(z^2)^2 = 9$.

Continuing our remarks prior to the above example, we may prefer to limit ourselves to substitutions which correspond to a transfer to another orthonormal set. For example, every quadric surface (just as in the above ellipse) determines at least one set of mutually perpendicular axes, along which it is customary to take the basis vectors. Substitutions which preserve orthonormal bases will be discussed in Chap. 8.

A general discussion of the contrast in the reductions in forms, etc., that are possible using nonsingular substitutions indiscriminately or, on the other hand, restricting oneself to those which preserve features of the original basis is best viewed from the standpoint of linear transformations of vectors. This will be developed in Sec. 6.5.

EXAMPLE 5.6

In mechanics, the moment of inertia of a system of n particles (with masses) m_i with respect to an axis is defined as the sum

$$\sum_{i=1}^{n} m_i r_i^2$$

where r_i denotes the distance of m_i from the axis. If $(x_i^1, x_i^2, x_i^3)^T$ is the column vector of coordinates of m_i with respect to a rectangular coordinate system with origin O, then the moments of inertia of the system about the coordinate axes are

$$A = \sum m_i((x_i^2)^2 + (x_i^3)^2)$$
$$B = \sum m_i((x_i^3)^2 + (x_i^1)^2)$$
$$C = \sum m_i((x_i^1)^2 + (x_i^2)^2)$$

The products of inertia of the system are defined as the sums

$$D = \sum m_i x_i^2 x_i^3 \qquad E = \sum m_i x_i^3 x_i^1 \qquad F = \sum m_i x_i^1 x_i^2$$

In terms of these six constants it is possible to evaluate the moment-of-inertia function I of the given system at a line through O and with direction cosines l, m, n. Indeed

$$I(l, m, n) = Al^2 + Bm^2 + Cn^2 - 2Dmn - 2Enl - 2Flm$$

so that I is a quadratic form. Let us measure off along a line through O with direction cosines l, m, n a distance OP such that $(OP)^2 I(l, m, n) = 1$. The locus of P has the equation

$$A(x^1)^2 + B(x^2)^2 + C(x^3)^2 - 2Dx^2x^3 - 2Ex^3x^1 - 2Fx^1x^2 = 1 \tag{8}$$

This is the equation of a quadric surface with center O. In general it is a closed surface, since I does not vanish for any line. Hence it is the equation of an ellipsoid, the ellipsoid of inertia of the system at O. Of course, with a change in coordinate axes, the equation (8) will be altered, but the quadric itself remains an invariant model of the inertial properties. Later we shall see that it is possible to rotate the given rectangular axes into coincidence with the principal axes of an ellipsoid, which will yield an equation for the ellipsoid of inertia free from cross-product terms.

EXAMPLE 5.7

Let φ denote a function on $V_n(R^\#)$ to $R^\#$ which possesses finite continuous partial derivatives in the neighborhood of a point $A = (a^1, a^2, \ldots, a^n)^T$.

We adopt the following notation:

$$\varphi_i = \left.\frac{\partial \varphi}{\partial x^i}\right|_A \qquad \varphi_{ij} = \left.\frac{\partial^2 \varphi}{\partial x^i\, \partial x^j}\right|_A = \varphi_{ji}$$

$$f = (\varphi_1, \varphi_2, \ldots, \varphi_n)^{\mathrm{T}} \qquad H = (h^1, h^2, \ldots, h^n)^{\mathrm{T}} \qquad F = (\varphi_{ij}) = F^{\mathrm{T}}$$

where H represents a column vector of increments. The Taylor series expansion of $\varphi(A + H)$ about the point A is

$$\varphi(A + H) = \varphi(A) + H^{\mathrm{T}}f + \tfrac{1}{2}H^{\mathrm{T}}FH + R$$

where R is the remainder. A necessary condition for a maximum or minimum of φ at A is that the vector f be the zero vector, since this ensures that φ is stationary at A. If this condition is fulfilled, A is called a critical point. Then we have

$$\varphi(A + H) - \varphi(A) = \tfrac{1}{2}H^{\mathrm{T}}FH + R$$

where for small values of the h^i's the controlling term on the right is the quadratic form $H^{\mathrm{T}}FH$. If this form has rank n, the behavior of φ can be described as follows:

(i) If $H^{\mathrm{T}}FH$ is positive definite, A is a minimum point.
(ii) If $H^{\mathrm{T}}FH$ is negative definite, A is a maximum point.
(iii) If $H^{\mathrm{T}}FH$ is indefinite, A is neither a maximum nor a minimum point.

If the rank of the form is less than n, a study of the remainder R is necessary in order to determine the behavior of φ at A.

PROBLEMS

1. If $(x^1, x^2)^{\mathrm{T}}$ are rectangular coordinates in the plane, the conic

$$\begin{pmatrix} x^1 \\ x^2 \end{pmatrix}^{\mathrm{T}} \begin{pmatrix} 6 & 12 \\ 12 & -1 \end{pmatrix} \begin{pmatrix} x^1 \\ x^2 \end{pmatrix} = -30$$

is a hyperbola. Reduce this equation using the substitution

$$\begin{pmatrix} x^1 \\ x^2 \end{pmatrix} = \begin{pmatrix} 0 & 1 \\ 1 & 12 \end{pmatrix} \begin{pmatrix} y^1 \\ y^2 \end{pmatrix}$$

and sketch the hyperbola relative to the new axes defined by the transformation.

2. Investigate each of the following functions for maximum and minimum points:

(a) $x^3 + y^3 - 3xy + 1$ (b) $x^2 + xy + y^2 - (ax + by)$

5.7. Hermitian Functions, Forms, and Matrices. In Chap. 8 we shall have occasion to consider a type of function on $V \times V$ to C, where V is an n-dimensional vector space over the complex field C, that is analogous to the bilinear functions of Sec. 5.2. The defining properties of such a function, f, which we shall call a *Hermitian conjugate bilinear function,* are the following:

(i) Linearity in the first argument:

$$f(a\xi + b\eta, \zeta) = af(\xi, \zeta) + bf(\eta, \zeta)$$

(ii) Hermitian symmetry:

$$f(\xi, \eta) = \overline{f(\eta, \xi)}$$

where the bar denotes complex conjugate.

Such a function is not linear, in general, in its second argument, since we can prove only that

$$\begin{aligned} f(\zeta, a\xi + b\eta) = \overline{f(a\xi + b\eta, \zeta)} &= \overline{af(\xi, \zeta) + bf(\eta, \zeta)} \\ &= \bar{a}\overline{f(\xi, \zeta)} + \bar{b}\overline{f(\eta, \zeta)} \\ &= \bar{a}f(\zeta, \xi) + \bar{b}f(\zeta, \eta) \end{aligned}$$

It is this property of f, together with (i), that is described by the phrase "conjugate bilinear." It is an easy matter to modify the proof of Theorem 5.1 for the case at hand to conclude that upon the selection of a basis for V, let us say the α-basis, a Hermitian conjugate bilinear function f is represented by a function on $V_n(C) \times V_n(C)$ to C, whose value at X, Y is

$$X^{\mathrm{T}}H\overline{Y} = \sum_{i,j=1}^{n} h_{ij}x^i\bar{y}^j \tag{9}$$

which we call a *Hermitian conjugate bilinear form.* Of course X and Y are the α-coordinates of the arguments ξ and η, respectively, of f; also $\overline{Y} = (\bar{y}_1, \bar{y}_2, \ldots, \bar{y}_n)^{\mathrm{T}}$, and $H = (h_{ij})$. Since

$$h_{ij} = f(\alpha_i, \alpha_j) = \overline{f(\alpha_j, \alpha_i)} = \bar{h}_{ji}$$

H satisfies the equation

$$H = \overline{H}^{\mathrm{T}} \tag{10}$$

This we take as the defining equation of a *Hermitian matrix* over the complex field. It is a generalization to the complex case of the notion of a real symmetric matrix.

As for the operation which consists in replacing each element of a matrix A over C by its conjugate, *i.e.*, replacing A by $\overline{A}$, observe that it has the following properties:

$$\overline{\overline{A}} = A, \ \overline{AB} = \overline{A}\,\overline{B}, \ (\overline{A})^{\mathrm{T}} = \overline{A^{\mathrm{T}}}$$

According to the last equation, (10) is unambiguous and also may be rewritten as $\overline{H} = H^{\mathrm{T}}$.

DEFINITION. *The conjugate transpose,* $\overline{A}^{T}$, *of a matrix* A *over* C *will be called the* associated adjoint *of* A *and written as* A^*.

In terms of this definition, the defining equation of a Hermitian matrix may be written as

$$H = H^*$$

At the moment, our interest lies in so-called "Hermitian functions." If in a Hermitian conjugate bilinear function on V to C we equate the two arguments, the resulting function of one variable is called a *Hermitian function.* It is an obvious (by virtue of the "self" Hermitian symmetry) yet important remark that *the values of a Hermitian function are real.* Corresponding to a basis for V, the α-basis for example, there is a representation h_α of a Hermitian function h whose value at X is [upon specializing (9)]

$$h_\alpha(X) = X^{T}H\overline{X} \qquad H = H^*$$

Such a function on $V_n(C)$ to C is called a *Hermitian form.* As a representation of a Hermitian function, the range of a Hermitian form is a set of real numbers. The analogue of Theorem 5.6 follows immediately:

Theorem 5.17. Two Hermitian forms are equivalent (*i.e.*, representations of one and the same Hermitian function) if and only if their respective matrices H, K are related by an equation of the form $K = PHP^*$, where P is nonsingular. Two matrices so related are called *conjunctive.* Conjunctivity is an equivalence relation over the set of $n \times n$ matrices over C.

The criterion for congruence (to which conjunctivity reduces over the real field) stated after the proof of Theorem 5.6 generalizes immediately to the current relation. We have: B is conjunctive to A if and only if

$$B = E_k(\cdots(E_2(E_1AE_1^*)E_2^*)\ \cdots)\ E_k^*$$

where each E_p is an elementary matrix (hence effects a row transformation) and E_p^* is its conjunctive column analogue. That is, if

$E_p = R_{ij}$, then E_p^* as a postmultiplier interchanges the ith and jth columns.

$E_p = R_i(c)$, then E_p^* as a postmultiplier multiplies the ith column by $\bar{c}$.

$E_p = R_{ij}(c)$, then E_p^* as a postmultiplier adds $\bar{c}$ times the jth column to the ith.

Analogous to our discussion of congruence, we shall limit that of conjunctivity to the set of Hermitian matrices. Since a matrix conjunctive to a Hermitian matrix is Hermitian, the set of $n \times n$ Hermitian matrices is a sum of certain conjunctivity classes, which implies that the conjunctivity relation restricted to the set of Hermitian matrices is simply conjunctivity. Next, with the definition of a C_{III} transformation modified in accordance with the effect of $R_{ij}^*(c)$ as a postmultiplier, Lemma 5.1 remains valid and the following analogue of Lemma 5.2 holds:

Lemma 5.3. A Hermitian matrix H of positive rank can be reduced with C transformations to a conjunctive matrix of the form

$$K = \text{diag}(K_1, K_2, \ldots, K_s, 0)$$

where each K_i has one of the following forms:

(a) where a is real and nonzero

$$K(a) = \begin{pmatrix} 0 & a \\ \bar{a} & 0 \end{pmatrix} \qquad \text{where } a \neq 0$$

Moreover, this reduction may be effected with a sequence of C_I transformations followed by a sequence of C_{III} transformations.

We shall not supply a proof of this or succeeding results, since they are immediate generalizations of the proofs of the earlier theorems that we are imitating. However, we feel obliged to comment on the form of K above. It stems from the fact that the determinant of a Hermitian matrix H (and hence each principal minor of H) is real. The chain of equalities below verifies the latter statement:

$$\det H = \det H^T = \det \bar{H} = \overline{\det H}$$

We may now apply Lemma 5.3 to prove the analogue of Theorem 5.7 (see the proof of Theorem 5.7, as well as Prob. 4 at the end of this section) where the diagonal entries of D are now all real. From this result follows, in turn, the analogue of Theorem 5.12, which we state next:

Theorem 5.18. Let V denote a vector space over C and h a Hermitian function of rank r on V to C. Then h can be represented by a form of the type

$$v^1\bar{v}^1 + v^2\bar{v}^2 + \cdots + v^p\bar{v}^p - v^{p+1}\bar{v}^{p+1} - \cdots - v^r\bar{v}^r \tag{11}$$

where the number p is uniquely determined by h. We call (11) the *canonical representation* of a Hermitian function. Stated for Hermitian forms, any Hermitian form can be reduced by a nonsingular substitution to a uniquely determined form (11). Finally, stated for Hermitian matrices, a Hermitian matrix of rank r is conjunctive to a uniquely determined diagonal matrix of the type

$$\text{diag}(I_p, -I_{r-p}, 0)$$

One notices that the earlier definition of signature is applicable to Hermitian functions and matrices. Thus, parallel to our earlier results, we infer from the preceding theorem that a Hermitian function on an n-dimensional space over C is characterized by its rank and signature. Alternatively, we can say that for the set of Hermitian forms in $x^1, x^2, \ldots, x^n$, rank and signature constitute a complete set of invariants under equivalence, and the forms (11) constitute a canonical set. Finally, the results imply that for the set of $n \times n$ Hermitian matrices, rank and signature constitute a complete set of invariants under conjunctivity and define a canonical set consisting of the matrices (6); in particular, two Hermitian matrices are conjunctive if and only if they have the same rank and signature.

Since the values of a Hermitian function (and form) are real, the classification as positive definite, etc., is possible. The analogue of Theorem 5.14 follows:

Theorem 5.19. A Hermitian function, or form, is positive definite, positive semidefinite, indefinite, respectively, if and only if its canonical form is

$$\sum_1^n v^i\bar{v}^i \qquad \sum_1^r v^i\bar{v}^i, \quad r < n \qquad \sum_1^p v^i\bar{v}^i - \sum_{p+1}^r v_i\bar{v}_i, \quad p < r \leqq n$$

Finally, concerning the analogue of the earlier results dealing with properties of symmetric matrices, we recall that we have available Lemma 5.1 and the analogue (Lemma 5.3) of Lemma 5.2. As such, *it is possible to deduce Theorems* 5.9, 5.10, 5.11, 5.13, 5.15, *and* 5.16 *with no changes whatsoever, apart from the replacement of "symmetric" and "quadratic" by "Hermitian."*

PROBLEMS

1. If G is a matrix of complex numbers, such that $X^T G\overline{X}$ is always real for complex X, prove that G is Hermitian.

2. Let A denote a real symmetric or complex Hermitian matrix. Prove that there exists a real number c, such that $cI + A$ is positive definite.

3. Show that if A is any matrix of complex numbers, then $\overline{A^T}A$ and $A\overline{A^T}$ are both Hermitian.

4. Show that the Hermitian matrix

$$H = \begin{pmatrix} 0 & a \\ \bar{a} & 0 \end{pmatrix} \qquad a \neq 0$$

is conjunctive to $D = \operatorname{diag}(2a\bar{a}, -\frac{1}{2})$, by applying to H the row transformations $R_{12}(a)$ and $R_{21}\left(-\dfrac{1}{2a}\right)$, each followed by its conjunctive column analogue. Determine the corresponding matrix P such that $PHP^* = D$.

5. Show directly that the values of a Hermitian form $X^T H \bar{X}$ are real numbers.

6. Supply a proof of Theorem 5.17.

7. Supply a proof of Lemma 5.3.

8. Construct a nonsingular substitution that will reduce the Hermitian form with matrix

$$H = \begin{pmatrix} 0 & -i & 1-i \\ i & 0 & 1+2i \\ i+1 & 1-2i & 0 \end{pmatrix}$$

to the canonical form (11) in Theorem 5.18. SUGGESTION: Imitate the method used in Example 5.2.

9. Determine the rank and signature of each of the following Hermitian matrices:

$$\begin{pmatrix} 5 & i & 1+2i \\ -i & 1 & -i \\ 1-2i & i & 1 \end{pmatrix} \qquad \begin{pmatrix} 2 & 1-i & -2i \\ 1+i & 0 & 1-i \\ 2i & 1+i & 0 \end{pmatrix}$$

CHAPTER 6

LINEAR TRANSFORMATIONS ON A VECTOR SPACE

6.1. Linear Functions and Matrices. Just as the preceding chapter, this one begins with a discussion of linear functions on a vector space V over a field F. However, we now assume that the values of such a function are in a vector space W over F. Such linear functions will always be denoted by capital letters in boldface type. Thus, typical notation for the functions we shall study is the following:

$$\mathbf{A}\xi = \eta \qquad \xi \text{ in } V,\ \eta \text{ in } W$$

where

$$\mathbf{A}(c_1\xi_1 + c_2\xi_2) = c_1\mathbf{A}\xi_1 + c_2\mathbf{A}\xi_2 \tag{1}$$

In general, in indicating a function value, the argument will be enclosed in parentheses only when a misinterpretation is possible.

If bases are introduced for V and W, such a function is displayed in a more familiar light. Indeed if $\{\alpha_1, \alpha_2, \ldots, \alpha_n\}$ is a basis for V and the vector ξ in V has the coordinates X relative to this basis, then

$$\mathbf{A}\xi = \mathbf{A}\left(\sum_1^n x^j\alpha_j\right) = \sum_1^n x^j\mathbf{A}\alpha_j$$

so that $\mathbf{A}\xi$ is known as soon as $\mathbf{A}\alpha_j$ is known for $j = 1, 2, \ldots, n$. If $\{\beta_1, \beta_2, \ldots, \beta_m\}$ is a basis for W, then $\mathbf{A}\alpha_j$ and η, as vectors in W, are linear combinations of the β-basis:

$$\mathbf{A}\alpha_j = \sum_{i=1}^m a_j^i\beta_i \qquad j = 1, 2, \ldots, n \qquad \eta = \sum_1^m y^i\beta_i \tag{2}$$

Thus we have

$$\eta = \sum_j x^j\mathbf{A}\alpha_j = \sum_j x^j\left(\sum_i a_j^i\beta_i\right) = \sum_i\left(\sum_j a_j^i x^j\right)\beta_i = \sum_i y^i\beta_i$$

Because of the uniqueness of the coordinates of a vector relative to a basis, we conclude that

$$\sum_{j=1}^n a_j^i x^j = y^i \qquad i = 1, 2, \ldots, m$$

Once again we encounter a system of linear equations, which of course we shall write in matrix notation as

$$AX = Y \tag{3}$$

Before commenting on (3), we digress briefly to indicate how (3) can be obtained directly, using matrix notation together with the following definition:

$$\mathbf{A}\alpha = (\mathbf{A}\alpha_1, \mathbf{A}\alpha_2, \ldots, \mathbf{A}\alpha_n)$$

where, as usual, α symbolizes the row matrix $(\alpha_1, \alpha_2, \ldots, \alpha_n)$ whose entries constitute a basis for V. Indeed, with (2) written as

$$\mathbf{A}\alpha_j = \beta A_j \qquad \text{where } A_j = (a_j^1, a_j^2, \ldots, a_j^m)^{\mathrm{T}}, j = 1, 2, \ldots, n$$

we have

$$\begin{aligned} \mathbf{A}\alpha = (\mathbf{A}\alpha_1, \mathbf{A}\alpha_2, \ldots, \mathbf{A}\alpha_n) &= (\beta A_1, \beta A_2, \ldots, \beta A_n) \\ &= \beta(A_1, A_2, \ldots, A_n) = \beta A \end{aligned} \tag{4}$$

Hence our earlier assumptions can be written as

$$\xi = \alpha X \qquad \eta = \beta Y \qquad \mathbf{A}\alpha = \beta A \tag{5}$$

to infer

$$\eta = \mathbf{A}\xi = \mathbf{A}(\alpha X) = (\mathbf{A}\alpha)X = (\beta A)X = \beta(AX) \qquad \text{or} \qquad AX = Y \tag{6}$$

where the third and fifth equality signs are justified by the linearity of **A** and the associativity law mentioned in the last paragraph of Sec. 3.6, respectively.

Returning to (3), our first comment is that in its present setting it has entirely different connotations from those it had before. It is a formula for the coordinates Y of the value η of the linear function **A** in terms of the coordinates X of the argument ξ. We observe that **A** is completely determined as soon as its values at n linearly independent vectors (where $n = d[V]$) are known, since the coordinates A_j of $\mathbf{A}\alpha_j$ relative to the β-basis are the columns of A and with A known, $Y = AX$ is known for every X. This is precisely the feature of **A** that is emphasized in (4), (5), and (6).

The above correspondence of linear function to matrix with respect to fixed bases for V and W is a one-one correspondence between the set of all linear functions on V to W and the set of all $m \times n$ matrices over F. Let us verify this. Relative to fixed bases for V and W, an $m \times n$ matrix A defines a function on V to W using (2), if it is assumed that the function is linear; A is then the correspondent of this function. Also, if two linear functions **A** and **B** define the same matrix, it is clear that $\mathbf{A} = \mathbf{B}$, that is, $\mathbf{A}\xi = \mathbf{B}\xi$ for all ξ in V. Indeed the definition of equality in Chap. 3 for matrices was chosen so that this latter statement would hold.

Next we contend that this one-one correspondence is preserved by operations already introduced for functions and matrices, respectively. In Sec. 3.7 it was mentioned that the set of all $m \times n$ matrices over F together with matrix addition and scalar multiplication is a vector space. On the other hand, the definitions of operations having these same names, defined in Sec. 1.2 for functions, apply to those at hand, and it is clear that the set of all linear functions on V to W together with these operations is a vector space. Finally it is easy to establish Theorem 6.1, which follows. Actually the definitions of equality, addition, and scalar multiplication of matrices were designed with this statement in mind.

Theorem 6.1. The one-one correspondence between the set of all linear functions on V to W (both vector spaces over F) and the set of all $m \times n$ matrices over F (where $m = d[W]$, $n = d[V]$) that is defined when bases are chosen for V and W is an isomorphism between the vector space of linear functions and that of matrices.

The analogue among linear functions of the product of two matrices is that of the product of two functions as defined in the section on notation and conventions. Thus, if $\mathbf{A}$ and $\mathbf{B}$ are linear functions on V to W and W to U, respectively, the product $\mathbf{BA}$ of $\mathbf{A}$ by $\mathbf{B}$ is the function on V to U defined by the equation

$$(\mathbf{BA})\xi = \mathbf{B}(\mathbf{A}\xi)$$

It is easily shown that $\mathbf{BA}$ is linear and, moreover, that if the matrices A and B correspond to $\mathbf{A}$ and $\mathbf{B}$, respectively, when bases are chosen for the spaces involved, then BA corresponds to $\mathbf{BA}$ relative to the same bases.

We have begun this chapter with an introduction to linear functions on one vector space to a second vector space in order to demonstrate the correlation between such functions and $m \times n$ matrices. Actually, our principal interest is in linear functions on a vector space V to itself. Such a linear function $\mathbf{A}$ is usually called a *linear transformation* or *linear operator on* V and the function value $\mathbf{A}\xi$ the *image* of ξ under $\mathbf{A}$. Regarding this terminology, the word "operator" has its origin in the realization $AX = Y$ of $\mathbf{A}$; A applied to (or operating on) X produces Y.† The use of the word "image" stems from the interpretation of a linear transformation as carrying the vector ξ into the vector $\mathbf{A}\xi$; in mechanics (*e.g.*, a study of shears, radial compressions, etc.) this is a natural viewpoint. In the next section we begin a systematic study of linear transformations. After investigating transformations singly, certain classes of transformations will be discussed; at every stage the correspondence of transformations with matrices will be stressed.

†Compare with the derivative operator in the calculus.

6.2. Nonsingular Linear Transformations. Specializing the results of the previous section to the case of a linear transformation **A** on an n-dimensional space V over F, we conclude that once a basis $\{\alpha_1, \alpha_2, \ldots, \alpha_n\}$ is chosen for V, **A** determines and is determined by a single matrix equation $AX = Y$ for the α-coordinates Y of the image η under **A** of the vector ξ with α-coordinates X. Moreover, the columns of A are the α-coordinates of the image of the α-basis. Using notation introduced in Sec. 6.1, X, Y, and A are defined by the relations

$$\xi = \alpha X,\ \eta = \alpha Y,\ \mathbf{A}\alpha = \alpha A, \text{ and } \mathbf{A}\xi = \eta$$

It will prove convenient to designate A by the symbol

$$(\mathbf{A}; \alpha)$$

which exhibits the transformation and basis that characterize A. Thus the equation $A = (\mathbf{A}; \alpha)$ is symbolism for "A is the matrix that corresponds to the linear transformation **A** on V, under the one-one correspondence between linear transformations on V and $n \times n$ matrices over F, relative to the α-basis."

The first topic for discussion in our exposition of linear transformations is the distinction between the interpretation of the matrix equation $AX = Y$ as a realization of a transformation and as a formula for a change of coordinates in a vector space. For the latter case, A is nonsingular by necessity, so that we shall make the same restriction for the former case.

Theorem 6.2. If A is nonsingular, the linear transformation, **A**, on V defined by $AX = Y$ relative to any basis is a one-one correspondence on V onto V. Conversely, if the linear transformation **A** on V is one-one, then $(\mathbf{A}; \alpha)$ is nonsingular. Thus it is appropriate to call the one-one linear transformations on V *nonsingular*.

Proof. If **A** is defined by $AX = Y$ relative to the α-basis for V, then $\mathbf{A}\xi = \eta$ if and only if $\xi = \alpha X$, $\eta = \alpha Y$. To show that **A** is one-one when A is nonsingular, we must prove that (i) given η in V there exists a ξ such that $\mathbf{A}\xi = \eta$ and (ii) $\xi_1 \neq \xi_2$ implies $\mathbf{A}\xi_1 \neq \mathbf{A}\xi_2$. For (i) suppose that $\eta = \alpha Y$. Define $\xi = \alpha A^{-1}Y$; then $\mathbf{A}\xi = \eta$. As for (ii), if $\mathbf{A}\xi_1 = \mathbf{A}\xi_2$, where $\xi_i = \alpha X_i$, $i = 1, 2$, then $AX_1 = AX_2$, which gives $X_1 = X_2$ or $\xi_1 = \xi_2$ upon multiplication by A^{-1}.

Conversely, if **A** is one-one and $(\mathbf{A}; \alpha) = A$, then $AX = Y$ has a unique solution for each choice of Y, which forces A to be nonsingular. This completes the proof.

As a one-one correspondence, a nonsingular linear transformation **A** on V has an inverse, $\mathbf{A}^{-1}$, defined by the equation

$$\mathbf{A}^{-1}\eta = \xi \qquad \text{if and only if } \mathbf{A}\xi = \eta \tag{7}$$

This transformation is linear, and if $\mathbf{I}$ denotes the *identity transformation* on V, that is,

$$\mathbf{I}\xi = \xi \qquad \text{all } \xi \text{ in } V$$

it is clear that $\mathbf{I}$ is linear and moreover that

$$\mathbf{A}\mathbf{A}^{-1} = \mathbf{A}^{-1}\mathbf{A} = \mathbf{I} \tag{8}$$

This property characterizes a one-one transformation according to the following theorem:

Theorem 6.3. A linear transformation $\mathbf{A}$ on V is one-one if and only if there exists a linear transformation $\mathbf{B}$ such that

$$\mathbf{AB} = \mathbf{BA} = \mathbf{I}$$

This formal inverse is then the inverse in the sense of equation (7).

Proof. If such a $\mathbf{B}$ exists, then for all ξ and η

$$\mathbf{A}(\mathbf{B}\eta) = (\mathbf{AB})\eta = \mathbf{I}\eta = \eta \qquad \mathbf{B}(\mathbf{A}\xi) = (\mathbf{BA})\xi = \mathbf{I}\xi = \xi$$

The first result asserts that any vector η is the image $\mathbf{A}\zeta$ of some vector $\zeta = \mathbf{B}\eta$. The second result asserts that if $\mathbf{A}\xi = \eta$ then $\mathbf{B}\eta = \xi$, which implies that η is the image of only one vector, *viz.*, the vector ξ such that $\mathbf{B}\eta = \xi$. Taken together, these statements mean that $\mathbf{A}$ is one-one.

Now for the discussion of the dual role of $AX = Y$ with A nonsingular. The use of this equation in Chap. 5 has its origin in Chap. 3, where it is used in conjunction with the equation $\alpha = \beta A$. If in a vector space the (old) α-basis is replaced by the (new) β-basis, the coordinates X of a vector ξ are changed to $Y = AX$. We call this the *alias* interpretation of $AX = Y$: upon a change of basis, the vector with coordinates X is *relabeled* with coordinates Y.

In the present chapter, when used in conjunction with a basis for V, $AX = Y$ defines a nonsingular linear transformation $\mathbf{A}$ where $(\mathbf{A}; \alpha) = A$, which associates with the vector αX the vector αY. We call this the *alibi* interpretation of $AX = Y$: the vector with coordinates X is *carried into* the vector with coordinates $Y = AX$.

Summarizing, $AX = Y$, with A nonsingular, may be considered as a transformation of coordinates (alias), or it may be considered as a transformation of vectors (alibi). Thus *every algebraic simplification of a given expression that can be accomplished by a change of coordinates in a vector space can be accomplished by a nonsingular linear transformation on the space, and conversely.* The implications of this statement are developed in Sec. 6.5.

As a final remark it should be observed that a change of basis in V defines a linear transformation on V, and conversely. For example, if the

α-basis and β-basis are given, a linear function **A** on V is defined by the equation

$$\mathbf{A}\alpha = \beta$$

The image of a vector αX is βX under **A**, since

$$\mathbf{A}(\alpha X) = (\mathbf{A}\alpha)X = \beta X$$

PROBLEMS

1. A linear transformation on $V_3(R)$ is defined by the equations

$$\mathbf{A}\epsilon_1 = \epsilon_1 + \epsilon_2 \qquad \mathbf{A}\epsilon_2 = \epsilon_1 + \epsilon_3 \qquad \mathbf{A}\epsilon_3 = \epsilon_2 + \epsilon_3$$

Determine $(\mathbf{A};\epsilon)$, and compute the coordinates of the image of $\beta = \epsilon_1 - \epsilon_2 + \epsilon_3$ under **A**. If $\alpha_i = \mathbf{A}\epsilon_i$, determine $(\mathbf{A};\alpha)$.

2. In Prob. 1 determine the coordinates of β relative to the α-basis.

3. A linear transformation **A** on $V_4(R^*)$ is defined by the following change of coordinates:

$$\begin{pmatrix} x^1 \\ x^2 \\ x^3 \\ x^4 \end{pmatrix} \rightarrow \begin{pmatrix} 2x^1 - x^2 + x^3 \\ -x^1 + x^2 - x^4 \\ x^1 + x^3 - x^4 \\ - x^2 - x^3 + 2x^4 \end{pmatrix}$$

Later we shall prove that the set, R, of all vectors which occur as images under **A**, as well as the set, N, of all vectors whose image is the zero vector, are subspaces. Determine a basis for R and one for N.

6.3. The Group $L_n(F)$. It is important to study not only individual nonsingular linear transformations on a vector space but in addition the totality of such. The latter is best viewed against the background of a type of abstract system which we have not yet directly encountered. We shall digress to discuss this new notion, *viz.*, that of a group.

DEFINITION. *A set $G = \{\alpha, \beta, \ldots\}$, together with a (binary) composition $\cdot$ in G, is called a group (in symbols, $G, \cdot$) if the following conditions hold:*

G_1. *Associative law: $\alpha(\beta\gamma) = (\alpha\beta)\gamma$ for all α, β, γ in G.*

G_2. *Unit element: G contains at least one element 1, called a right unit element, with the property $\alpha 1 = \alpha$ for all α in G.*

G_3. *Inverse element: For each unit element 1 and every element α in G, the equation $\alpha\xi = 1$ has a solution in G.*

If the composition in G is commutative, so that

$$\alpha\beta = \beta\alpha \qquad \text{all } \alpha, \beta \text{ in } G$$

the group is called *commutative* or *abelian*. As in the case of fields and vector spaces, we shall often use the term "group G" for the set part of the group.

We have used multiplicative notation in the definition of a group, so that it is appropriate to call the composition multiplication and $\alpha\beta$ the product of α and β. However, we could just as well indicate the composition by $+$, call it addition, and use additive notation throughout. In that event, in order to achieve consistency in the notation, one speaks of a *zero element* 0 in place of a unit element 1 and a *negative* in place of an inverse.

If the postulates for a field stated in Chap. 1 are reviewed, it is seen that now a field can be described in the following elegant fashion. A field is a system consisting of a set F, together with two binary compositions in F, addition and multiplication, such that (i) F and addition is a commutative group with zero element 0, (ii) F, with 0 omitted, and multiplication is a commutative group, and (iii) multiplication distributes over addition. Also, we observe that those postulates for a vector space pertaining to addition can be summarized by the statement that, relative to addition, a vector space is an abelian group. In this connection, the properties of a vector space which are consequences of the addition postulates (rules 1 to 5 in Sec. 2.2) are valid for groups in general. Thus, in particular, stated in multiplicative notation, a group G has a unique unit element 1,

$$\alpha 1 = 1\alpha = \alpha \qquad \text{all } \alpha \text{ in } G$$

relative to which each element has a unique inverse α^{-1}

$$\alpha\alpha^{-1} = \alpha^{-1}\alpha = 1$$

To prove these statements without the use of the commutative law, we multiply $\alpha\xi = 1$ on the left by ξ to obtain $\xi(\alpha\xi) = \xi$ or $(\xi\alpha)\xi = \xi$ (using G_1) and then multiply this equation on the right by a right inverse of ξ with respect to 1 (using G_3) to obtain $\xi\alpha 1 = 1$ or $\xi\alpha = 1$, which shows that ξ is a left inverse of α with respect to 1. Furthermore,

$$1\alpha = (\alpha\xi)\alpha = \alpha(\xi\alpha) = \alpha 1 = \alpha$$

so that 1 is also a left unit for all α. Now we can follow the method of Sec. 2.2 to show that the right and left cancellation laws for multiplication hold, that division is possible and unique, and, finally, that the inverse and unit are unique. The reader should verify these statements.

Various primitive notions discussed for vector spaces are meaningful when scalar multiplication is ignored and hence are basically group-theory notions. We shall enumerate several of these now; examples appear later. First there is the notion of a *subgroup* of a group G as a nonempty subset H, closed under the composition in G, which, together with the thereby induced composition in H, is a group. Necessary and sufficient conditions on a subset H that $H, \cdot$ be a subgroup of $G, \cdot$ are that

(i) α, β in H imply that $\alpha\beta$ is in H.
(ii) 1 is in H.
(iii) α in H implies that α^{-1} is in H.

Before we describe the next concept, which is the analogue of the subspace generated by a set of vectors, the *integral powers* α^m of a group element α must be defined. If $m > 0$, we define

$$\alpha^m = \alpha \cdot \alpha \cdots \alpha (m \text{ factors}) \qquad \alpha^0 = 1 \qquad \alpha^{-m} = (\alpha^{-1})^m$$

Two of the familiar laws of exponents hold:

$$\alpha^m\alpha^n = \alpha^{m+n} \qquad (\alpha^m)^n = \alpha^{mn}$$

It is then clear what is meant by a power product with a finite number of factors, with or without repetitions. The set of all such power products that can be constructed using the members of a subset S of elements in a group G is seen to be (using the above criterion for a subgroup) a subgroup of G, which is called *the subgroup of G generated by the elements of S.*

Finally there is the concept of a quotient group which underlies our earlier discussion of a quotient space (Sec. 2.7). For this a more detailed study is necessary since our previous exposition relies heavily on the commutativity of the addition operation in the vector space.

A review of Theorem 2.17 will reveal that, with scalar multiplication discarded, we proved that, starting with an additive abelian group V and a subgroup S, the set V/S of residue classes $\bar{\xi}$ modulo S, together with the composition

$$\bar{\xi} + \bar{\eta} = \overline{\xi + \eta}$$

is a group, indeed a homomorphic image of V. A similar theorem is true for noncommutative groups provided we restrict ourselves to subgroups which commute with the group elements in the sense of the following definition:

DEFINITION. *A subgroup N of a (multiplicative) group G is* normal *in G if and only if $\xi N = N\xi$ for all ξ in G. Here ξN, for a fixed ξ in G, denotes the set of all elements $\xi\alpha$ where α ranges over N. Such a subset of G is called a* (left) coset *of N.*

To construct a quotient group starting with a given normal subgroup N of a group G, we imitate the additive case by defining the elements ξ and η as being congruent modulo N, in symbols $\xi \equiv \eta (\text{mod } N)$, if and only if $\xi\eta^{-1}$ is in N. This is easily shown to be an equivalence relation over G with the property that modulo N,

$$(9) \qquad \xi \equiv \eta \text{ implies } \xi\zeta \equiv \eta\zeta \text{ and } \zeta\xi \equiv \zeta\eta \qquad \text{for all } \zeta \text{ in } G$$

To prove (9), observe that

$$\xi\zeta(\eta\zeta)^{-1} = \xi\zeta(\zeta^{-1}\eta^{-1}) = \xi\eta^{-1}$$

is in N whenever $\xi \equiv \eta$. Also

$$\zeta\xi(\zeta\eta)^{-1} = \zeta(\xi\eta^{-1})\zeta^{-1}$$

is in N whenever $\xi\eta^{-1}$ is in N, because of the normality of N. Using (9), it follows that a multiplication can be defined in the set G/N of equivalence classes $\bar{\xi}$ modulo N as follows:

$$\bar{\xi} \cdot \bar{\eta} = \overline{\xi\eta} \tag{10}$$

Then $G/N, \cdot$ is a group, *the quotient group of G modulo N*, and indeed a homomorphic image of G under the correspondence $\xi \to \bar{\xi}$, according to (10).

To grasp the essence of the normality condition in this connection, it is helpful to notice that a residue class of G modulo N is a coset ξN (the analogue of $\xi + S$ in Theorem 2.17). The multiplication rule (10) for residue classes gives

$$\xi N \cdot \eta N = \xi\eta N$$

If we define the *product of two left cosets* ξN and ηN of an arbitrary subgroup N as the set of all products $\sigma\tau$ with σ and τ in ξN and ηN, respectively, then when N is normal, we obtain the above formula. For $N\eta$ may be replaced by ηN, which gives $\xi N \cdot \eta N = \xi\eta NN$, and finally $N \cdot N = N$, since, as a group, N is closed under multiplication and contains a unit element.

In addition, it is clear that the coset N is a unit element in the set of cosets and that $\xi^{-1}N$ is an inverse of ξN.

Theorem 6.4. If N is a normal subgroup of G, the set G/N of residue classes ξN modulo N with the composition $\xi M \cdot \eta N = \xi\eta N$ is a group. This quotient group is a homomorphic image of G. All homomorphic images of G may be obtained (within an isomorphism) in this way.

Proof. The first two statements summarize the preceding construction. To prove the final one, suppose that $G \xrightarrow{\varphi} G'$. If $1'$ is the unit element in G', it is easily shown that $N = \varphi^{-1}1'$, that is, the set of those elements of G which correspond to $1'$, determines a normal subgroup. More generally, if

$$\xi \to \xi'$$

then

$$\varphi^{-1}\xi' = \xi N$$

so that the partition of G defined by the homomorphism agrees with that defined by N (*i.e.*, the cosets of N in G). Moreover, since

$$\xi \to \xi' \text{ and } \eta \to \eta' \text{ imply } \xi\eta \to \xi'\eta'$$

it follows that $\xi N \cdot \eta N = \xi\eta N$. In other words, G/N is isomorphic with G'. This completes the proof.

We turn now to a study of several examples to illustrate the foregoing concepts. In Sec. 1.9 the four elements $\bar{1}$, $\bar{2}$, $\bar{3}$, $\bar{4}$ of J_5 determine a multiplicative group of four elements. If, in the multiplication table there for J_5, $\bar{0}$ is discarded, the multiplication table for this group results. In the case of a *finite group*, *i.e.*, one containing a finite number of elements, a multiplication table can always be constructed and all properties of the group deduced from it. In the example at hand notice that all group elements can be expressed as powers of the element $\bar{2}$:

$$\bar{2}^0 = \bar{1},\ \bar{2}^1 = \bar{2},\ \bar{2}^2 = \bar{4},\ \bar{2}^3 = \bar{3}$$

A group all of whose elements can be expressed as powers of a single element is called *cyclic*. Every cyclic group is commutative.

EXAMPLE 6.1

Consider the set of six functions $\{f_1, f_2, \ldots, f_6\}$ of a complex variable z where

$$f_1(z) = z,\ f_2(z) = \frac{1}{1-z},\ f_3(z) = \frac{z-1}{z},$$

$$f_4(z) = \frac{1}{z},\ f_5(z) = 1 - z,\ f_6(z) = \frac{z}{z-1}$$

with the composition f_if_j introduced earlier for functions, *viz.*,

$$(f_if_j)(z) = f_i(f_j(z))$$

The multiplication table for the set follows:

	f_1	f_2	f_3	f_4	f_5	f_6
f_1	f_1	f_2	f_3	f_4	f_5	f_6
f_3	f_3	f_1	f_2	f_5	f_6	f_4
f_2	f_2	f_3	f_1	f_6	f_4	f_5
f_4	f_4	f_5	f_6	f_1	f_2	f_3
f_5	f_5	f_6	f_4	f_3	f_1	f_2
f_6	f_6	f_4	f_5	f_2	f_3	f_1

The table demonstrates that all of the group postulates, except possibly the associative law, are valid. The validity of the associative law for the operation used here will be shown, in general, in the next section. We shall accept this fact now to conclude that the system is a group G. The group is noncommutative and has

$$N = \{f_1, f_2, f_3\}$$

as its only normal subgroup, apart from the full group and the subgroup consisting of the unit element f_1 alone. The cosets of N in G are N and f_4N, which form a cyclic group of two elements.

EXAMPLE 6.2

The set of all nonsingular nth-order matrices with entries in a field F and matrix multiplication constitute a group. The identity matrix I_n is the unit element and A^{-1} the inverse of A.

This example is, in effect, equivalent to the culminating result of this section, *viz.*, that *the set of all nonsingular linear transformations on a vector space and multiplication* (*i.e.* function of a function) *is a group.* It was the desire to make known this important property of the nonsingular linear transformations that motivated the introduction of the concept of group at this time. A proof of this result can be based on the correspondence between nonsingular linear transformations and nonsingular matrices, together with the assertion of Example 6.2. For since this correspondence is preserved by multiplication in the respective systems (see the paragraph after Theorem 6.1), we can deduce that since the matrices determine a group, the transformations do likewise. As for a direct proof, the group postulates G_2 and G_3 have been verified. The validity of the associative law G_1 remains to be confirmed; this is done in the next section for a more general class of transformations. The group of all nonsingular linear transformations (and sometimes its matrix analogue) is called the *full linear group* and denoted by $L_n(F)$, where n is the dimension of the vector space and F its field of scalars.

PROBLEMS

1. In a group the equations $\alpha\xi = \beta$ and $\eta\alpha = \beta$ have (unique) solutions for fixed α, β. Show that a system G with a composition $\alpha\beta$ satisfying G_1 and the following postulate G_4 is a group:

G_4 : All equations $\alpha\xi = \beta$ and $\eta\alpha = \beta$ have solutions ξ and η in G.

2. In Prob. 1 show that G_4 can be replaced by the weaker assumption G_5 in the event G is finite.

G_5 : The cancellation rules hold in G: $\alpha\gamma = \beta\gamma$ and $\gamma\alpha = \gamma\beta$ each implies $\alpha = \beta$.

3. Use Prob. 2 to prove that if G is a finite group and H a subset closed under multiplication, then H is a subgroup.

4. Let H denote a subgroup of the group G. Verify that the decomposition of G into left cosets ξH can be carried out as in the case of any normal subgroup. Use this fact to deduce that the *order of H* (*i.e.*, the number of elements in H) divides the order of G, assuming the latter is finite.

5. If α is an element in a finite group G, there is a least positive integer n such that $\alpha^n = 1$. Show that n divides the order of G.

6. Consider all triangular matrices

$$M(p, q, r) = \begin{pmatrix} 1 & p & q \\ 0 & 1 & r \\ 0 & 0 & 1 \end{pmatrix}$$

where p, q, r are integers. Show that this set, together with matrix multiplication, is a group G. Determine whether or not G is abelian. Determine whether $M(p, q, r) \rightarrow p$ is a homomorphism on G onto the additive group of integers.

7. Imagine that a regular polygon P_n of n sides is cut out of cardboard and laid in the plane of this page and its outline traced on the page. A rotation through $2\pi k/n$ radians, $k = 0, 1, \ldots, n - 1$, of P_n about its center brings P_n into coincidence with its original trace, as does each of the n reflections (out of the plane) of P_n into itself about each of its axes of symmetry. If a composition M_1M_2 of two such motions M_1, M_2 is defined by the equation $(M_1M_2)P_n = M_1(M_2P_n)$, show that the totality of $2n$ motions with this composition is a group. This is the *group of symmetries of* P_n.

8. Show that the set of all matrices

$$A(v) = \left(1 - \frac{v^2}{c^2}\right)^{-\frac{1}{2}} \begin{pmatrix} 1 & -v \\ -v/c^2 & 1 \end{pmatrix}$$

where v ranges over the interval $-c < v < c$, c a positive constant, together with matrix multiplication is a group. This is the *Lorentz group.*

9. What is the order of the full linear group $L_2(J_3)$?

10. A *translation* $\mathbf{T}$ on a vector space V is a transformation such that

$$\mathbf{T}\xi = \xi + \alpha \qquad \alpha \text{ a fixed vector in } V$$

Since $\mathbf{T}$ is determined by α, a suitable notation for this transformation is $\mathbf{T}_\alpha$. Show that the set of all translations and multiplication constitute an abelian group which is isomorphic to the additive group of the vectors of V under the correspondence $\mathbf{T}_\alpha \rightarrow \alpha$.

11. An *affine transformation* $\mathbf{S}$ on a vector space V over F is a linear transformation followed by a translation,

$$\mathbf{S}\xi = \mathbf{A}\xi + \alpha$$

where $\mathbf{A}$ is a linear transformation and α a fixed vector. Show that $A_n(F)$, the set of all affine transformations on V with $\mathbf{A}$ nonsingular together with multiplication is a group. Show also that the group T of all translations is a normal subgroup of $A_n(F)$. Describe $A_n(F)/T$.

6.4. Groups of Transformations in General. The group $L_n(F)$ provides an instance of one of the simplest ways to construct groups. Namely, let $S = \{a, b, c, \ldots\}$ denote an arbitrary set of objects and $G(S)$ the set of all one-one correspondences φ on S onto itself. As functions on S to S, equality and multiplication (*viz.*, function of a function) have been defined for these correspondences, and we contend that $G(S)$, $\cdot$ is a group. The closure property is clear. To prove that the associative law holds for multi plication, we observe that for every a in S

$$[\varphi_1(\varphi_2\varphi_3)]a = \varphi_1[(\varphi_2\varphi_3)a] = \varphi_1[\varphi_2(\varphi_3 a)]$$

and

$$[(\varphi_1\varphi_2)\varphi_3]a = (\varphi_1\varphi_2)(\varphi_3 a) = \varphi_1[\varphi_2(\varphi_3 a)]$$

Since the function values of $\varphi_1(\varphi_2\varphi_3)$ and $(\varphi_1\varphi_2)\varphi_3$ are equal for every a, the functions are equal. Next there is an identity correspondence 1 in $G(S)$ defined by

$$1a = a \qquad \text{all } a \text{ in } S$$

satisfying G_2 and relative to which each φ has a formal inverse, namely, φ^{-1}. This completes the proof of the following result:

Theorem 6.5. The set $G(S)$ of all one-one correspondences on an arbitrary set S onto itself, together with multiplication, is a group.

In general $G(S)$ has many subgroups, *e.g.*, the set of all φ in G for which $\varphi a = a$, where a is fixed. A subgroup of $G(S)$ is frequently called a *group of correspondences* on S. In a certain sense this type of group is most general, according to the following result:

Theorem 6.6 (Cayley). For every group G there is an isomorphic group of correspondences.

Proof. We take as the set S of objects upon which the correspondences shall be defined the set G of group elements $\{a, b, \ldots\}$ and shall show that the group G is isomorphic to a group of correspondences on the set G. Consider the correspondence φ_a on G defined in terms of the group element a as follows:

$$\varphi_a x = ax \text{ (group multiplication)} \qquad \text{all } x \text{ in } G$$

Using postulates G_2 and G_3 for a group, φ_a is seen to be a one-one correspondence on G onto itself. To show that the set Γ of all φ_a's is a group, notice first that

$$(\varphi_a\varphi_b)x = \varphi_a(\varphi_b x) = \varphi_a(bx) = a(bx) = (ab)x = \varphi_{ab}x \tag{11}$$

so that the closure property holds. The unit element e of G ensures the presence of an identity correspondence in Γ since

$$\varphi_e x = ex = x \qquad \text{all } x \text{ in } G$$

The presence in G of an inverse a^{-1} for each a in G ensures the presence in Γ of an inverse for φ_a since from (11)

$$(\varphi_a\varphi_{a^{-1}})x = (aa^{-1})x = x = \varphi_e x \qquad \text{all } x \text{ in } G$$

Thus Γ is a group, and we contend that $G \cong \Gamma$ under the correspondence

$$a \to \varphi_a$$

First, the correspondence is one-one since $\varphi_a = \varphi_b$ implies that $ae = be$ or $a = b$. Moreover (11) shows that the correspondence is preserved under multiplication:

$$\varphi_a \varphi_b = \varphi_{ab}$$

This completes the proof.

Since, in the set of all groups, isomorphism is an equivalence relation, the above theorem asserts that a group of correspondences can be chosen as a representative in each equivalence class of isomorphic groups.

A group of one-one correspondences on a finite set N, or, as it is usually called, a *permutation group*, can be indicated in a simple way which is worthy of mention. With N finite it can be identified with a segment of the positive integers, for example $\{1, 2, \ldots, n\}$, and a correspondence (permutation) π described by a symbol

$$\begin{pmatrix} 1 & 2 & \cdots & n \\ \pi 1 & \pi 2 & \cdots & \pi n \end{pmatrix}$$

where under k is written the image πk of k, $k = 1, 2, \ldots, n$. The group S_n of all permutations of the set $1, 2, \ldots, n$ is called the *symmetric group on n letters*. One important subgroup, the *alternating group A_n on n letters*, consists of all those permutations π for which

$$e_\pi = \prod_{i<k} \frac{\pi k - \pi i}{k - i}$$

is 1. For any permutation this number is ± 1; those permutations for which it is $+1$ are called *even* and the others *odd*. It is obvious that S_n contains $n!$ elements, and it is not difficult to see that A_n contains half this number of elements.

EXAMPLE 6.3

The symmetric group S_3 on $\{1, 2, 3\}$ consists of the six permutations

$$\varphi_1 = \begin{pmatrix} 1 & 2 & 3 \\ 1 & 2 & 3 \end{pmatrix}, \varphi_2 = \begin{pmatrix} 1 & 2 & 3 \\ 2 & 3 & 1 \end{pmatrix}, \varphi_3 = \begin{pmatrix} 1 & 2 & 3 \\ 3 & 1 & 2 \end{pmatrix},$$

$$\varphi_4 = \begin{pmatrix} 1 & 2 & 3 \\ 2 & 1 & 3 \end{pmatrix}, \varphi_5 = \begin{pmatrix} 1 & 2 & 3 \\ 1 & 3 & 2 \end{pmatrix}, \varphi_6 = \begin{pmatrix} 1 & 2 & 3 \\ 3 & 2 & 1 \end{pmatrix}$$

It is isomorphic to the group of Example 6.1 under the correspondence

$f_i \to \varphi_i$, $i = 1, 2, \ldots, 6$. This representation of the group of functions as a permutation group is not one of the type constructed in Theorem 6.6, since that representation would be written on six letters.

PROBLEMS

1. Let S denote the set of all real numbers and G the set of all correspondences $[a, b]$ on S defined by the equation $[a, b]\,x = ax + b$, where $a = \pm 1$ and $b = 0, \pm 1, \pm 2, \ldots$.

(a) Show that $G, \cdot$ is a group.

(b) Is G abelian? Is G cyclic?

(c) Determine the orders of the transformations $[1, 1]$ and $[-1, -1]$.

(d) Specify all values of a and b for which $[a, b]$ belongs to the subgroup H generated by the particular transformations $[1, 2]$ and $[-1, 0]$.

(e) Specify two transformations that together generate G itself.

(f) Show that G is isomorphic to the subgroup H of (d).

2. By numbering the vertices of the regular polygon P_n of Prob. 7, Sec. 6.3, show that the group of symmetries of P_n is isomorphic to a subgroup of S_n.

3. Let G be the group of symmetries of the octahedron $|x| + |y| + |z| = 1$. Each symmetry involves a rearrangement of x, y, z accompanied by changes of sign [for example, $(x, y, z) \to (-y, x, -z)$] and so can be briefly indicated by an expression [for example, $(-y, x, -z)$] that specifies the rearrangement and changes of sign.

(a) Exhibit G as a rectangular array whose columns are the cosets of a normal subgroup and whose rows are left cosets of a subgroup isomorphic to S_3.

(b) Determine a cyclic subgroup of G of order 6.

(c) If possible, determine subgroups of G isomorphic to the alternating groups A_3 and A_4.

4. Show that A_n is a normal subgroup of S_n.

5. Let H denote the subgroup consisting of f_1 and f_4 in the group G of Example 6.1. Form the three left cosets of H in G (see Prob. 4, Sec. 6.3), and let this collection of cosets play the role of the set used in the proof of Theorem 6.6 to obtain, as an isomorphic image of G, the permutation group of Example 6.2, assuming that the cosets are designated by 1, 2, 3 in some order.

6.5. A Reinterpretation of Chapter 5. In the terminology of Sec. 6.2, Chap. 5, as written, is primarily a study of forms from the alias point of view. We now know that the results obtained can also be described from the alibi viewpoint, and it is instructive to do so. First, however, in order to assist the reader in obtaining a perspective of the topic at hand, let us define the notion of the equivalence of two forms under a group of transformations. For simplicity, we shall phrase our definition and ensuing remarks in terms of quadratic forms and ask the reader to rephrase these items for Hermitian forms. No difficulties arise in this connection. Let $f = X^T AX$ and $g = Y^T BY$ denote two quadratic forms on $V_n(F)$ to F, and let G denote a group of linear transformations $X \to Y$ on $V_n(F)$, that is, a subgroup of $L_n(F)$. We say that f *is equivalent to* g *under the group* G if and only if there exists a transformation $X \to Y = PX$ in G

which carries f into g. The phrase "a transformation $X \to Y = PX$ which carries f into g" we adopt as the alias terminology for the alibi expression "f reduces to g by the nonsingular substitution $Y = PX$," used earlier. It is clear that equivalence of forms under a given G is an equivalence relation over the set of all quadratic forms on $V_n(F)$ to F. Since this equivalence relation has its origin in the group G, it is appropriate that the customary terminology in connection with an equivalence relation be modified to read "canonical set under G" and "canonical form under G."

Let us now discuss Chap. 5 in terms of the above concepts. Since there we assumed that all nonsingular substitutions were at our disposal, we may say that Chap. 5 is a study of quadratic forms under the full linear group $L_n(F)$ [and of Hermitian forms under $L_n(C)$]. The earlier definition of equivalent forms (that is, f and g are equivalent if and only if they represent the same quadratic function) we may now interpret as meaning equivalence under $L_n(F)$. Then our most notable result, namely Theorem 5.12, states that for the set of real quadratic forms in $x^1, x^2, \ldots, x^n$, the forms of the type $\sum_1^p (v^i)^2 - \sum_{p+1}^r (v^i)^2$ where $p \leqq r \leqq n$ constitute a canonical set under $L_n(R^\#)$, which is determined by the complete set of invariants under $L_n(R^\#)$ consisting of the rank function and signature function.

In contrast to the real (and complex) case, for forms over an arbitrary scalar field F, we have only the weak result labeled Theorem 5.7: An equivalence class under $L_n(F)$ contains a form of the type $\sum d_i(u^i)^2$.

Returning to the general notion of the equivalence of two forms under a group G of transformations, it is clear that this equivalence induces an equivalence relation over the set of figures defined by quadratic forms (see Example 5.4): the figure $L_1 : f(X) = a$ *is equivalent to the figure* $L_2 : g(Y) = a$ *under* G if and only if f is equivalent to g under G. Thus, the equivalence of L_1 and L_2 under G means that there exists a transformation $X \to Y = PX$ in G which carries the equation $f(X) = a$ into the equation $g(Y) = a$; simultaneously, the same transformation carries the points X of L_1 into the points $Y = PX$ of L_2, or, as we shall say, carries L_1 into L_2.

For example, the ellipse and circle defined by $f(X) = 4(x^1)^2 + (x^2)^2 = 1$ and $g(Y) = (y^1)^2 + (y^2)^2 = 1$, respectively, are equivalent under any group G which contains the transformation

$$Y = \begin{pmatrix} 2 & 0 \\ 0 & 1 \end{pmatrix} X$$

since this transformation carries f into g. Simultaneously it carries the points of the ellipse into points of the circle.

A much broader example is provided by Example 5.4, where it is indicated that, under the full linear group $L_n(R^\#)$, any central quadric surface in $V_n(R^\#)$ is equivalent to one of the type

$$\sum_{i=1}^{p} (y^i)^2 - \sum_{i=p+1}^{r} (y^i)^2 = 0 \quad \text{or} \quad 1$$

In particular, in the two-dimensional case, this means that every pair of ellipses, both with center at the origin, are equivalent under $L_n(R^\#)$, since both are equivalent to the unit circle. This last remark is illustrated by Example 5.5, where a substitution

$$X = A^{-1}Y \qquad \text{or} \qquad Y = AX$$

was presented which reduces the equation

$$5(x^1)^2 + 4x^1x^2 + 8(x^2)^2 = 9 \qquad \text{to} \qquad (y^1)^2 + (y^2)^2 = 1$$

Changing to the alibi viewpoint, we interpret the above matrix equation as the linear transformation $X \to Y = AX$ on the plane and conclude that it carries the given ellipse into the unit circle.

It will do much to establish the value of the alibi viewpoint in geometry if we analyze this transformation further. Accompanying a factorization of the matrix A into elementary matrices

$$A = \begin{pmatrix} \frac{1}{3} & -\frac{2}{3} \\ \frac{2}{3} & \frac{2}{3} \end{pmatrix} = \begin{pmatrix} 1 & -1 \\ 0 & 1 \end{pmatrix} \begin{pmatrix} 1 & 0 \\ 0 & \frac{2}{3} \end{pmatrix} \begin{pmatrix} 1 & 0 \\ 1 & 1 \end{pmatrix} = A_3A_2A_1$$

let us say, there is a decomposition of the transformation $X \to Y$ into the product of three transformations

$$A_1X = U \qquad A_2U = V \qquad A_3V = Y$$

In expanded form these read as follows:

$$\begin{array}{lll} u^1 = x^1 & v^1 = u^1 & y^1 = v^1 - v^2 \\ u^2 = x^1 + x^2 & v^2 = \frac{2}{3}u^2 & y^2 = v^2 \end{array}$$

The first of these is a shear parallel to the x^2 axis, the second a uniform compression toward the u^1 axis, and the third a shear parallel to the v^1 axis. We conclude that these operations, carried out upon the plane, distort the original ellipse into the unit circle.

It is easily seen how the analysis of this example can be generalized to produce the following result, which we state for the two-dimensional case, although it may be extended to higher dimensions:

Theorem 6.7. A nonsingular linear transformation on a Euclidean plane may be decomposed into a product of transformations, each of which corresponds to a shear, a stretching (or compression), or a reflection.

The final type of transformation does not appear in the above example. It corresponds to an elementary matrix of type I.

The preceding theorem gives a precise description of the distortions that are possible with the full linear group $L_2(R^{\#})$ and, consequently, of the variety of figures that are equivalent to a given figure under this group. Simultaneously, we acquire a description of the alterations that are possible in the equation of a figure when all possible bases for the space $V_2(R^{\#})$ are at our disposal.

We are now in a position where we can indicate the usefulness of the concept of equivalence of forms under a group. We have seen that the alibi analogue of the alias situation where all possible bases are available, is equivalence under the full linear group. At the other extreme, an agreement never to change the basis in a space corresponds to equivalence under the identity group. *It is equivalence under groups between these two extremes that is the counterpart of the restriction to a limited class of bases in a space.* For example (relying once again on the reader's knowledge of analytic geometry; see Example 5.5), in three-dimensional Euclidean space, one may wish to restrict himself to orthonormal bases; this corresponds to equivalence under a subgroup that we shall later define as the orthogonal group. These matters are taken up in Chap. 8.

6.6. Singular Linear Transformations. Those linear transformations on a vector space V which are not nonsingular are called *singular*, since by Theorem 6.2 they correspond to singular matrices. Thus, if $\mathbf{A}$ is singular, $A = (\mathbf{A}; \alpha)$ is singular, and hence the set of images

$$\mathbf{A}\alpha_j = \alpha A_j \qquad j = 1, 2, \ldots, n \tag{12}$$

of the members of the α-basis is dependent (Theorem 3.13). Precisely, if $r(A) = r$, then r is the rank of the set $\{\mathbf{A}\alpha_1, \mathbf{A}\alpha_2, \ldots, \mathbf{A}\alpha_n\}$, in other words, r is the dimension of the space spanned by this set, since (12) establishes an isomorphism between this space and the column space of A. To expedite further discussion, several definitions are now introduced.

DEFINITION. *If* $\mathbf{A}$ *is a linear transformation on* V *and* S *is a subspace of* V, *the set of all vectors of the form* $\mathbf{A}\xi$, *where* ξ *is in* S, *is denoted by* $\mathbf{A}S$. *The particular set,* $\mathbf{A}V$, *of this type is called the* range of $\mathbf{A}$ *and denoted by* $R(\mathbf{A})$. *The set of all vectors* ξ *in* V *such that* $\mathbf{A}\xi = 0$ *is called the* null space of $\mathbf{A}$ *and denoted by* $N(\mathbf{A})$.

With the aid of Theorem 2.1, a direct computation shows that $\mathbf{A}S$ [hence $R(\mathbf{A})$] and $N(\mathbf{A})$ are subspaces of V.

DEFINITION. *The dimension of* $R(\mathbf{A})$ *is called the* rank of $\mathbf{A}$ *and denoted by* $r(\mathbf{A})$. *The dimension of* $N(\mathbf{A})$ *is called the* nullity of $\mathbf{A}$ *and denoted by* $n(\mathbf{A})$.

Since above we proved that the rank of **A** is the rank of A, the use of the notation $r(\mathbf{A})$ is justified. As for imitating the notation $n(A)$ for the nullity of a matrix (see Prob. 6, Sec. 3.8) in the present connection, the following theorem supplies the reason:

Theorem 6.8. The sum of the rank and nullity of a linear transformation **A** on V is equal to the dimension of V:

$$r(\mathbf{A}) + n(\mathbf{A}) = d[V]$$

Proof. Let s denote the nullity of **A**, so that $N(\mathbf{A})$ has a basis $\{\alpha_1, \alpha_2, \ldots, \alpha_s\}$ of s elements. Extend this set to a basis for V by adjoining $\{\beta_1, \beta_2, \ldots, \beta_t\}$. Then

$$s + t = d[V]$$

Since $\mathbf{A}\alpha_i = 0$, the vectors $\mathbf{A}\beta_i$ span the range $R(\mathbf{A})$ of **A**. Actually $\{\mathbf{A}\beta_1, \mathbf{A}\beta_2, \ldots, \mathbf{A}\beta_t\}$ is a basis for $R(\mathbf{A})$ since $\sum c^i(\mathbf{A}\beta_i) = 0$ implies $\mathbf{A}\sum c^i\beta_i = 0$ or $\sum c^i\beta_i$ is in $N(\mathbf{A})$. Hence $c^1 = c^2 = \cdots = c^t = 0$. It follows that $s + t$ is the nullity plus the rank of **A**. This completes the proof.

If **A** is nonsingular, it is clear that $R(\mathbf{A}) = V$ and $N(\mathbf{A}) = O$, the subspace consisting of the zero vector alone. According to the preceding theorem, each of these conditions implies the other. In turn, the assertion of the next theorem that each implies the nonsingularity of **A** is of importance. This means that if a correspondence on a vector space V to itself is a linear transformation on V, then each of the two conditions necessary in general for a one-one correspondence implies the other.

Theorem 6.9. If **A** is a linear transformation on V, the following three statements are equivalent:

(i) **A** is nonsingular
(ii) $R(\mathbf{A}) = V$
(iii) $N(\mathbf{A}) = O$

Proof. In view of the above remarks, it is sufficient to show that (ii) implies (i). If (ii) holds, then every element of V occurs as an image under **A**. Moreover, distinct elements have distinct images since $\mathbf{A}\xi_1 = \mathbf{A}\xi_2$ implies $\mathbf{A}(\xi_1 - \xi_2) = 0$ and consequently, since (ii) implies (iii), $\xi_1 - \xi_2 = 0$ or $\xi_1 = \xi_2$. Hence **A** is a one-one or nonsingular transformation on V.

To obtain our first example of a class of singular transformations, we imitate the familiar notion of the projection of Euclidean space upon a line or plane through the origin. Suppose that the vector space V is decomposed into a direct sum

$$V = S_1 \oplus S_2$$

of two subspaces, so that every vector ξ has a unique representation in the form

$$\xi = \xi_1 + \xi_2 \qquad \xi_i \text{ in } S_i$$

We define a transformation $\mathbf{E}$ on V by the rule

$$\mathbf{E}\xi = \xi_1$$

and call $\mathbf{E}$ the *projection on* S_1 *along* S_2. A direct computation shows that $\mathbf{E}$ is linear. Moreover, it is clear that

$$R(\mathbf{E}) = S_1 \qquad \text{and} \qquad N(\mathbf{E}) = S_2$$

so that V is the direct sum of the range and nullity of $\mathbf{E}$. Thus Theorem 6.8 is trivial for a projection. It is accidental that $V = R(\mathbf{E}) \oplus N(\mathbf{E})$; in general it need not even be true that these subspaces are disjoint, as the following example shows:

EXAMPLE 6.4

Let $V = P_2(C)$, the space of polynomials of degree less than 2 of Example 2.3, and $\mathbf{A} = \mathbf{D}$, the derivative operator. Obviously $\mathbf{D}$ is linear. In this case

$$R(\mathbf{D}) = N(\mathbf{D}) = P_1(C)$$

as one sees immediately from the definitions of the range and null spaces of $\mathbf{D}$.

It is of interest to know how projections are characterized among the singular transformations. A moment's reflection upon the following result will show that it is the obvious way to describe a projection geometrically:

Theorem 6.10. A linear transformation $\mathbf{A}$ on V is a projection on some subspace S_1 if and only if it is *idempotent*: $\mathbf{A}^2 = \mathbf{A}$ (where $\mathbf{A}^2$ denotes $\mathbf{AA}$).

Proof. If $\mathbf{A}$ is a projection on S_1 along S_2 and $\xi = \xi_1 + \xi_2$ is the decomposition of ξ with ξ_i in S_i, the decomposition of ξ_1 is $\xi_1 + 0$ so that

$$\mathbf{A}^2\xi = \mathbf{A}(\mathbf{A}\xi) = \mathbf{A}\xi_1 = \xi_1 = \mathbf{A}\xi$$

and $\mathbf{A}$ is idempotent.

Conversely, suppose $\mathbf{A}^2 = \mathbf{A}$. We define S_1 to be the set of all vectors ξ such that $\mathbf{A}\xi = \xi$ and $S_2 = N(\mathbf{A})$. Now S_1 and S_2 are subspaces, and we shall prove that $V = S_1 \oplus S_2$. First, $S_1 \cap S_2 = O$ from the definition of S_1 and S_2. Next, writing ξ in the form

$$\xi = \mathbf{A}\xi + (\mathbf{I} - \mathbf{A})\xi = \xi_1 + \xi_2$$

let us say, we have

$$\mathbf{A}\xi_1 = \mathbf{A}^2\xi = \mathbf{A}\xi = \xi_1 \qquad \mathbf{A}\xi_2 = \mathbf{A}(\mathbf{I} - \mathbf{A})\xi = (\mathbf{A} - \mathbf{A}^2)\xi = 0$$

Thus ξ_i is in S_i, and $V = S_1 + S_2$. Hence $V = S_1 \oplus S_2$, and $\mathbf{A}$ is a projection on S_1 along S_2.

The next result describes the structure of a singular transformation in terms of a projection and a nonsingular transformation.

Theorem 6.11. Let $\mathbf{A}$ denote a linear transformation on V. There exists a nonsingular linear transformation $\mathbf{A}_1$ on V for which $\mathbf{A}_1\mathbf{A}$ is a projection $\mathbf{E}$. Hence $\mathbf{A}$ is a projection followed by a nonsingular transformation.

Proof. Let $r = r(\mathbf{A})$ and $\{\alpha_1, \alpha_2, \ldots, \alpha_r\}$ be a basis for $R(\mathbf{A})$. Extend this to a basis $\{\alpha_1, \alpha_2, \ldots, \alpha_n\}$ for V. Since $\alpha_1, \alpha_2, \ldots, \alpha_r$ are in $R(\mathbf{A})$, there exist vectors $\beta_1, \beta_2, \ldots, \beta_r$ such that $\mathbf{A}\beta_i = \alpha_i$, $i = 1, 2, \ldots, r$. Finally, we choose a basis which we may denote by $\{\beta_{r+1}, \beta_{r+2}, \ldots, \beta_n\}$, for $N(\mathbf{A})$. We shall prove that $\{\beta_1, \beta_2, \ldots, \beta_n\}$ is a basis for V. It is sufficient to prove that this set is linearly independent. If $\sum_1^n c^i\beta_i = 0$, then $\mathbf{A}\sum_1^n c^i\beta_i = \sum_1^r c^i\alpha_i = 0$ so that $c^1 = c^2 = \cdots = c^r = 0$. Consequently $\sum_{r+1}^n c^i\beta_i = 0$, which forces the remaining c^i's to be zero.

The transformation $\mathbf{A}_1$ referred to in the theorem is now defined as follows:

$$\mathbf{A}_1\alpha_i = \beta_i \qquad i = 1, 2, \ldots, n$$

Then

$$\begin{aligned}\mathbf{A}_1\mathbf{A}\beta_i &= \mathbf{A}_1\alpha_i = \beta_i & i &= 1, 2, \ldots, r\\ &= \mathbf{A}_1 0 = 0 & i &= r+1, r+2, \ldots, n\end{aligned}$$

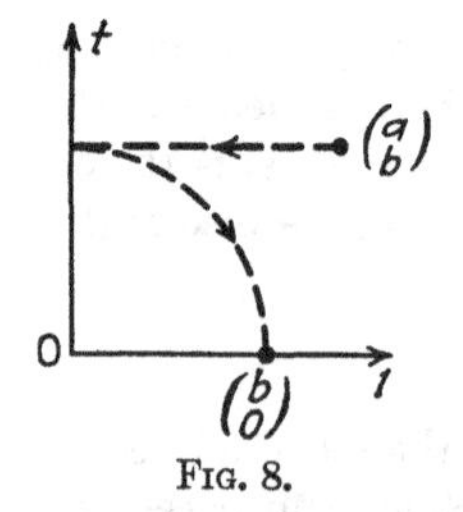

FIG. 8.

so that $\mathbf{A}_1\mathbf{A}$ is a projection, $\mathbf{E}$. Since $\mathbf{A}_1$ is nonsingular, we obtain, finally, that $\mathbf{A} = \mathbf{A}_1^{-1}\mathbf{E}$.

It will be instructive for the reader to imitate the above proof with V as the space $P_n(C)$ and $\mathbf{A}$ as the derivative operator $\mathbf{D}$. For $n = 2$, an α-basis is $\{1, t\}$ and a β-basis may be chosen to be $\{t, 1\}$. Thus $\mathbf{A}_1$ is an interchange of the two basis vectors, and $\mathbf{A}_1\mathbf{D}$ is the projection of the space on $[t]$ along $[1]$. Consequently $\mathbf{D}$ is this projection followed by $\mathbf{A}_1^{-1} = \mathbf{A}_1$. The transformation is illustrated in Fig. 8, where $(a, b)^T$ represents the vector $a + bt$.

PROBLEMS

1. Referring to the proof of Theorem 6.10, show that if **E** is the projection on S_1 along S_2, then S_1 is the set of all solutions of the equation $\mathbf{E}\xi = \xi$ and S_2 is the set of all solutions of $\mathbf{E}\xi = 0$.

2. Using the preceding problem and Theorem 6.10, show that the linear transformation **E** on V is a projection if and only if $\mathbf{I} - \mathbf{E}$ is a projection; if **E** is the projection on S_1 along S_2, then $\mathbf{I} - \mathbf{E}$ is the projection on S_2 along S_1.

3. Let **E** denote the projection on S_1 along S_2 and $\{\alpha_1, \alpha_2, \ldots, \alpha_r\}$ be a basis for S_1. Extend this set to a basis $\{\alpha_1, \alpha_2, \ldots, \alpha_n\}$ for V, and compute $(\mathbf{E}; \alpha)$.

4. A linear transformation **A** on $V_4(R^{\#})$ is defined by the matrix

$$A = \frac{1}{3}\begin{pmatrix} 2 & -1 & 1 & 0 \\ -1 & 1 & 0 & -1 \\ 1 & 0 & 1 & -1 \\ 0 & -1 & -1 & 2 \end{pmatrix}$$

relative to the ϵ-basis. Show that **A** is a projection, and determine subspaces S_1 and S_2 such that **A** is a projection on S_1 along S_2. Choose a basis for S_1 and S_2, and determine the matrix corresponding to **A** relative to the resulting basis for V.

6.7. The Ring of All Linear Transformations on a Vector Space. In this section we introduce a further type of algebraic system in order to classify the set of all linear transformations on a vector space.

DEFINITION. *A* ring *is a system consisting of a set D and two binary compositions in D called addition* (+) *and multiplication* (·) *such that*

(*i*) *D together with addition is a commutative group*

(*ii*) *Multiplication is associative*: $a(bc) = (ab)c$

(*iii*) *The following distributive laws hold*:

$$a(b + c) = ab + ac \qquad (a + b)c = ac + bc$$

If multiplication in D is commutative, D is called a commutative ring. *If D contains a multiplicative unit* 1 (*thus* $a \cdot 1 = 1 \cdot a = a$ *for all a*), *D is called a* ring with unit element.

It is clear upon a comparison of the ring definition with that of a field, as stated in Sec. 6.3, that a fortiori every field is a ring. That the converse is false is demonstrated by the commutative ring with unit element consisting of the set of integers $\{0, \pm 1, \pm 2, \ldots\}$ and the usual compositions. The subset of even integers of this ring determines a commutative ring without unit element. Again, whereas the set J_p, together with the compositions defined for this set in Sec. 1.9, is a field when p is a prime, the system is merely a commutative ring with unit element when p is composite.

The simplest example of a noncommutative ring with unit element is the system consisting of the set F_n of all $n \times n$ matrices with entries in a field F and matrix addition and multiplication. In effect, this statement summarizes the important properties of these two operations when restricted to the case of square matrices. The zero matrix 0 is the zero element of the abelian group F_n, $+$, and the identity matrix I is the unit element. The example following rule 2, Sec. 3.7, shows that multiplication is not commutative in general.

Turning next to the set of all linear transformations on an n-dimensional vector space V over F, we assert that this set, together with addition and multiplication is likewise a ring with unit element. A proof of this statement may be based upon the following three remarks: Upon the selection of a basis for V, the set of linear transformations V is in one-one correspondence with the set of matrices F_n (Sec. 6.2); this correspondence is preserved by addition and multiplication of transformations and matrices, respectively; F_n is a ring. Of course, a direct proof can be given. Since it is entirely straightforward, we shall omit it. The relation between transformations and matrices is summarized in the following theorem (see Sec. 2.5):

Theorem 6.12. The ring of linear transformations on an n-dimensional vector space V over F is isomorphic to the ring of $n \times n$ matrices over F under the natural correspondence of transformation to matrix that is determined upon the choice of a basis for V.

If φ_α denotes the isomorphism which is determined by the α-basis for V, it is immediately seen that the transformation $\mathbf{0}$ such that $\varphi_\alpha(\mathbf{0}) = 0$ has the property

$$\mathbf{0}\xi = 0 \qquad \text{all } \xi \text{ in } V$$

and that this characterizes $\mathbf{0}$. We call $\mathbf{0}$ the *zero transformation* on V. Just as the zero (linear) transformation corresponds to the zero matrix in any isomorphism, so does the identity transformation $\mathbf{I}$ correspond to the identity matrix. Finally, recalling the relation between nonsingular transformations and matrices, it is worth while mentioning that under any φ_α the nonsingular transformations correspond to the nonsingular matrices.

PROBLEMS

1. Let $D, +, \cdot$ be a ring with unit element 1. Show that the system $D, \oplus, \odot$, where $a \oplus b = a + b - 1$ and $a \odot b = a + b - ab$, is likewise a ring with unit element. Moreover, show that the new ring bears the same relationship to the old ring as the old ring does to the new.

2. Let D denote a ring with the property that every element satisfies the equation $x^2 = x$. Show that every element satisfies the equation $x + x = 0$ and, moreover, that D is commutative.

3. Let $[0, 1]$ denote the unit interval on the x axis, *i.e.*, the set of all points with abscissa x such that $0 \leqq x \leqq 1$, and D the set of all subsets of $[0, 1]$. We regard the set of all points, as well as the set of no points (the void set), as members of D. It may be emphasized that an element a of D is a subset of $[0, 1]$. In D we define $a + b$ to be the set of all points in a or b, but not in both, and ab to be the set of points in both a and b. Show that $D, +, \cdot$ is a ring.

6.8. The Polynomial Ring $F[\lambda]$. We extend our list of examples of rings by constructing a type of ring that will have indirect applications in the next chapter. Let F denote a field, λ a symbol, or *indeterminate* (not an element of F), and consider all finite expressions

$$f(\lambda) = a_{i_1}\lambda^{i_1} + a_{i_2}\lambda^{i_2} + \cdots + a_{i_k}\lambda^{i_k} \tag{13}$$

where $i_1, i_2, \ldots, i_k$ are distinct, nonnegative integers and $a_{i_1}, a_{i_2}, \ldots, a_{i_k}$ are elements of F. Such an expression is called a *polynomial in* (*the indeterminate*) λ with *coefficients* a_{i_j} and *terms* $a_{i_j}\lambda^{i_j}$ or, briefly, a polynomial in λ over F. For example,

$$2\lambda^0 + 3\lambda^2 + \tfrac{1}{2}\lambda^5 \qquad \text{and} \qquad 2\lambda^3 + 4\lambda^7 + 0\lambda^6 + 1\lambda^1$$

are polynomials in λ over the field of rational numbers. The reader is cautioned against thinking of λ as a variable and f as the polynomial function whose value at λ is given by (13); rather, λ is only a symbol whose powers, together with field elements, are used to construct expressions like (13). We shall define two binary operations in the set $F[\lambda]$ of all polynomials in λ such that the resulting system is a commutative ring with unit element.

First, there is the definition of equality: Two polynomials in λ are called *equal* if and only if they contain exactly the same terms apart from terms with zero coefficients. It follows that a polynomial (13) is equal to one of the type

$$f(\lambda) = a_0\lambda^0 + a_1\lambda^1 + \cdots + a_n\lambda^n \tag{13$'$}$$

which is the representation that we shall use in order to define compositions in $F[\lambda]$. The *sum* $f(\lambda) + g(\lambda)$ of $f(\lambda)$ in (13)$'$ and

$$g(\lambda) = b_0\lambda^0 + b_1\lambda^1 + \cdots + b_m\lambda^m$$

is defined to be the polynomial

$$(a_0 + b_0)\lambda^0 + (a_1 + b_1)\lambda^1 + \cdots + (a_m + b_m)\lambda^m + a_{m+1}\lambda^{m+1} + \cdots + a_n\lambda^n$$

if $m \leqq n$; if $m > n$ the obvious modification is made. According to this definition it is clear that a polynomial is actually the sum of its monomial constituents, which gives the addition indicated in (13) the meaning of actual addition. Since addition of polynomials reduces to addition of their coefficients, it is obvious that the system $F[\lambda], +$ is an abelian group with zero element $0\lambda^0$, which is called the *zero polynomial.*

The *product* $f(\lambda)g(\lambda)$ is defined to be the polynomial

$$c_0\lambda^0 + c_1\lambda^1 + \cdots + c_{m+n}\lambda^{m+n} \qquad \text{where } c_k = \sum_{i+j=k} a_i b_j \qquad k = 0, 1, \ldots, m+n \tag{14}$$

which is obtained by multiplying $f(\lambda)$ and $g(\lambda)$ according to the rules of elementary algebra. To prove that multiplication is associative, we introduce a third polynomial $h(\lambda) = \sum c_k\lambda^k$ and observe that the coefficient of λ^l in $[f(\lambda)g(\lambda)]h(\lambda)$ is

$$\sum_{r+k=l} (\sum_{i+j=r} a_i b_j) c_k = \sum_{i+j+k=l} a_i b_j c_k$$

Similarly the corresponding term of $f(\lambda)[g(\lambda)h(\lambda)]$ is

$$\sum_{r+i=l} a_i (\sum_{j+k=r} b_j c_k) = \sum_{i+j+k=l} a_i b_j c_k$$

Hence, the associative law holds. Similarly we can verify the distributive laws. Hence, the system $F[\lambda]$, $+$, $\cdot$ is a ring. The commutativity of multiplication follows from that of F. Finally, the polynomial $1\lambda^0$, where 1 is the unit element in F, is the unit element for $F[\lambda]$.

The *degree* of a polynomial different from the zero polynomial $0\lambda^0$ is the exponent of the highest power of λ which has a nonzero coefficient, and this coefficient is called the *leading coefficient*. A polynomial whose leading coefficient is 1 may be conveniently called a *monic polynomial*. Polynomials of degree zero are those of the form $a_0\lambda^0$, $a_0 \neq 0$, and may be identified with the elements $a_0 \neq 0$ of F. This is permissible since these polynomials, together with the zero polynomial, which we identify with 0 in F, constitute a field isomorphic with F. Such polynomials will be called *constants*.

From the definition of $F[\lambda]$ it is clear that λ itself is not an element of $F[\lambda]$. However $1\lambda^1$ is, and we may identify it with λ. This is justified since $(1\lambda^1)^k = 1\lambda^k$ and, therefore, $a(1\lambda^1)^k = a\lambda^k$. With λ an element of $F[\lambda]$, $a\lambda^1 = (a\lambda^0)(1\lambda^1)$ is identified with $a\lambda$, and the latter becomes the actual product of the elements a and λ of $F[\lambda]$.

Having seen the expression that we labeled "a polynomial in the indeterminate λ" made to look exactly like the value of a polynomial function, the reader may feel entitled to an explanation of the difference between these two possible interpretations of an expression such as the following:

$$a_0 + a_1\lambda + \cdots + a_n\lambda^n \qquad a_i \text{ in } F \tag{15}$$

With polynomial functions in mind, (15) is the value of a polynomial function at λ, an interpretation which carries the connotation that λ and hence (15) are field elements. From the standpoint of a polynomial in an indeterminate, (15) is a symbol, an element of a ring $F[\lambda]$. It is clear that from the latter point of view (15) defines a unique polynomial function, *viz.*, the function f whose value at λ in F is $f(\lambda) = \sum a_i\lambda^i$. However, different polynomials in λ can define the same function over F. For example, in $J_5[\lambda]$ the unequal polynomials $\lambda^2 + \lambda^5$ and $\lambda + \lambda^6$ define the same function over J_5.

Not only is (15) meaningful when the indeterminate is replaced by an element in F, but more generally, when λ is replaced by an element in any ring D which includes F since only addition and multiplication are involved in the evaluation. *If, moreover, the element μ which is substituted for λ commutes with all elements of F, all relations among polynomials in λ which are expressible in terms of the operations of addition and multiplication remain valid when λ is replaced by μ.* The proof of this *substitution principle* follows. If $f(\lambda) = \sum a_i\lambda^i$ and $g(\lambda) = \sum b_j\lambda^j$ are two polynomials and $s(\lambda)$ and $p(\lambda)$ denote their sum and product, respectively, thus

$$f(\lambda) + g(\lambda) = s(\lambda) \qquad f(\lambda)g(\lambda) = p(\lambda)$$

we must prove that

$$f(\mu) + g(\mu) = s(\mu) \qquad f(\mu)g(\mu) = p(\mu)$$

where $f(\mu)$, for example, denotes the ring element obtained when λ is replaced by μ in $f(\lambda)$. The first formula follows upon an obvious regrouping of the powers of μ which occur in $f(\mu) + g(\mu)$. For the second formula we use the definition of $p(\lambda)$ in (14) to conclude that

$$\begin{aligned} p(\mu) = \sum c_k\mu^k &= \sum_k \sum_{i+j=k} a_i b_j \mu^k = \sum_i \sum_j a_i b_j \mu^{i+j} \\ &= (\sum a_i\mu^i)(\sum b_j\mu^j) = f(\mu)g(\mu) \end{aligned}$$

The extreme importance of the concept of polynomials in an indeterminate lies in the validity of this substitution principle. In order to substantiate this statement, let us first indicate the types of relations that can exist among polynomials. These are presented in the following theorems and definitions, all of which, with λ regarded as a variable over the (real) coefficient field, can be found in an elementary algebra text. There, the relations are shown to be "identities in λ," that is, valid for all values of λ, by, in effect, treating λ as an indeterminate. Thus the proofs carry over to the more general case. As such we shall omit the majority of proofs.

The *division algorithm*, which is merely a statement of what can be accomplished by dividing one polynomial by a polynomial different from zero, is the starting point of any discussion of polynomials. We state this as our first result, where, as in most of the theorems to follow, polynomials are indicated by letters.

Theorem 6.13. Let f and $g \neq 0$ denote elements of $F[\lambda]$. Then there exist unique elements q and r in $F[\lambda]$ such that

$$f = qg + r$$

where the remainder r is zero or of degree less than the degree of g.

Upon setting $g = \lambda - a$, a in F, it is then easy to establish in succession the remainder and factor theorems as corollaries of this theorem.

COROLLARY 1. When $f(\lambda)$ in $F[\lambda]$ is divided by $\lambda - a$, the remainder is $f(a)$, the field element obtained when λ is replaced by a in $f(\lambda)$; that is, $f(\lambda) = q(\lambda)(\lambda - a) + f(a)$.

COROLLARY 2. An element a in F is a root of the polynomial $f(\lambda)$, that is, $f(a) = 0$, if and only if $f(\lambda) = q(\lambda)(\lambda - a)$.

DEFINITION. *If f and g denote elements of $F[\lambda]$, then f is said to be* divisible by g, *or g is a* divisor of f, *in symbols $g|f$, if there exists an element h in $F[\lambda]$ such that*

$$f = gh \tag{16}$$

If $g|f$ and $f|g$, then f and g are called associates. *If $g|f$ and g has lower degree than f, then g is called a* proper divisor *of f.*

Since the degree of the product of two polynomials is the sum of the degrees of the factors, it is easily shown that two polynomials are associates if and only if each is a constant multiple of the other. If in (16) g and h are polynomials of positive degree and have leading coefficients a and b, respectively, we may rewrite (16) in the form

$$f = ab(a^{-1}g)(b^{-1}h)$$

where $a^{-1}g$ and $b^{-1}h$ are monic polynomials. It is clear, therefore, that in factoring a polynomial we may always replace any factor of positive degree by a monic associate.

DEFINITION. *A monic polynomial d is called a* greatest common divisor (g.c.d.) *of the polynomials f and g if (i) d is a common divisor of f and g and (ii) every common divisor of f and g divides d.*

The familiar result that any two polynomials, not both zero, have a unique g.c.d. may be derived by repeated use of the division algorithm. However, a more elegant proof is possible using the following notion, which is of extreme importance in the study of polynomials:

DEFINITION. *An* ideal *of the ring* $F[\lambda]$ *is a nonempty subset of* $F[\lambda]$ *possessing the following two properties*:

(*i*) *If* f *and* g *belong to the set, so does* $f + g$.

(*ii*) *If* f *belongs to the set, so does every multiple* hf, h *in* $F[\lambda]$.

The set consisting of the zero polynomial alone is an ideal, called the *zero ideal.* The set consisting of all multiples of a fixed polynomial f is easily seen to be an ideal; this ideal we denote by (f) and call the *principal ideal generated by* f. In this notation the zero ideal can be written as (0). If $\{f_1, f_2, \ldots, f_m\}$ is any finite set of polynomials, the set of all linear combinations $g_1f_1 + g_2f_2 + \cdots + g_mf_m$ with g_i in $F[\lambda]$, $i = 1, 2, \ldots, m$, is an ideal which we denote by $(f_1, f_2, \ldots, f_m)$ and call the *ideal generated by* $\{f_1, f_2, \ldots, f_m\}$. The fundamental theorem concerning ideals in $F[\lambda]$ follows:

Theorem 6.14. Every ideal E in $F[\lambda]$ is a principal ideal. If $E \neq (0)$, the generator of E may be characterized within associates as a polynomial of least degree in E.

Proof. If E consists of the zero polynomial, then $E = (0)$. Otherwise, E contains a polynomial $f \neq 0$ of least degree. Then $E = (f)$. For on the one hand, every multiple of f is in E; on the other hand, if g is in E, then, using Theorem 6.13, we can write $g = qf + r$, where $r = g - qf$ is 0, or a polynomial of degree less than the degree of f. The latter possibility leads to a contradiction, since with g and f in E, so is r, contrary to our choice of f. Hence $r = 0$, and $g = qf$. Finally, if f^* also generates E, then $f|f^*$ and $f^*|f$ so that f and f^* are associates. This completes the proof.

Theorem 6.15. Any two polynomials f and g, not both zero, of $F[\lambda]$ have a unique g.c.d. which may be characterized as the monic polynomial of minimum degree which can be expressed in the form

$$sf + tg$$

where s and t are elements of $F[\lambda]$.

Proof. We apply the preceding result to the ideal $(f, g) \neq (0)$ to conclude that $(f, g) = (d)$, where, if d is chosen as monic, it is uniquely determined. The equality $(f, g) = (d)$ implies that both f and g are multiples of d, hence that d is a common divisor. Moreover, it implies that $d = sf + tg$ for polynomials s, t; hence any common divisor of f and g divides d. Thus d is a g.c.d. of f and g. The uniqueness of the g.c.d. follows from the easily established fact that if each of two monic polynomials is divisible by the other, they are identical.

In passing, we note that the notion of a *least common multiple* (l.c.m.) of two polynomials can be handled efficiently using the concept of ideal. Once it is observed that the set of all common multiples of f and g is an ideal, Theorem 6.14 implies that the ideal is principal. A l.c.m. of f and g may then be defined as a generator of this ideal and the following result established: If d is the g.c.d. of f and g, then fg/d is a l.c.m.

We next sketch the results pertaining to factorization of elements of $F[\lambda]$.

DEFINITION. *If f is a polynomial of positive degree and there exist polynomials g and h of positive degree of $F[\lambda]$ such that $f = gh$, then f is called* reducible over F, *otherwise* irreducible, *or* prime, over F.

The concept of reducibility is relative to the given field F. For example, $\lambda^2 + 1$ as an element of $R^{\#}[\lambda]$ is irreducible but as an element of $C[\lambda]$ is reducible; thus $\lambda^2 + 1$ is irreducible over the real field but reducible over the complex field. Since each factor in a product which demonstrates the reducibility of a polynomial f in $F[\lambda]$ has lower degree than the degree of f, it is quickly shown that every polynomial can be written in at least one way as a finite product of irreducible factors. For example, we can state that a polynomial $f \neq 0$ can be written as

$$f = ap_1p_2 \cdots p_s$$

where a is the leading coefficient of f and each p_i is a monic, irreducible polynomial over F. The final result in this direction is our next theorem.

Theorem 6.16. An element f of $F[\lambda]$ of positive degree can be expressed uniquely, apart from the order of the factors, in the form

$$f = ap_1^{k_1}p_2^{k_2} \cdots p_r^{k_r} \tag{17}$$

where a is the leading coefficient of f, the p_i are distinct irreducible monic polynomials, and the k_i are positive integers.

The key to the proof of this theorem is the possibility of expressing the g.c.d. of two polynomials f and g as a linear combination of f and g (Theorem 6.15). For this implies that if a product of polynomials is divisible by an irreducible polynomial p, then at least one factor is divisible by p. It is then possible, starting with two factorizations of a polynomial into irreducible factors, to deduce their agreement.

Referring to (17), for $i \neq j$ the polynomials p_i and p_j or, more generally, $p_i^{k_i}$ and $p_j^{k_j}$ have no common factor of positive degree. Two polynomials with this property are called *relatively prime*. Thus (17) when written in the form

$$f = af_1f_2 \cdots f_r \qquad f_i = p_i^{k_i} \qquad i = 1, 2, \ldots, r$$

is a decomposition of f into factors f_i relatively prime in pairs.

In the event F is the field of complex numbers C, the only irreducible monic polynomials have the form $\lambda - a$ (this is assumed to be known) and the preceding theorem reads as follows:

Theorem 6.17. An element f of $C[\lambda]$ of positive degree can be expressed uniquely, apart from the order of the factors, in the form

$$f = a(\lambda - c_1)^{k_1}(\lambda - c_2)^{k_2} \cdots (\lambda - c_r)^{k_r} \tag{18}$$

where a is the leading coefficient of f, the c_i are distinct complex numbers, and the k_i are positive integers.

The above factorization of f implies that the distinct roots of $f(\lambda)$ are $c_1, c_2, \ldots, c_r$, where c_i is a root of multiplicity k_i, $i = 1, 2, \ldots, r$. A root of multiplicity 1 is called a *simple root.* When F is the real field $R^{\#}$, the irreducible polynomials over $R^{\#}$ are the first degree polynomials, together with those quadratic factors which have negative discriminants (again, this is assumed to be known). Thus the decomposition of a polynomial involves both linear and quadratic factors in general.

PROBLEMS

1. Prove that any finite set of elements, not all zero, in $F[\lambda]$ have a g.c.d. which is a linear combination of the given polynomials.

2. Show that the set of all common multiples of $(\lambda - 1)^2$ and $\lambda^3 - 3\lambda + 2$, polynomials in $R[\lambda]$, is an ideal, and find a single generator for this ideal. Show that the generator is a l.c.m. of the given polynomials. Recall that R denotes the rational field.

3. Each of the statements below describes a set of elements in the ring $R[\lambda]$. Determine which sets are ideals. For each set which is an ideal show that it is principal by finding a generator.

(*a*) All polynomials whose leading coefficient is an even integer together with the zero polynomial

(*b*) All polynomials which have 1 and 2 as roots

(*c*) All polynomials such that some power is divisible by $(\lambda - 1)(\lambda - 2)^3$

(*d*) All polynomials $f(\lambda)$ such that $f(-1) = f(1)$ and $f(4) = 0$

(*e*) All polynomials of even degree

(*f*) All polynomials of degree greater than 1, together with the zero polynomial

6.9. Polynomials in a Linear Transformation. In the ring of all linear transformations on a vector space V over F occur the transformations $c\mathbf{I}$ which multiply every vector by the scalar c. These determine a field that is isomorphic to F under the correspondence $c\mathbf{I} \to c$. Moreover, they commute with every linear transformation $\mathbf{A}$ on V:

$$(c\mathbf{I})\mathbf{A} = \mathbf{A}(c\mathbf{I}) \tag{19}$$

since, with $\mathbf{A}$ linear, it is immaterial whether one multiplies by c before or after determining the image of a vector under $\mathbf{A}$.†

Consider now the polynomials $f(\lambda)$ with coefficients in the field of all linear transformations of the form $c\mathbf{I}$. Since this field is included in the ring of all linear transformations $\mathbf{A}$ on V and the commutativity condition (19) holds, we are in a position to apply the substitution principle announced in the preceding section. Namely, we can replace the indeterminate λ by $\mathbf{A}$ in polynomials $f(\lambda)$ to obtain linear transformations, $f(\mathbf{A})$, for which persist all relations expressible in terms of addition and multiplication that are valid for the corresponding polynomials in λ. For example, since for all polynomials f and g, $fg = gf$, it follows that for all polynomials f and g and every linear transformation $\mathbf{A}$

$$f(\mathbf{A})g(\mathbf{A}) = g(\mathbf{A})f(\mathbf{A})$$

In the future, in order to simplify the notation in explicit computations, we propose to do the following: We shall calculate with polynomials having coefficients in F (rather than in the isomorphic field of transformations $c\mathbf{I}$) and, whenever we replace an indeterminate λ by a transformation $\mathbf{A}$, shall simultaneously replace any field element c that occurs by $c\mathbf{I}$, the transformation to which it corresponds. For example, if the decomposition of a polynomial $f(\lambda)$ into prime powers is

$$f(\lambda) = \lambda^n + a_1\lambda^{n-1} + \cdots + a_{n-1}\lambda + a_n = (\lambda - c_1)^{k_1} \cdots (\lambda - c_r)^{k_r}$$

we shall infer that

$$f(\mathbf{A}) = \mathbf{A}^n + a_1\mathbf{A}^{n-1} + \cdots + a_{n-1}\mathbf{A} + a\mathbf{I} = (\mathbf{A} - c_1\mathbf{I})^{k_1} \cdots (\mathbf{A} - c_r\mathbf{I})^{k_r}$$

where in writing $a_1\mathbf{A}^{n-1}$ [rather than $(a_1\mathbf{I})\mathbf{A}^{n-1}$] in place of $a_1\lambda^{n-1}$, etc., we have used the fact that the scalar multiple $a_1\mathbf{A}^{n-1}$ of $\mathbf{A}^{n-1}$ is equal to the product $(a_1\mathbf{I})\mathbf{A}^{n-1}$.

We conclude this section with a request to the reader that he reread the section substituting matrices for linear transformations throughout. It will be found that all assertions remain valid. As such, we shall feel free to carry out computations with matrices as outlined above for transformations.

†These remarks are equivalent to our earlier results (Sec. 3.7) that scalar matrices commute with all matrices and determine a field isomorphic to F, since in the correspondence of transformations to matrices that is determined upon the selection of a basis for V, $c\mathbf{I} \rightarrow \operatorname{diag}(c, c, \ldots, c)$.

CHAPTER 7

CANONICAL REPRESENTATIONS OF A LINEAR TRANSFORMATION

7.1. Introduction. The observations made at the beginning of Sec. 6.2 may be summarized by the remark that upon the selection of a basis $\{\alpha_1, \alpha_2, \ldots, \alpha_n\}$ for a vector space V over F, and the designation of vectors by their α-coordinates, the linear transformation $\xi \to \mathbf{A}\xi$ on V assumes the concrete form $X \to AX$ where $A = (\mathbf{A}; \alpha)$. Now $X \to AX$ may be interpreted as a linear transformation on $V_n(F)$. Indeed, using the isomorphism φ between V and $V_n(F)$ determined by the α-basis (thus $\varphi\xi = X$ if $\xi = \alpha X$), $X \to AX$ is simply $\varphi\mathbf{A}\varphi^{-1}$, the transformation on $V_n(F)$ induced by $\mathbf{A}$, relative to the α-basis. This transformation is indicated in the diagram of Fig. 9, which is the analogue of Fig. 6 in Chap. 5 for the case at hand. Recalling the terminology introduced in Chap. 5, we label the transformation $X \to AX$ on $V_n(F)$ a *representation of* $\mathbf{A}$; it is a copy of $\mathbf{A}$ in the sense that if $\varphi\xi = X$, then $\varphi(\mathbf{A}\xi) = AX$. Since $X \to AX$ is determined by A we shall often call A a representation of $\mathbf{A}$.

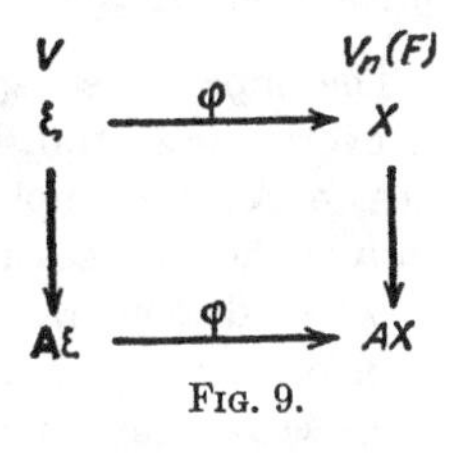

FIG. 9.

It is clear that representations are of practical importance for analyzing the structure of a linear transformation, and that some representations will be more helpful than others in this respect. One criterion for evaluating the usefulness of a representation is in terms of its nearness to diagonal form, since, as the following example suggests, a diagonal representation is the essence of simplicity for the analysis of a transformation.

EXAMPLE 7.1

Let $\mathbf{A}$ be the linear transformation on $V_3(R^{\#})$ such that

$$(\mathbf{A}; \alpha) = \begin{bmatrix} 0 & 2 & -1 \\ -2 & 5 & -2 \\ -4 & 8 & -3 \end{bmatrix}$$

Later we shall verify that $\mathbf{A}$ is also represented by diag(1, 1, 0) which reveals the structure of **A** immediately. Indeed, we may conclude that relative to a suitable basis for the space, **A** is a projection of the space on the plane determined by the first two basis vectors along the line generated by the third basis vector.

The purpose of this chapter is to discuss such properties of a linear transformation as enable one to write down a simple representation of the transformation. Ultimately a characterization of a transformation will be derived from which the simplest possible representation can be computed. However, the chapter is so arranged that the results of primary interest for some applications (*e.g.*, the existence of a superdiagonal representation and a criterion for the existence of a diagonal representation) are concluded in Sec. 7.5. An understanding of the chapter through this section is all that is presupposed for reading Chap. 8.

7.2. Invariant Spaces. The principal tool for investigating linear transformations is introduced next.

DEFINITION. *A subspace S of a vector space V is called an* invariant space *of the linear transformation* **A** *on V if and only if* $\mathbf{A}S \subseteq S$.

The improper subspaces V and O of V are (improper) invariant spaces of every linear transformation on V. The null space $N(\mathbf{A})$ and range $R(\mathbf{A})$ of **A** are examples of invariant spaces of **A**; if **A** is singular, but not **0**, both of these spaces are proper invariant spaces of **A**.

Let us demonstrate the usefulness of such spaces both for a study of a transformation as well as for the determination of simple representations. Suppose S is an r-dimensional invariant space of **A**, $0 < r < n = d[V]$, and that we select a basis for V by first choosing a basis $\{\alpha_1, \alpha_2, \ldots, \alpha_r\}$ for S and then extending this set to a basis $\{\alpha_1, \alpha_2, \ldots, \alpha_n\}$ for V. Recalling the rule for finding the representation $(\mathbf{A}; \alpha)$, we see that

$$(1) \qquad (\mathbf{A}; \alpha) = \begin{bmatrix} A_1 & A_3 \\ 0 & A_2 \end{bmatrix}$$

where A_1 is of order r and A_2 is of order $n - r$. The form of the above matrix motivates the terminology "S *reduces* **A**" to express the invariance of S under **A**. If we restrict our attention to the space S, it is clear that **A** induces a linear transformation $\mathbf{A}_1$ on S if we set

$$\mathbf{A}_1\xi = \mathbf{A}\xi \qquad \text{all } \xi \text{ in } S$$

Then A_1 is the representation of $\mathbf{A}_1$ relative to the α-basis for S.

The matrix A_2 can be given an analogous interpretation in terms of the quotient space V/S. Indeed, denoting elements of V/S in the usual way, we observe first that $\{\bar{\alpha}_{r+1}, \bar{\alpha}_{r+2}, \ldots, \bar{\alpha}_n\}$ is a basis for V/S, in other words, that $\{\bar{\alpha}_{r+1}, \bar{\alpha}_{r+2}, \ldots, \bar{\alpha}_n\}$ is a linearly independent set modulo S. Next, $\mathbf{A}$ induces a linear transformation $\mathbf{A}_2$ on V/S according to the definition

$$\mathbf{A}_2\bar{\alpha}_k = \overline{\mathbf{A}\alpha_k} \qquad k = r+1, r+2, \ldots, n$$

The correctness of this statement is assured once we verify that the definition is independent of the representative chosen for a residue class. But since $\alpha_k \equiv \alpha_k'(\text{mod } S)$ implies $\alpha_k - \alpha_k'$ is in S, $\mathbf{A}(\alpha_k - \alpha_k') = \mathbf{A}\alpha_k - \mathbf{A}\alpha_k'$ is in S and hence $\mathbf{A}\alpha_k \equiv \mathbf{A}\alpha_k'(\text{mod } S)$ as desired. Relative to the $\bar{\alpha}$-basis for V/S, A_2 represents $\mathbf{A}_2$ since

$$\begin{aligned}\mathbf{A}\alpha_k &= a_k^1\alpha_1 + \cdots + a_k^r\alpha_r + a_k^{r+1}\alpha_{r+1} + \cdots + a_k^n\alpha_n \\ &\equiv a_k^{r+1}\alpha_{r+1} + \cdots + a_k^n\alpha_n(\text{mod } S)\end{aligned}$$

and hence

$$\mathbf{A}_2\bar{\alpha}_k = a_k^{r+1}\bar{\alpha}_{r+1} + \cdots + a_k^n\bar{\alpha}_n \qquad k = r+1, r+2, \ldots, n$$

The only interest in the submatrix A_3 is whether it is the zero matrix. To show that this need not be the case, as well as to illustrate the foregoing, we present the following example:

EXAMPLE 7.2

Consider the derivative operator $\mathbf{D}$ on the space $P_4(R^\#)$ of polynomial functions of degree at most 3 in t. The subspace $S = [1, t]$ is an invariant space of $\mathbf{D}$. If we extend this basis for S to the basis $\{1, t, t^2, t^3\}$ for P_4, the following representation of D is obtained:

$$D = \begin{bmatrix} 0 & 1 & 0 & 0 \\ 0 & 0 & 2 & 0 \\ 0 & 0 & 0 & 3 \\ 0 & 0 & 0 & 0 \end{bmatrix} \quad \text{where } D_1 = \begin{bmatrix} 0 & 1 \\ 0 & 0 \end{bmatrix}, D_2 = \begin{bmatrix} 0 & 3 \\ 0 & 0 \end{bmatrix}, D_3 = \begin{bmatrix} 0 & 0 \\ 2 & 0 \end{bmatrix}$$

Then $P_4/S = [\bar{t}^2, \bar{t}^3]$, and $\mathbf{D}_2$ is defined by the equations

$$\mathbf{D}_2\bar{t}^2 = \bar{0} \qquad \mathbf{D}_2\bar{t}^3 = 3\bar{t}^2$$

Returning to the general discussion, if in (1) $A_3 = 0$, then $T = [\alpha_{r+1}, \alpha_{r+2}, \ldots, \alpha_n]$, a complement of S in V, as well as S, is an invariant space of $\mathbf{A}$. Then V is a *direct sum of two invariant spaces* S and T, and we say that $\mathbf{A}$ is *completely reduced* by the pair $\{S, T\}$. In the corresponding representation

$$\begin{bmatrix} A_1 & 0 \\ 0 & A_2 \end{bmatrix}$$

of $\mathbf{A}$, A_2 may be given the simpler interpretation of a representation of the transformation $\mathbf{A}_2$ induced on T by $\mathbf{A}$. Since each element of V is uniquely expressible as the sum of an element of S and one of T, if $\mathbf{A}_1$ and $\mathbf{A}_2$ on S and T, respectively, are known, $\mathbf{A}$ on V is known. Thus, when $\{S, T\}$ completely reduces $\mathbf{A}$, it decomposes into two transformations $\mathbf{A}_1$ and $\mathbf{A}_2$ which determine $\mathbf{A}$. We describe this situation by calling $\mathbf{A}$ the *direct sum* of $\mathbf{A}_1$ and $\mathbf{A}_2$; in symbols,

$$\mathbf{A} = \mathbf{A}_1 \oplus \mathbf{A}_2$$

Of course the foregoing can be extended to the case of several summands. If V can be decomposed into a direct sum of invariant spaces V_i of $\mathbf{A}$,

$$V = V_1 \oplus V_2 \oplus \cdots \oplus V_r \tag{2}$$

that is, $\mathbf{A}$ is completely reduced by the set $\{V_1, V_2, \ldots, V_r\}$, then $\mathbf{A}$ induces a transformation $\mathbf{A}_i$ on V_i, $i = 1, 2, \ldots, r$. Collectively the $\mathbf{A}_i$'s determine $\mathbf{A}$ by virtue of the characteristic property (see Sec. 2.7) of the direct sum decomposition (2), *viz.*, the uniqueness of the representation of a vector ξ in V as a sum $\sum \xi_i$ with ξ_i in V_i. Thus we say that $\mathbf{A}$ is the *direct sum* of $\mathbf{A}_1, \mathbf{A}_2, \ldots, \mathbf{A}_r$ and write

$$\mathbf{A} = \mathbf{A}_1 \oplus \mathbf{A}_2 \oplus \cdots \oplus \mathbf{A}_r \tag{3}$$

Relative to a basis for V consisting of a basis for V_1, together with one for V_2, etc., or, as we shall say, *a basis adapted to the spaces* V_i, $\mathbf{A}$ is represented by a matrix

$$A = \begin{bmatrix} A_1 & 0 & \cdots & 0 \\ 0 & A_2 & \cdots & 0 \\ \cdot & \cdot & \cdots & \cdot \\ 0 & 0 & \cdots & A_r \end{bmatrix}$$

of diagonal blocks A_i such that A_i represents $\mathbf{A}_i$ on V_i relative to the basis chosen for V_i. In view of the notation used in (2) and (3) it is convenient to rewrite the above matrix A as

$$A = A_1 \oplus A_2 \oplus \cdots \oplus A_r \tag{4}$$

and call it the *direct sum* of $A_1, A_2, \ldots, A_r$.

The form of this representation (4) of $\mathbf{A}$, together with the corresponding decomposition (3) of $\mathbf{A}$, both of which accompany a decomposition of V into a direct sum of invariant spaces of $\mathbf{A}$, clearly justifies the study of such spaces.

In the next section we shall describe a systematic method for finding invariant spaces.

PROBLEM

1. Show that it is impossible to find a pair of subspaces $\{S, T\}$ of $P_4(R^*)$ which completely reduce the derivative operator **D** on this space.

7.3. Null Spaces. To expedite the discussion in this section, let us agree that **A** denotes a fixed linear transformation on a vector space V over F. We direct our attention now to polynomials in **A**, that is, linear transformations $f(\mathbf{A})$ which result upon replacing the indeterminate λ in polynomials $f(\lambda)$ from $F[\lambda]$ by **A** (see Sec. 6.9) and assert that the null space of $f(\mathbf{A})$, which we may denote by $N(f)$ without ambiguity, is an invariant space of **A**. This is a consequence of the relation (see Sec. 6.9)

$$\mathbf{A} \cdot f(\mathbf{A}) = f(\mathbf{A}) \cdot \mathbf{A} \qquad \text{all } f \text{ in } F[\lambda]$$

For if ξ is in $N(f)$,

$$f(\mathbf{A})(\mathbf{A}\xi) = [f(\mathbf{A}) \cdot \mathbf{A}]\xi = [\mathbf{A} \cdot f(\mathbf{A})]\xi = \mathbf{A}[f(\mathbf{A})\xi] = 0$$

and hence the image $\mathbf{A}\xi$ of ξ is also in $N(f)$. We state this conclusion as our first result.

Lemma 7.1. The null space, $N(f)$, of the linear transformation $f(\mathbf{A})$ on V is an invariant space of **A**.

One might anticipate that relations among polynomials in λ would be reflected in the null spaces of the corresponding polynomials in **A**. We prove three such results in succession. The latter two make use of the notions of g.c.d. and l.c.m. for a set of polynomials; these are immediate extensions of the corresponding notions for a pair of polynomials as explained in Sec. 6.8.

Lemma 7.2. If f and g are any two members of $F[\lambda]$ such that $g|f$, then $N(g) \subseteq N(f)$.

Proof. By assumption there exists a polynomial h in $F[\lambda]$ such that $f = gh$. Hence, for any vector ξ in $N(g)$,

$$f(\mathbf{A})\xi = [h(\mathbf{A})g(\mathbf{A})]\xi = h(\mathbf{A})[g(\mathbf{A})\xi] = 0$$

so that ξ is in $N(f)$.

Lemma 7.3. Let $\{f_1, f_2, \ldots, f_r\}$ denote a finite set of polynomials in $F[\lambda]$ and d their g.c.d. Then

$$N(d) = N(f_1) \cap N(f_2) \cap \cdots \cap N(f_r)$$

Proof. Let D denote the right-hand member. Then $N(d) \subseteq D$ since $d|f_i$ implies $N(d) \subseteq N(f_i)$, $i = 1, 2, \ldots, r$ (Lemma 7.2). To establish the reverse inequality, we recall (Theorem 6.15) that there exist polynomials $g_1, g_2, \ldots, g_r$ such that $d = \sum g_i f_i$. If ξ is any vector in D [hence $f_i(\mathbf{A})\xi = 0$, $i = 1, 2, \ldots, r$], then

$$d(\mathbf{A})\xi = [g_1(\mathbf{A})f_1(\mathbf{A}) + \cdots + g_r(\mathbf{A})f_r(\mathbf{A})]\xi = 0$$

so that ξ is in $N(d)$. This completes the proof.

Lemma 7.4. Let $\{f_1, f_2, \ldots, f_r\}$ denote a finite set of polynomials in $F[\lambda]$ and h a l.c.m. Then

$$N(h) = N(f_1) + N(f_2) + \cdots + N(f_r)$$

Proof. Let S denote the right-hand member, which is the subspace of V consisting of all sums

$$\xi = \sum_1^r \xi_i \qquad \xi_i \text{ in } N(f_i) \tag{5}$$

Then $N(h) \supseteq S$. For since $f_i | h$, $N(f_i) \subseteq N(h)$, $i = 1, 2, \ldots, r$, and hence $N(h)$ contains all sums of the form (5). To establish the reverse inequality, we show that a vector in $N(h)$ admits a decomposition of the form (5). Since $f_i | h$, $h = h_i f_i$, $i = 1, 2, \ldots, r$, where it is clear that the g.c.d. of the set $\{h_1, h_2, \ldots, h_r\}$ is 1. Hence (Theorem 6.15) there exist polynomials s_i in $F[\lambda]$ such that $\sum s_i h_i = 1$. Replacing λ by $\mathbf{A}$ in this equation gives the following representation of the identity transformation, $\mathbf{I}$, on V:

$$\sum_1^r s_i(\mathbf{A})h_i(\mathbf{A}) = \mathbf{I}$$

This implies that for all ξ in V

$$\mathbf{I}\xi = \xi = \sum_1^r [s_i(\mathbf{A})h_i(\mathbf{A})]\xi$$

which we claim is of the same type as (5) if ξ is a member of $N(h)$. To verify this, observe that

$$f_i(\mathbf{A})[s_i(\mathbf{A})h_i(\mathbf{A})]\xi = s_i(\mathbf{A})[h_i(\mathbf{A})f_i(\mathbf{A})]\xi = s_i(\mathbf{A})[h(\mathbf{A})]\xi = 0$$

Thus $N(h) \subseteq S$, and equality follows.

The two preceding results may be combined to deduce the following fundamental theorem:

Theorem 7.1. Let

$$f = f_1 f_2 \cdots f_r$$

be a decomposition of the polynomial f in $F[\lambda]$ into factors f_i relatively prime in pairs. Then $N(f)$ is the direct sum of the null spaces $N(f_i)$:

$$N(f) = N(f_1) \oplus N(f_2) \oplus \cdots \oplus N(f_r) \tag{6}$$

Proof. The assumptions imply that f is a l.c.m. of $\{f_1, f_2, \ldots, f_r\}$ so that $N(f)$ is the sum of the various null spaces $N(f_i)$ by Lemma 7.4. To show that the sum is direct, we use the following implication of the assumptions: The g.c.d. of f_k and $\prod_{i \neq k} f_i$ is 1, $k = 1, 2, \ldots, r$. Since the polynomial in **A** corresponding to the polynomial 1 in λ is **I** and $N(\mathbf{I}) = O$, we conclude from Lemma 7.3 that

$$O = N(f_k) \cap N(\prod_{i \neq k} f_i) \quad \text{or} \quad O = N(f_k) \cap \sum_{i \neq k} N(f_i) \tag{7}$$

where, in rewriting $N(\prod_{i \neq k} f_i)$ as $\sum_{i \neq k} N(f_i)$, we have used Lemma 7.4 again. Equation (7) is the defining equation for the directness of the sum (6).

If there can be found a polynomial f, such that $N(f) = V$, equation (6) determines a decomposition of V into a direct sum of invariant spaces of **A**, that is, the spaces $N(f_i)$ completely reduce V. Of course the only transformation whose null space is V is **0**. The existence of a polynomial f such that $f(\mathbf{A}) = \mathbf{0}$ can be shown as follows: If α is a nonzero vector, there exists a least positive integer $m \leqq d[V]$ such that $\{\alpha, \mathbf{A}\alpha, \ldots, \mathbf{A}^m\alpha\}$ is a linearly dependent set, let us say $\sum_0^m c_i\mathbf{A}^i\alpha = 0$. Consequently the polynomial $g(\lambda) = \sum c_i\lambda^i$ has the property that $g(\mathbf{A})\alpha = 0$. If such a polynomial is found for each member α_i of a basis of V and f denotes a l.c.m. of this set, then $f(\mathbf{A})\alpha_i = 0$, $i = 1, 2, \ldots, n$ and since $f(\mathbf{A})$ is linear, $f(\mathbf{A})\sum a^i\alpha_i = 0$, that is, $f(\mathbf{A}) = \mathbf{0}$.

Theorem 7.2. There exists a unique monic polynomial of least degree in $F[\lambda]$, called the *minimum function* of **A** and designated by $m(\lambda)$, such that $m(\mathbf{A}) = \mathbf{0}$. If $f(\lambda)$ is any member of $F[\lambda]$, $f(\mathbf{A}) = \mathbf{0}$ if and only if $m|f$.

Proof. We have just shown that the set of polynomials $f(\lambda)$ in $F[\lambda]$ such that $f(\mathbf{A}) = \mathbf{0}$ is nonempty. The remaining requirements for an ideal are quickly verified so that Theorem 6.14 is applicable. The assertions follow immediately from this theorem.

The application of Theorem 7.1 to the minimum functions of **A** gives the following important result, apart from the last sentence, which is left as an exercise:

Theorem 7.3. If $m = p_1^{k_1}p_2^{k_2} \cdots p_r^{k_r}$ is the decomposition of the minimum function of **A** into powers of distinct, irreducible monic factors p_i

and if V_i denotes the null space of $[p_i(\mathbf{A})]^{k_i}$, $i = 1, 2, \ldots, r$, then $\{V_1, V_2, \ldots, V_r\}$ completely reduces $\mathbf{A}$. That is, V is the direct sum of the invariant spaces V_i, and $\mathbf{A}$ is the direct sum of the transformations $\mathbf{A}_i$ induced on these spaces by $\mathbf{A}$. Relative to a basis for V adapted to the V_i's, $\mathbf{A}$ is represented by the direct sum of matrices $A_1, A_2, \ldots, A_r$, where A_i represents $\mathbf{A}_i$. The order of A_i is equal to $d[V_i]$, which in turn is equal to $n[p_i(\mathbf{A})]^{k_i}$. Finally, the transformation $\mathbf{A}_i$ on V_i has $p_i^{k_i}$ as its minimum function.

The concept of minimum function is applicable to square matrices by virtue of the isomorphism between linear transformations and matrices; the minimum function of a linear transformation $\mathbf{A}$ coincides with that of any representation matrix. Hence if $\mathbf{A}$ is defined by a representation, the minimum function can be determined from the matrix; the same is true for invariant spaces of $\mathbf{A}$. We illustrate the foregoing by justifying the assertions made in Example 7.1.

EXAMPLE 7.3

Let us regard the matrix

$$A = \begin{bmatrix} 0 & 2 & -1 \\ -2 & 5 & -2 \\ -4 & 8 & -3 \end{bmatrix}$$

of Example 7.1 as the representation of $\mathbf{A}$ relative to the α-basis for $V_3(R^*)$. In order to find the minimum function, $m(\lambda)$, of A (and $\mathbf{A}$), we look for a linear dependence among the powers of A. Since $A^2 = A$, we conclude that A is a root of the polynomial $\lambda^2 - \lambda$ and consequently (Theorem 7.2) $m(\lambda)$ is one of $\lambda - 1, \lambda, \lambda^2 - \lambda$. Since $\lambda - 1$ and λ are the minimum functions of the unit and zero matrices, respectively, it follows that $m(\lambda) = \lambda^2 - \lambda$. Next let us determine the null spaces $N(\lambda - 1)$ and $N(\lambda)$ which accompany the irreducible factors $\lambda - 1$ and λ, respectively, of $m(\lambda)$. The first space, which consists of all vectors with α-coordinates X, such that $(A - I)X = 0$, has dimension 2. The second space, which consists of all vectors with α-coordinates X such that $AX = 0$ has dimension 1. Now $\{N(\lambda - 1), N(\lambda)\}$ completely reduces $\mathbf{A}$, and relative to a basis for $V_3(R^*)$ adapted to these spaces, $\mathbf{A}$ is represented by a matrix of the form $A_1 \oplus A_2$. Since the transformation $\mathbf{A}_1$ induced by $\mathbf{A}$ on $N(\lambda - 1)$ is the identity transformation, A_1 is the identity matrix relative to any basis chosen for the space. Likewise, the transformation $\mathbf{A}_2$ induced by $\mathbf{A}$ on $N(\lambda)$ is the zero transformation so that $A_2 = 0$. Thus

A is represented by

$$\text{diag}(1, 1, 0) = A_1 \oplus A_2$$

The next example introduces a transformation which has been constructed with the idea of serving as a master example for the entire chapter. As methods are developed for determining representations which are progressively simpler, they will be illustrated in connection with this transformation. In this way the reader will see its simplest possible representation emerge.

EXAMPLE 7.4

The decomposition into irreducible factors over the real field of the minimum function of the matrix

$$A = \begin{bmatrix} 0 & 2 & 4 & -2 & -2 & 1 & 4 & -2 \\ -3 & -2 & 7 & 2 & 1 & 2 & 6 & 7 \\ 0 & 4 & 0 & -3 & -4 & 1 & 0 & -4 \\ 1 & 2 & -2 & 1 & -2 & 0 & -2 & -2 \\ -2 & -1 & 5 & 1 & 0 & 2 & 4 & 4 \\ -1 & 2 & 3 & 0 & -3 & 2 & 2 & -1 \\ 0 & -4 & 2 & 3 & 4 & -1 & 2 & 4 \\ -1 & -3 & 2 & 1 & 3 & 0 & 2 & 5 \end{bmatrix}$$

is

$$\lambda(\lambda - 2)^2(\lambda^2 - \lambda + 1)^2$$

We shall regard A as the representation of the linear transformation **A** on $V_8(R^*)$ relative to the ϵ-basis. The three matrices A,

$$(A - 2I)^2 = \begin{bmatrix} 1 & -6 & -1 & 6 & 5 & -2 & -2 & 7 \\ 9 & 0 & -21 & 1 & 7 & -8 & -18 & -7 \\ -4 & -12 & 13 & 9 & 11 & -2 & 8 & 13 \\ -3 & 0 & 6 & -1 & 0 & 1 & 6 & 0 \\ 6 & 0 & -15 & 0 & 7 & -7 & -12 & -7 \\ 3 & 0 & -9 & -1 & 3 & -2 & -6 & -3 \\ 4 & 12 & -13 & -9 & -11 & 2 & -8 & -13 \\ 3 & 0 & -6 & 1 & 0 & -1 & -6 & 0 \end{bmatrix}$$

$$(A^2 - A + I)^2 = \begin{bmatrix} 0 & 0 & 18 & 0 & 0 & 0 & 18 & 0 \\ 0 & -45 & 0 & 18 & 46 & -1 & 0 & 62 \\ 0 & 0 & 0 & 0 & 1 & -1 & 0 & -1 \\ 0 & 18 & 0 & 0 & -18 & 0 & 0 & -18 \\ 0 & -18 & 0 & 9 & 19 & -1 & 0 & 26 \\ 0 & 18 & 0 & 0 & -18 & 0 & 0 & -18 \\ 0 & 0 & 9 & 0 & -1 & 1 & 9 & 1 \\ 0 & -36 & 0 & 9 & 36 & 0 & 0 & 45 \end{bmatrix}$$

have ranks 7, 5, 4 and nullities 1, 3, 4, respectively. It follows that the null spaces

$$V_1 = N(\lambda),\ V_2 = N(\lambda - 2)^2,\ V_3 = N(\lambda^2 - \lambda + 1)^2$$

have dimensions 1, 3, 4, respectively, so that, relative to a basis for $V_8(R^{\#})$ adapted to these invariant spaces, **A** is represented by a direct sum

$$A_1 \oplus A_2 \oplus A_3$$

of matrices A_1, A_2, A_3 of orders 1, 3, 4, respectively. Since a variety of bases can be found for V_2 and V_3, there are many choices for A_2 and A_3, and hence the possibility of finding simple forms for them.

PROBLEMS

1. Show that the transformation **A** on $V_3(R^{\#})$ which is represented by the matrix

$$\begin{bmatrix} 0 & 1 & 1 \\ 1 & 0 & 1 \\ 1 & 1 & 0 \end{bmatrix}$$

is also represented by the matrix diag$(-1, -1, 2)$.

2. Show that $\lambda^2(\lambda - 3)(\lambda + 1)$ is the minimum function of the matrix

$$A = \begin{bmatrix} -1 & 0 & 0 & 0 \\ 1 & 2 & 1 & 0 \\ 0 & 3 & 2 & -1 \\ 0 & 5 & 3 & -1 \end{bmatrix}$$

If A is assumed to be the representation of **A** relative to the α-basis of $V_4(R^{\#})$, find the null spaces $N(\lambda^2)$, $N(\lambda - 3)$, $N(\lambda + 1)$ and then a representation of **A** relative to a basis for $V_4(R^{\#})$ adapted to these null spaces.

3. Prove the final statement in Theorem 7.3.

4. Prove that if $g(\lambda)$ is any element of $F[\lambda]$, $m(\lambda)$ the minimum function of a linear transformation **A**, and $d(\lambda)$ the g.c.d. of g and m, then $N(g) = N(d)$.

5. Describe the linear transformations **A** such that $\mathbf{A}^2 = \mathbf{A}$, using Theorem 7.3. Theorem 6.10 should be recalled in this connection.

7.4. Similarity and Canonical Representations. The problem of finding a simple representation of a linear transformation **A** on a vector space V is that of making a strategic choice of basis for V, or, what amounts to the same, a selection from the set of all matrices which represent **A**. Let us determine the relationship among the members of this set. Suppose that $(\mathbf{A}; \alpha)$ and $(\mathbf{A}; \beta)$ are two representations of **A**, where the α- and β-bases are related by the equation $\alpha = \beta P$. Then the defining equation for $(\mathbf{A}; \alpha)$,

$$\mathbf{A}\alpha = \alpha(\mathbf{A}; \alpha)$$

can be rewritten as†

$$\mathbf{A}\beta = \beta P(\mathbf{A}; \alpha)P^{-1}$$

to conclude that

$$(\mathbf{A}; \beta) = P(\mathbf{A}; \alpha)P^{-1}$$

Conversely, it is seen immediately that the matrices A and PAP^{-1}, for any nonsingular P, determine the same transformation relative to the α- and αP^{-1}-bases, respectively. It follows that the totality of representations of **A** is the set of distinct matrices of the form PAP^{-1}, where A is any one representation and P is any nonsingular matrix.

DEFINITION. *The matrix PAP^{-1} is called the* transform of A by P. *A matrix B so related to A for some P is called* similar *to A; in symbols, $A \overset{S}{\sim} B$.*

It is clear that similarity is an equivalence relation for $n \times n$ matrices so that we may regard the proof of the next result as complete.

Theorem 7.4. Similarity is an equivalence relation over the set of $n \times n$ matrices in a field F. Two $n \times n$ matrices over F represent one and the same linear transformation if and only if they are similar.

It may have occurred to the reader that the rule which associates a matrix A over F with an ordered pair $(\mathbf{A}, \alpha)$, whose first member is a linear transformation on V over F and whose second member denotes a basis for V, is simply a function, which we have denoted by $(\ ;\ )$, on the product set of linear transformations on V and bases for V to the set of

†This step includes rewriting $\mathbf{A}(\beta P)$ as $(\mathbf{A}\beta)P$. The following computation, wherein $P_1, P_2, \ldots, P_n$ denote the columns of P, justifies this:

$$\begin{aligned}\mathbf{A}(\beta P) = \mathbf{A}(\beta P_1, \beta P_2, \ldots, \beta P_n) &= (\mathbf{A}(\beta P_1), \mathbf{A}(\beta P_2), \ldots, \mathbf{A}(\beta P_n)) \\ &= ((\mathbf{A}\beta)P_1, (\mathbf{A}\beta)P_2, \ldots, (\mathbf{A}\beta)P_n) = (\mathbf{A}\beta)P\end{aligned}$$

$n \times n$ matrices over F. In terms of this function the above notion of similarity describes the relation among function values for a fixed first argument **A**. Again, our earlier result concerning the one-one correspondence between transformations and matrices, relative to a fixed basis, is a property of the same function for a fixed second argument (*i.e.*, a fixed basis for V).

In terms of the function (;), the next question that we shall discuss can be phrased as follows: What is the relationship among the solutions of the equation $A = (\mathbf{A}; \alpha)$ for a fixed A? This is the question of the relation between two linear transformations which can be defined by a single matrix. If $(\mathbf{A}; \alpha) = (\mathbf{B}; \beta) = A$, then

$$\mathbf{A}\alpha = \alpha A \qquad \text{and} \qquad \mathbf{B}\beta = \beta A$$

Defining a third (nonsingular) linear transformation **C** by the equation $\mathbf{C}\alpha = \beta$, we substitute above for β to conclude that $\mathbf{BC}\alpha = \mathbf{C}\alpha A$, or $\mathbf{C}^{-1}\mathbf{BC}\alpha = \alpha A$. By comparison with the relation $\mathbf{A}\alpha = \alpha A$ we conclude that

$$\mathbf{A} = \mathbf{C}^{-1}\mathbf{BC}$$

Conversely, if **A** and **B** are so related for some **C**, that is, are so-called *similar transformations*, then $(\mathbf{A}; \alpha) = (\mathbf{B}; \mathbf{C}\alpha)$.

The bond between the similar transformations $\mathbf{A} = \mathbf{C}^{-1}\mathbf{BC}$ and **B** may be described thus: **B** transforms the space $\mathbf{C}V$ (which is simply V with its members relabeled) in the same way (as suggested by the diagram in Fig. 10) as **A** transforms V. To verify this, we deduce from $\mathbf{A}\xi = \eta$ that $\mathbf{CA}\xi = \mathbf{C}\eta$ and hence, replacing **CA** by **BC**, that $\mathbf{B}(\mathbf{C}\xi) = \mathbf{C}\eta$. If we generalize the notion of a representation as defined in Sec. 7.1, we can classify **B** as a representation of **A**; the likeness of Fig. 10 to Fig. 9 should be noted in this connection.

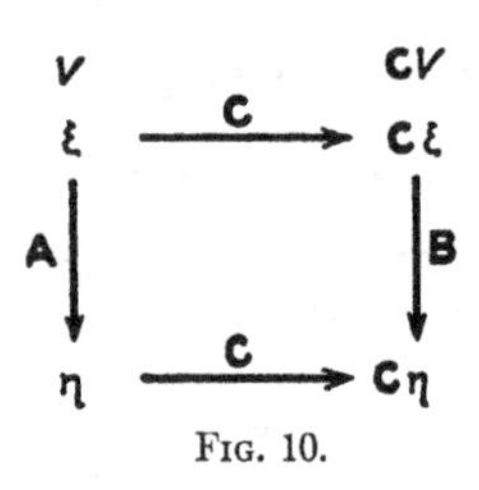

FIG. 10.

Similarity is an equivalence relation over the set of all linear transformations on a space V and the similarity classes of transformations are in one-one correspondence with the similarity classes of $n \times n$ matrices. Because the members of a similarity class of linear transformations are simply copies of each other, we shall not distinguish among them; however, we shall carefully distinguish among the members of the similarity class of matrices which represent a transformation. Indeed, beginning in Sec. 7.6, the necessary background is supplied for describing several canonical sets under similarity. The members of these canonical sets (*i.e.*, the canonical forms under similarity) will be found to have definite preference in the analysis of the corresponding linear transformations.

Our agreement to not distinguish among similar transformations has the consequence that a linear transformation **A** can be defined by the statement of a representing matrix A for **A**. However, in this event, before a property of, or a definition pertaining to, A can be assigned to **A**, one must verify that it is invariant under similarity, *i.e.*, is unchanged when A is replaced by any similar matrix, for then and only then is the property, or definition, independent of the choice of basis for the space, and hence applicable to **A**. As an example of this principle, we point out that the rank and nullity of **A** (see Chap. 6) could have been defined as the rank and nullity, respectively, of a representation. In the next section we shall have use for a polynomial in λ,

$$\det(\lambda I - A) = \det \begin{bmatrix} \lambda - a_1^1 & -a_2^1 & \cdots & -a_n^1 \\ -a_1^2 & \lambda - a_2^2 & \cdots & -a_n^2 \\ \cdot & \cdot & \cdots & \cdot \\ -a_1^n & -a_2^n & \cdots & \lambda - a_n^n \end{bmatrix} = \lambda^n + \cdots \pm \det A$$

the so-called *characteristic function* of the matrix A. Since

$$\det(\lambda I - PAP^{-1}) = \det P(\lambda I - A)P^{-1} = \det(\lambda I - A)$$

it is permissible to label this the characteristic function of the transformation **A** defined by A.

To study the linear transformation **A** defined by an $n \times n$ matrix A over F, there is no loss of generality if we assume that **A** is the transformation $X \to AX$ on $V_n(F)$ or, in other words, that $A = (\mathbf{A}; \epsilon)$. According to the definition of a representation, **A** is thereby a representation of itself! If a new basis $\{\rho_1, \rho_2, \ldots, \rho_n\}$ is chosen for $V_n(F)$, where

$$\rho_i = (p_i^1, p_i^2, \ldots, p_i^n)^T \qquad \text{or} \qquad \rho = \epsilon P,\ P = (p_i^j)$$

then **A** is represented by $P^{-1}X \to (P^{-1}AP)P^{-1}X$ or $Y \to (P^{-1}AP)Y$ if we define $Y = P^{-1}X$. The following example illustrates these remarks:

EXAMPLE 7.5

Let

$$A = \begin{bmatrix} -1 & -2 & 3 & 2 \\ 0 & 1 & 0 & 1 \\ -2 & -2 & 4 & 2 \\ 0 & 0 & 0 & 2 \end{bmatrix}$$

define the transformation **A**: $X \to AX$ on $V = V_4(R^*)$. We shall investigate **A** with the aid of Theorem 7.3. The minimum function of A

is $(\lambda - 1)(\lambda - 2)$. Solving in turn the homogeneous systems $(A - I)X = 0$ and $(A - 2I)X = 0$, we find that

$$\{(3, 0, 2, 0)^T, (2, 1, 2, 0)^T\} \qquad \text{and} \qquad \{(1, 0, 1, 0)^T, (0, 1, 0, 1)^T\}$$

are bases for $N(\lambda - 1)$ and $N(\lambda - 2)$, respectively. The composite set is a basis $\{\rho_1, \rho_2, \rho_3, \rho_4\}$ for V, such that $\rho = \epsilon P$ where

$$P = \begin{bmatrix} 3 & 2 & 1 & 0 \\ 0 & 1 & 0 & 1 \\ 2 & 2 & 1 & 0 \\ 0 & 0 & 0 & 1 \end{bmatrix} \qquad P^{-1} = \begin{bmatrix} 1 & 0 & -1 & 0 \\ 0 & 1 & 0 & -1 \\ -2 & -2 & 3 & 2 \\ 0 & 0 & 0 & 1 \end{bmatrix}$$

It is clear that $P^{-1}AP = \text{diag}(1, 1, 2, 2)$; this matrix then represents **A** relative to the ρ-basis for V.

It should not escape the reader's attention that the columns of P are the members of the basis relative to which A is represented by $P^{-1}AP$. The validity of this rule, in general, has already been established in the paragraph preceding this example.

We conclude this section with several comments on the various equivalence relations for matrices encountered so far. It is appropriate to do so at this time since similarity completes our enumeration of fundamental equivalence relations for the set of $n \times n$ matrices over a field.† Our purpose is to correlate these relations with the main topics that we have studied as well as to point out a pattern into which they fall. We point out first that each of the principal topics discussed has led us to an equivalence relation for matrices. Indeed, the study of systems of linear equations reduced to one of *row equivalence*; bilinear functions and forms led us to examine *equivalent* matrices; the investigation of quadratic forms became that of *congruent* symmetric matrices (or *conjunctive* Hermitian matrices in the complex case); finally, the topic of linear transformations has directed us to the *similarity* relation. A complete analysis of each of these topics would include, recalling the discussion in Sec. 2.6, the determination of a canonical set of matrices under the associated equivalence relation and a practical set of necessary and sufficient conditions that two matrices are in the specified relation. To date we have done this in full generality for only row equivalence, equivalence, and for restricted classes of matrices in the case of congruence and conjunctivity. Later in this chapter similarity is treated exhaustively. For this relation the rank and signature functions on matrices, which were found to be useful invariants for equivalence,

†Uniformity is gained with no essential loss of generality by restricting the discussion to square matrices.

congruence and conjunctivity, are inadequate so that further functions will be introduced. It is the setting in which a matrix appears that dictates which equivalence relation is applicable and hence what properties of the matrix are pertinent.

Below, the definitions of the above relations have been arranged in a table to indicate how they stem from the basic one of equivalence:

Equivalence	PAQ	
Congruence	PAQ	with $Q = P^T$
Conjunctivity	PAQ	with $Q = P^*$
Similarity	PAQ	with $Q = P^{-1}$

The likeness of the similarity and congruence relation suggests that a technique analogous to that used to study congruent matrices might be applied to similar matrices, *i.e.*, that "elementary S transformations" be defined for a matrix, etc. This can be done, and, once it is known what to expect in the way of canonical forms, elementary operations can be used to determine them.† However, this approach has little value in suggesting normal forms and determining whether two matrices are similar.

PROBLEMS

1. If $g(\lambda)$ denotes any member of $F[\lambda]$ and A is any $n \times n$ matrix, show that $Pg(A)P^{-1} = g(PAP^{-1})$.

2. Show that the transformation $X \rightarrow AX$ on $V_3(R^*)$, where

$$A = \begin{bmatrix} 1 & -1 & 1 \\ 4 & 0 & -1 \\ 4 & -2 & 1 \end{bmatrix}$$

has a diagonal representation D. Find a basis for the space relative to which the transformation is represented by D, and use it to determine a matrix P such that $P^{-1}AP = D$.

3. Show that $m(\lambda) = (\lambda - 1)^2(\lambda - 4)$ is both the minimum and characteristic function of the matrix

$$A = \begin{bmatrix} -3 & -6 & 7 \\ 1 & 1 & -1 \\ -4 & -6 & 8 \end{bmatrix}$$

Determine (i) a basis for V adapted to the null spaces of $(\lambda - 1)^2$ and $\lambda - 4$, (ii) the representation, B, of the transformation $X \rightarrow AX$ relative to this basis, and (iii) a matrix P such that $P^{-1}AP = B$.

†Two papers by M. F. Smiley [*Am. Math. Monthly*, vol. 49, pp. 451–454 (1942); vol. 56, pp. 542–545 (1949)] should be consulted in this connection.

4. Show that $a_0 + a_1\lambda + \cdots + a_{n-1}\lambda^{n-1} + \lambda^n$ is the characteristic function of the matrix

$$\begin{bmatrix} 0 & 0 & \cdots & 0 & -a_0 \\ 1 & 0 & \cdots & 0 & -a_1 \\ 0 & 1 & \cdots & 0 & -a_2 \\ \cdot & \cdot & \cdots & \cdot & \cdot \\ 0 & 0 & \cdots & 1 & -a_{n-1} \end{bmatrix}$$

5. Show that $\det(\lambda I - A) = \lambda^n - p_1\lambda^{n-1} + p_2\lambda^{n-2} - \cdots + (-1)^n p_n$, where p_k is the sum of the principal minors of order k of A.

6. If $A = (a_{rs})$ is an $n \times n$ matrix, where $a_{rs} = c^{(r-1)(s-1)}$ and $c = e^{2\pi i/n}$, determine A^2 and show that the distinct roots of the characteristic function of A^2 are $\pm n$. Determine their multiplicities.

7. Let A denote a nilpotent matrix: $A^k = 0, k \geqq 1$. Using the characteristic equation of A, show that $\det(A + I) = 1$. This is a special case of the result: if $AB = BA$ and $A^k = 0$, then $\det(A + B) = \det B$.

8. Prove that if A and B are $n \times n$ matrices, such that at least one is nonsingular, then $AB \overset{S}{\sim} BA$ (and hence these products have the same characteristic function). Show that in all cases AB and BA have the same characteristic function. HINT: If $r(A) = r$, select nonsingular P and Q, such that $PAQ = \operatorname{diag}(I_r, 0)$, and show that $PABP^{-1}$ and $Q^{-1}BAQ$, which are similar to AB and BA, respectively, have the same characteristic function.

7.5. Diagonal and Superdiagonal Representations. For the convenience of the reader whose interest lies primarily in linear transformations on vector spaces over the complex field, we shall discuss in this section the representations of a class of transformations which include these. Regardless of whether or not we favor the complex field, our starting point is plausible; we ask for a characterization of those transformations which admit the simplest type of representation—in other words, a diagonal representation. One such characterization can be established immediately.

Theorem 7.5. A linear transformation $\mathbf{A}$ on V over F admits a diagonal representation if and only if the minimum function of $\mathbf{A}$ factors over F into a product of distinct linear factors. The multiplicity of a scalar c in a diagonal representation is equal to the nullity of the transformation $\mathbf{A} - c\mathbf{I}$.

Proof. If the minimum function, $m(\lambda)$, of $\mathbf{A}$ decomposes into linear factors over F

$$m(\lambda) = \prod_{i=1}^{r} (\lambda - c_i) \qquad c_i \neq c_j \text{ if } i \neq j$$

then, according to Theorem 7.3, V can be decomposed into a direct sum of the null spaces V_i of the transformations $\mathbf{A} - c_i\mathbf{I}$, $i = 1, 2, \ldots, r$. Since, for every vector ξ in V_i, $(\mathbf{A} - c_i\mathbf{I})\xi = 0$ or $\mathbf{A}\xi = c_i\xi$, this equation holds for each member of a basis for V_i. Hence, relative to any basis for V adapted to the V_i's, $\mathbf{A}$ is represented by the matrix

$$A_1 \oplus A_2 \oplus \cdots \oplus A_r \qquad A_i = c_i I_{n_i}, \; n_i = d[V_i] \qquad i = 1, 2, \ldots, r$$

For the converse assume that

$$(\mathbf{A}; \alpha) = \operatorname{diag}(c_1, c_2, \ldots, c_n) \qquad c_i \text{ in } F$$

where $\{c_{i_1}, c_{i_2}, \ldots, c_{i_r}\}$ is a subset of $\{c_1, c_2, \ldots, c_n\}$ obtained by discarding all repetitions. We shall ask the reader to verify that if

$$p(\lambda) = \prod_{j=1}^{r} (\lambda - c_{i_j})$$

then $p(\mathbf{A}) = \mathbf{0}$, by proving that $p(\mathbf{A})\alpha_k = 0$, $k = 1, 2, \ldots, n$. It follows that $m(\lambda)$ is a product of distinct linear factors since $m|p$.

The reader should always bear in mind (since in many cases we shall not mention it) that every theorem which asserts the existence, for a type of linear transformation, of a representation with a property, p, can be restated as a theorem of the following type about matrices: If a matrix A over the field F is of a specified type, then there exists a matrix P over F such that PAP^{-1} has property p. If the matrix P is unimportant, the conclusion can be stated more simply as: There exists a matrix with property p similar to A, or A can be transformed to a matrix with property p in the field F. The latter formulation is appropriate when the result depends upon the choice of coefficient field. The proof of the foregoing remark, that a theorem concerning representations of a linear transformation can be reformulated as a theorem about matrices, is immediate—the given matrix A defines a linear transformation $\mathbf{A}$ on some vector space, and any other representation of $\mathbf{A}$ is similar to A.

To illustrate the foregoing, we restate Theorem 7.5 in matrix language.

COROLLARY. A square matrix A over F can be transformed to diagonal form in the field F if and only if the minimum function of A decomposes in $F[\lambda]$ into simple linear factors.

We refer the reader to Example 7.5 for an illustration of the above theorem and its corollary, as well as the computations involved in finding a basis relative to which a transformation, whose minimum function decomposes into simple linear factors, has a diagonal representation.

Since there may be difficulty in determining the minimum function of a linear transformation $\mathbf{A}$, the preceding theorem may not always be helpful. However, it will emerge from the following discussion that the minimum function of **A** can be studied by means of the characteristic function of **A** (see Sec. 7.4). Since the latter can be found from any representation of **A**, the characteristic function becomes of great practical importance.

It is obvious that the existence of a diagonal representation for a linear transformation **A** is contingent upon the existence of one-dimensional invariant spaces for **A**, in other words, nonzero vectors α such that $\mathbf{A}\alpha = c\alpha$.

DEFINITION. *A vector* $\alpha \neq 0$ *is called a* characteristic vector *of the linear transformation* **A** *if there exists a scalar c such that* $\mathbf{A}\alpha = c\alpha$. *A scalar c is called a* characteristic value *of* **A** *if there exists a vector* $\alpha \neq 0$ *such that* $\mathbf{A}\alpha = c\alpha$.

We have introduced the notion of characteristic value, along with that of characteristic vector, since a knowledge of the characteristic values of **A** facilitates the determination of its characteristic vectors, as we shall show. In passing, we observe that the transformations which admit a diagonal representation are precisely those whose characteristic vectors span V.

Theorem 7.6. The set of characteristic values of **A** coincides with the set of distinct roots in F of the minimum function, $m(\lambda)$, of **A**.

Proof. Let c denote a characteristic value of **A** and α a corresponding characteristic vector. Then $\mathbf{A}\alpha = c\alpha$, or $(\mathbf{A} - c\mathbf{I})\alpha = 0$. Writing $m(\lambda)$ in the form $m(\lambda) = q(\lambda)(\lambda - c) + m(c)$ implies that $0 = m(\mathbf{A})\alpha = m(c)\alpha$ and hence (since $\alpha \neq 0$) that $m(c) = 0$.

Conversely, if c is a root in F of $m(\lambda)$, $m(\lambda) = (\lambda - c)q(\lambda)$, where there exists a vector β such that $\alpha = q(\mathbf{A})\beta \neq 0$, by virtue of the definition of $m(\lambda)$. The equation

$$0 = m(\mathbf{A})\beta = (\mathbf{A} - c\mathbf{I})q(\mathbf{A})\beta = (\mathbf{A} - c\mathbf{I})\alpha$$

then demonstrates that c is a characteristic value.

COROLLARY. A characteristic vector of **A** exists if and only if $m(\lambda)$ has a root in F.

The reader might find it profitable to reexamine Examples 7.3 to 7.5 in the light of this theorem. In the first and last of these examples there are sufficient characteristic vectors accompanying the various characteristic values to construct a basis of such vectors and, as a consequence, a diagonal representation exists. In Example 7.4, 0 and 2 are the only characteristic values. Since $A = A - 0I$ and $A - 2I$ have ranks 7 and 6, respectively, there is a one-dimensional space (less the zero vector) of characteristic vectors accompanying 0 and a two-dimensional space (less zero) accompanying 2.

If a representation A of $\mathbf{A}$ is known, a characteristic value of $\mathbf{A}$ is a scalar c such that the matrix equation

$$AX = cX \qquad \text{or} \qquad (cI - A)X = 0$$

has a nontrivial solution. This condition holds, in turn, if and only if $\det(cI - A) = 0$. Thus the characteristic values of $\mathbf{A}$ are the roots in F of the characteristic function $\det(\lambda I - A)$ of $\mathbf{A}$. It follows that (see Theorem 7.6) *the minimum function, $m(\lambda)$, and the characteristic function, $f(\lambda)$, of* $\mathbf{A}$ *have the same distinct roots in F.* Later the existence of a much stronger relation between m and f will be shown: $m|f$, and there exists a positive integer k such that $f|m^k$. For example, if $\mathbf{A} = \mathbf{I}$, then $m(\lambda) = \lambda - 1$ and $f(\lambda) = (\lambda - 1)^n$ so that $m|f$ and $f|m^n$. Again, in Example 7.4 one finds that $f(\lambda) = \lambda(\lambda - 2)^3(\lambda^2 - \lambda + 1)^2$ so that $f|m^2$.

Not only can one determine the roots in F of $m(\lambda)$ from $f(\lambda)$, but, in addition, the simplicity of these roots can be examined without a knowledge of $m(\lambda)$. The next theorem demonstrates this.

First we prove a preliminary result.

Lemma 7.5. Suppose that $g(\lambda)$ is a proper divisor of a divisor $f(\lambda)$ of the minimum function, $m(\lambda)$, of a linear transformation $\mathbf{A}$. Then $N(g)$ is properly contained in $N(f)$.

Proof. By assumption, $m = fh$, and $k = gh$ is a proper divisor of m. Consequently, there exists a vector α such that $k(\mathbf{A})\alpha = g(\mathbf{A})[h(\mathbf{A})]\alpha \neq 0$. But then $h(\mathbf{A})\alpha$, which is a member of $N(f)$, is not contained in $N(g)$.

Theorem 7.7. A characteristic value, c, of $\mathbf{A}$, that is, a root in F of the minimum function, $m(\lambda)$, of A, is a simple root if and only if $N(\lambda - c) = N(\lambda - c)^2$.

Proof. If c is a simple root of $m(\lambda)$, the g.c.d. of $(\lambda - c)^2$ and $m(\lambda)$ is $\lambda - c$, so that (Lemma 7.3)

$$N(\lambda - c) = N(\lambda - c)^2 \cap N(m) = N(\lambda - c)^2$$

since $N(m) = V$.

For the converse, we remark first that, since the inclusion $N(\lambda - c) \subseteq N(\lambda - c)^2$ is always true, it is sufficient to prove that if $\lambda - c$ is a multiple factor of $m(\lambda)$, then the inclusion is proper. This is a consequence of Lemma 7.5 with $g = \lambda - c$ and $f = (\lambda - c)^2$.

Another criterion for the simplicity of a root c in F of $m(\lambda)$ may be expressed in terms of the following two numbers associated with c: the *geometric multiplicity*, g, of c, which is equal to $d[N(\lambda - c)]$, and the *algebraic multiplicity*, a, of c, which is equal to the multiplicity of $\lambda - c$ in $f(\lambda)$.

The inequality $g \leqq a$ is valid in general. To prove this, we construct a basis for V by extending a basis for the invariant space $N(\lambda - c)$ for **A**. Relative to such a basis, **A** is represented by a matrix of the form

$$A = \begin{bmatrix} cI_g & P \\ 0 & Q \end{bmatrix}$$

Then the characteristic function of A is seen to have the form $f(\lambda) = (\lambda - c)^g \det(\lambda I - Q)$, which implies that $g \leqq a$. If $g = a$, then $\det(cI - Q) \neq 0$, or $cI - Q$ and hence $Q - cI$ are nonsingular. With $Q - cI$ nonsingular, $A - cI$ and $(A - cI)^2$ are seen to have the same rank, and hence (Theorem 7.7) c is a simple root of $m(\lambda)$. If, conversely, c is a simple root of $m(\lambda)$, then $g = a$. This is left as an exercise for the reader. The foregoing is summarized next.

Theorem 7.8. The characteristic value c of **A** is a simple root of the minimum function if and only if the algebraic and geometric multiplicities of c are equal.

EXAMPLE 7.6

From the characteristic function $f(\lambda) = \lambda(\lambda - 2)^3(\lambda^2 - \lambda + 1)^2$ of the transformation in Example 7.4 it follows that the algebraic multiplicity of the characteristic value 2 is 3. Since the geometric multiplicity of 2 is $n(A - 2I) = 2$, it is a repeated root of the minimum function. As a simple root of $f(\lambda)$, the characteristic value 0 is necessarily a simple root of $m(\lambda)$. The derivative operator **D** of Example 7.2 has λ^4 as its characteristic function. Obviously the geometric multiplicity of 0 is less than the algebraic multiplicity, 4, since otherwise **D** would equal **0**.

We conclude this section with several remarks concerning linear transformations on vector spaces over the field of complex numbers C. Since every polynomial in λ with coefficients in C can be decomposed into a product of linear factors (Theorem 6.17), this is true, in particular, for the characteristic and minimum function of a linear transformation **A** on V over C. Thus characteristic values and vectors always exist for such a transformation. Although we know that **A** does not necessarily admit a diagonal representation, the existence of a representation by a superdiagonal matrix is easily shown. Such a matrix, the construction of which is described in the next theorem, exhibits the characteristic values of **A** and is useful in many applications. For the simplest possible representation of a transformation on V over C the reader is referred to Sec. 7.9.

Theorem 7.9. For a linear transformation **A** on V over C there exists a superdiagonal representation:

$$\begin{bmatrix} c_1 & c_2^1 & c_3^1 & \cdots & c_n^1 \\ 0 & c_2 & c_3^2 & \cdots & c_n^2 \\ 0 & 0 & c_3 & \cdots & c_n^3 \\ \cdot & \cdot & \cdot & \cdots & \cdot \\ 0 & 0 & 0 & \cdots & c_n \end{bmatrix}$$

Any such representation exhibits the characteristic values of **A**, each repeated a number of times equal to its algebraic multiplicity.

Proof. Let c_1 denote a characteristic value and α_1 a corresponding characteristic vector for **A**. Relative to any basis $\{\alpha_1, \alpha_2, \ldots, \alpha_n\}$ for V whose first member is α_1, **A** is represented by a matrix of the form

$$\begin{bmatrix} c_1 & B_1 \\ 0 & A_1 \end{bmatrix}$$

where (see Sec. 7.2) A_1 is the representation of $\mathbf{A}_1$, the transformation induced in $V_1 = V/[\alpha_1]$ by **A**, relative to the basis $\{\bar{\alpha}_2, \bar{\alpha}_3, \ldots, \bar{\alpha}_n\}$ for V_1. Let c_2 denote a characteristic value and $\bar{\beta}_2$ a corresponding characteristic vector for $\mathbf{A}_1$; thus $\mathbf{A}_1\bar{\beta}_2 = c_2\bar{\beta}_2$, where $\bar{\beta}_2 \neq \bar{0}$. Then $\overline{\mathbf{A}\beta_2} = \overline{c_2\beta_2}$, or $\mathbf{A}\beta_2 = c_2^1\alpha_1 + c_2\beta_2$. Moreover, $\bar{\beta}_2 \neq \bar{0}$ implies that β_2 is not a member of $[\alpha_1]$, and consequently there exists a basis for V of the form $\{\alpha_1, \beta_2, \ldots\}$. Relative to such a basis, **A** is represented by a matrix of the form

$$\begin{bmatrix} C_2 & B_2 \\ 0 & A_2 \end{bmatrix} \quad \text{where } C_2 = \begin{bmatrix} c_1 & c_2^1 \\ 0 & c_2 \end{bmatrix}$$

Here, A_2 represents the transformation induced in $V_2 = V/[\alpha_1, \beta_2]$ by **A**. Next, the step of selecting a characteristic value and vector is repeated. Continuing in this way will eventually produce a representation with all zeros below the principal diagonal. The final assertion of the theorem is obvious when the appropriate matrix is written down.

The method of proof used above provides a practical means for deriving a superdiagonal representation. The following example illustrates this:

EXAMPLE 7.7

The characteristic function of the transformation **A** on $V_4(C)$, defined by the matrix

$$A = \begin{bmatrix} 8 & 9 & -9 & 0 \\ 0 & 2 & 0 & 2 \\ 4 & 6 & -4 & 0 \\ 0 & 0 & 0 & 3 \end{bmatrix}$$

relative to the ϵ-basis, is $(\lambda - 2)^3(\lambda - 3)$. Beginning with the characteristic value 2, any nonzero solution of $(A - 2I)X = 0$ is a characteristic vector corresponding to 2. One such solution has been chosen as the first member of the following basis, whose remaining members may be selected in any manner:

$$\alpha_1 = \begin{bmatrix} 0 \\ 1 \\ 1 \\ 0 \end{bmatrix} \qquad \alpha_2 = \begin{bmatrix} 1 \\ 0 \\ 0 \\ 0 \end{bmatrix} \qquad \alpha_3 = \begin{bmatrix} 0 \\ 1 \\ 0 \\ 0 \end{bmatrix} \qquad \alpha_4 = \begin{bmatrix} 0 \\ 0 \\ 0 \\ 1 \end{bmatrix}$$

The transformation is represented by the matrix

$$P^{-1}AP = \begin{bmatrix} 2 & 4 & 6 & 0 \\ 0 & 8 & 9 & 0 \\ 0 & -4 & -4 & 2 \\ 0 & 0 & 0 & 3 \end{bmatrix} \qquad \text{where } P = \begin{bmatrix} 0 & 1 & 0 & 0 \\ 1 & 0 & 1 & 0 \\ 1 & 0 & 0 & 0 \\ 0 & 0 & 0 & 1 \end{bmatrix}$$

relative to the α-basis (see Example 7.5). The vector $\bar{\beta}_2 = 3\bar{\alpha}_2 - 2\bar{\alpha}_3$ is a characteristic vector corresponding to the characteristic value 2 of the transformation $\mathbf{A}_1$ induced by $\mathbf{A}$ on $V/[\alpha_1]$. As such we choose a basis for V whose first two members are α_1 and β_2 :

$$\alpha_1 = \begin{bmatrix} 0 \\ 1 \\ 1 \\ 0 \end{bmatrix} \qquad \beta_2 = \begin{bmatrix} 3 \\ -2 \\ 0 \\ 0 \end{bmatrix} \qquad \beta_3 = \begin{bmatrix} 0 \\ 0 \\ 1 \\ 0 \end{bmatrix} \qquad \beta_4 = \begin{bmatrix} 0 \\ 0 \\ 0 \\ 1 \end{bmatrix}$$

This basis is defined in terms of the previous one by the matrix Q below. Because of the simplicity of the example, the representation of $\mathbf{A}$ relative to this new basis, $Q^{-1}(P^{-1}AP)Q$, is in superdiagonal form.

$$Q = \begin{bmatrix} 1 & 0 & 1 & 0 \\ 0 & 3 & 0 & 0 \\ 0 & -2 & -1 & 0 \\ 0 & 0 & 0 & 1 \end{bmatrix} \qquad Q^{-1}(P^{-1}AP)Q = \begin{bmatrix} 2 & 0 & -6 & 2 \\ 0 & 2 & -3 & 0 \\ 0 & 0 & 2 & -2 \\ 0 & 0 & 0 & 3 \end{bmatrix}$$

PROBLEMS

1. In the proof of the second half of Theorem 7.5 show that $m(\lambda) = p(\lambda)$.

2. Verify the statement: $\mathbf{A}$ admits a diagonal representation if and only if its characteristic vectors span V.

3. Complete the proof of Theorem 7.8.

4. Let **A** denote a linear transformation on V over C such that $\mathbf{A}^k = \mathbf{I}$ for a positive integer k. Show that **A** can be represented by a diagonal matrix.

5. Let **A** denote a linear transformation on V over C such that $\mathbf{A}^k = \mathbf{0}$ for a positive integer k. Show that **A** can never be represented by a diagonal matrix, assuming that $\mathbf{A} \neq \mathbf{0}$.

6. Consider the linear transformation **A** on an n-dimensional space over C defined by the following system of equations: $y^k = c_k x^{n-k+1}$, $k = 1, 2, \ldots, n$. Determine necessary and sufficient conditions that **A** admit a diagonal representation.

7. Show that the transformation on $V_4(C)$ represented by the matrix below admits no diagonal representation by comparing the geometric and algebraic multiplicities of each of the distinct characteristic roots (2 and -1). Find a superdiagonal representation for the transformation and the accompanying basis.

$$A = \begin{bmatrix} 2 & -8 & -8 & -9 \\ -6 & -1 & 0 & 17 \\ 6 & 3 & 2 & -17 \\ 0 & -3 & -3 & -1 \end{bmatrix}$$

8. Show that a diagonal representation exists for the transformation on $V_4(C)$ defined by the matrix

$$A = \begin{bmatrix} -2 & -7 & 22 & 33 \\ 1 & 3 & -11 & -22 \\ 0 & 0 & -2 & -7 \\ 0 & 0 & 1 & 3 \end{bmatrix}$$

9. Find a diagonal or superdiagonal matrix similar to each of the following matrices:

$$\begin{bmatrix} 2 & -1 & 1 \\ 2 & 2 & -1 \\ 1 & 2 & -1 \end{bmatrix} \qquad \begin{bmatrix} 0 & 0 & 1 \\ 1 & 0 & -3 \\ 0 & 1 & 3 \end{bmatrix}$$

10. Let **A** denote a transformation on $V_n(C)$ and $g(\lambda)$ a polynomial in λ with coefficients in C. Use a superdiagonal representation for **A** to prove that the characteristic values of $g(\mathbf{A})$ are precisely the numbers $g(c)$, where c ranges over the characteristic values of **A**.

7.6. A-bases for a Vector Space. The remainder of the chapter will be devoted to the study of an alternative method for determining invariant spaces of a linear transformation. This study will culminate in a practical means for computing the simplest representations of those transformations which do not admit a diagonal representation. Our starting point is to be found in the paragraph preceding Theorem 7.2. There it is shown that if **A** is a linear transformation on the n-dimensional space V over F, and if α is a (fixed) nonzero vector, there exists a polynomial $g(\lambda)$ in $F[\lambda]$ associated with each α such that $g(\mathbf{A})\alpha = 0$. The set of all such polynomials is an ideal, and consequently there exists a unique monic polynomial (which depends upon α)

$$m_\alpha(\lambda) = c_0 + c_1\lambda + \cdots + c_{m-1}\lambda^{m-1} + \lambda^m$$

of least degree, the so-called *index polynomial* of α, such that $m_\alpha(\mathbf{A})\alpha = 0$. Then $g(\mathbf{A})\alpha = 0$ if and only if $m_\alpha | g$. It follows immediately, upon examination of an arbitrary linear combination

$$a_0\alpha + a_1\mathbf{A}\alpha + \cdots + a_{m-1}\mathbf{A}^{m-1}\alpha \tag{8}$$

of the vectors α, $\mathbf{A}\alpha, \ldots, \mathbf{A}^{m-1}\alpha$, that $\{\alpha, \mathbf{A}\alpha, \ldots, \mathbf{A}^{m-1}\alpha\}$ is a linearly independent set and that

$$S_\alpha = [\alpha, \mathbf{A}\alpha, \ldots, \mathbf{A}^{m-1}\alpha] \tag{9}$$

is an invariant space of $\mathbf{A}$ [which, incidentally, is a subspace of $N(m_\alpha)$]. We now ask whether V can be decomposed into a direct sum of such spaces:

$$V = S_{\alpha_1} \oplus S_{\alpha_2} \oplus \cdots \oplus S_{\alpha_t} \tag{10}$$

The existence of a diagonal representation is coextensive with an extreme case of this type, *viz.*, that of n one-dimensional invariant spaces (in other words, n characteristic vectors). This explains our interest in decompositions of the form (10).

The existence of a decomposition of V of the type (10) accompanying an arbitrary $\mathbf{A}$ can be reformulated as:

(i) There exists a set of vectors $\{\alpha_1, \alpha_2, \ldots, \alpha_t\}$ in V such that every vector ξ in V has a representation

$$\xi = \sum_1^t f^i(\mathbf{A})\alpha_i \tag{11}$$

where $f^i(\lambda)$ is a polynomial of degree less than that of the index polynomial of α_i.

(ii) This representation satisfies the uniqueness condition

$$\xi = \sum_1^t f^i(\mathbf{A})\alpha_i = \sum_1^t g^i(\mathbf{A})\alpha_i \text{ implies } f^i(\mathbf{A})\alpha_i = g^i(\mathbf{A})\alpha_i \qquad i = 1, 2, \ldots, t \tag{12}$$

or the equivalent one

$$\sum_1^t f^i(\mathbf{A})\alpha_i = 0 \text{ implies } f^i(\mathbf{A})\alpha_i = 0 \qquad i = 1, 2, \ldots, t \tag{13}$$

To verify the assertion, we remark that (10) is equivalent to the condition that each ξ have a unique representation in the form $\xi = \sum \xi_i$, ξ_i in S_{α_i}.

Now (i) asserts the existence of such a decomposition. Indeed, since a summand S_{α_i} has the form (9), each ξ_i has the form (8), which clearly can be written as $f(\mathbf{A})\alpha$, where $f(\lambda) = \sum a_i\lambda^i$. Finally (12) asserts the uniqueness of (11). For this it is always sufficient to require merely the uniqueness of the zero vector, that is, (13), as a direct computation shows.

A set of vectors $\{\alpha_1, \alpha_2, \ldots, \alpha_t\}$ satisfying (i) above is called an **A**-*generating system* for V; if in addition (ii) is satisfied, so that a decomposition (10) exists, the set is called an **A**-*basis* for V. These notions are analogous, respectively, to a spanning set and a basis for V in the sense of Chap. 2. It is clear that an ordinary basis for V is a trivial example of an **A**-generating system for any **A**.

In order to obtain a reformulation of the properties of an **A**-basis, further definitions are needed.

DEFINITION. *A* relation *for an* **A**-*generating system* $\{\alpha_1, \alpha_2, \ldots, \alpha_t\}$ *is a column matrix*

$$(f^1(\lambda), f^2(\lambda), \ldots, f^t(\lambda))^T \qquad f^i(\lambda) \text{ in } F[\lambda]$$

such that

$$f^1(\mathbf{A})\alpha_1 + f^2(\mathbf{A})\alpha_2 + \cdots + f^t(\mathbf{A})\alpha_t = 0$$

A matrix $M = (f^i_j(\lambda))$ *such that each column is a relation for* $\{\alpha_1, \alpha_2, \ldots, \alpha_t\}$ *is called a* relation matrix *for* $\{\alpha_1, \alpha_2, \ldots, \alpha_t\}$.

It is clear that the sum of two relations and any polynomial multiple of a relation are relations. A relation matrix M for $\{\alpha_1, \alpha_2, \ldots, \alpha_t\}$ is said to be *complete* when every relation can be obtained from the columns of M by a finite sequence of these two operations. The reader may verify that *every* **A**-*basis is precisely an* **A**-*generating system* $\{\alpha_1, \alpha_2, \ldots, \alpha_t\}$ *for which there exists a complete diagonal relation matrix*

$$\begin{bmatrix} h_1(\lambda) & & & & \\ & h_2(\lambda) & & & \\ & & \cdot & & \\ & & & \cdot & \\ & & & & \cdot \\ & & & & & h_t(\lambda) \end{bmatrix}$$

where $h_i(\lambda)$ *is the index polynomial of* α_i, $i = 1, 2, \ldots, t$.

A complete relation matrix is easily found for the type of **A**-generating system mentioned earlier, *viz.*, an ordinary basis $\{\alpha_1, \alpha_2, \ldots, \alpha_n\}$ for V, as soon as a representation $A = (\mathbf{A}; \alpha)$ is known. Indeed, when the defining equation for A, $\mathbf{A}\alpha = \alpha A$, is written in expanded form as

$$\sum_{i=1}^{n} (\delta_j^i \mathbf{A} - a_j^i \mathbf{I})\alpha_i = 0 \qquad j = 1, 2, \ldots, n$$

it is observed that the so-called *characteristic matrix* of A (or the characteristic matrix of $\mathbf{A}$ relative to the α-basis)

$$\lambda I - A$$

is a relation matrix for $\{\alpha_1, \alpha_2, \ldots, \alpha_n\}$. We shall merely outline a proof of the completeness of $\lambda I - A$. If $r = (f^1, f^2, \ldots, f^n)^{\mathrm{T}}$ is a relation for $\{\alpha_1, \alpha_2, \ldots, \alpha_n\}$, then by subtracting multiples of the columns of $\lambda I - A$ from r it is possible to deduce a relation $r' = (c^1, c^2, \ldots, c^n)^{\mathrm{T}}$, where each c_i is a constant. Recalling that the α_i's form a basis for V, we may conclude that $r' = (0, 0, \ldots, 0)^{\mathrm{T}}$. Finally, since each step of the reduction of r to r' is reversible, it follows that r is a linear combination of the relations in $\lambda I - A$.

Since it suffices for our purposes to develop a method for deriving an $\mathbf{A}$-basis for V from an ordinary basis, and for the latter a square complete relation matrix can always be found, we shall confine our attention henceforth to finite relation matrices. The results obtained so far suggest that the problem of deriving an $\mathbf{A}$-basis from an $\mathbf{A}$-generating system $\{\alpha_1, \alpha_2, \ldots, \alpha_n\}$ be studied in terms of the reduction of a complete relation matrix

$$M = (f_j^i) \qquad i, j = 1, 2, \ldots, n$$

for $\{\alpha_1, \alpha_2, \ldots, \alpha_n\}$ to diagonal form. Since in Chap. 3 this problem was studied for matrices over a field using elementary row and column transformations (Sec. 3.5), we propose to investigate the applicability of the same tool to the current problem. We may use the symbols R_{ij}, $R_i(c)$, and $R_{ij}(c)$ introduced in Sec. 3.5 for elementary row transformations of types I, II, and III, respectively, and R_{ij}^*, etc., for the corresponding column transformations. First it is clear that a column transformation on M replaces it by a relation matrix N for the same $\mathbf{A}$-generating system and that the completeness of N is assured if the transformation is reversible.† Thus an $R_{ij}^*(g)$ transformation, where g is an arbitrary polynomial, is admissible, but in an $R_i^*(g)$ transformation we must restrict g to the set of nonzero constants. Next we assert that a row transformation on M yields a complete relation matrix N for an $\mathbf{A}$-generating system which can be obtained

†In this connection, the reader should convince himself that each elementary column transformation is expressible in terms of the two operations pertinent to the definition of a complete relation matrix.

from $\{\alpha_1, \alpha_2, \ldots, \alpha_n\}$ in a simple way. Below, opposite each type of row transform of M is indicated the appropriate alteration in the α_i's.

$$
\begin{aligned}
N &= R_{ij}M && \text{interchange } \alpha_i \text{ and } \alpha_j \text{ in } \{\alpha_1, \alpha_2, \ldots, \alpha_n\} \\
&= R_i(c)M && \text{replace } \alpha_i \text{ by } c^{-1}\alpha_i \text{ in } \{\alpha_1, \alpha_2, \ldots, \alpha_n\} \\
&= R_{ij}(g)M && \text{replace } \alpha_j \text{ by } \alpha_j - g(\mathbf{A})\alpha_i \text{ in } \{\alpha_1, \alpha_2, \ldots, \alpha_n\}
\end{aligned}
$$

In the way of proof of these statements we first remark that obviously each of the indicated types of operations (where, of course, in the second type c is a nonzero constant) on $\{\alpha_1, \alpha_2, \ldots, \alpha_n\}$ yields an **A**-generating system. A straightforward computation in each case verifies that N is a relation matrix for the corresponding generating system; the completeness of N follows from that of M together with the reversibility of the operations used.

As the foregoing results suggest, a systematic use of elementary transformations is the key to the solution of our problem. Since, however, further developments utilize properties of $n \times n$ matrices over $F[\lambda]$ that are independent of their possible interpretation as relation matrices, we interrupt our discussion to present these properties in the next section.

PROBLEMS

1. Verify that the space S_α in (9) is an invariant space for **A**.

2. Demonstrate that an **A**-basis is an **A**-generating system for which there exists a complete diagonal relation matrix.

3. Describe an iterative scheme for showing that $\lambda I - A$ is a complete relation matrix for $\{\alpha_1, \alpha_2, \ldots, \alpha_n\}$, where $A = (\mathbf{A}; \alpha)$.

4. Let $m(\lambda)$ denote the minimum function of **A**, a linear transformation on V, and assume that $p_1^{e_1}p_2^{e_2} \cdots p_s^{e_s}$ is the decomposition of m into a product of powers of distinct polynomials, irreducible over F. Below is outlined a construction for (i) a vector γ_i whose index polynomial is $p_i^{e_i}$ and (ii) a vector whose index polynomial is m. Fill in the details.

If the polynomial P_i is defined by the equation $m = p_i^{e_i}P_i$, then $p_i^{e_i-1}(\mathbf{A})P_i(\mathbf{A}) \neq \mathbf{0}$. Hence there exists a basis vector, for example α_k, whose image under the last transformation is not zero. Then $\gamma_i = P_i(\mathbf{A})\alpha_k$ has $p_i^{e_i}$ as its index polynomial, and $\gamma = \sum_1^s \gamma_i$ has m as its index polynomial.

7.7. Polynomial Matrices. This section is devoted primarily to the derivation of a sequence of results for $n \times n$ matrices over $F[\lambda]$ or, as we prefer to say, *polynomial matrices* that correspond to those obtained in Sec. 3.5 for matrices over a field. We carry over the definition of equivalent matrices to the current case (where it is now understood that in a type II transformation the multiplier is restricted to the set of nonzero constants) and demonstrate the existence of a diagonal matrix in each equivalence class.

First it is necessary to say a word about the notion of the rank of a polynomial matrix, since this has been defined only for matrices over a field. Just as the ring of integers (see Sec. 6.7) can be regarded as a subset of the field of rational numbers, so can the ring $F[\lambda]$ be regarded as a subset of the field $F(\lambda)$, whose elements are of the form f/g, where f and g are polynomials and $g \neq 0$, and where equality, addition, and multiplication are defined exactly as for rational numbers. Thus a polynomial matrix M (*i.e.*, a matrix over $F[\lambda]$) is a matrix over the field $F(\lambda)$, and, consequently, its rank is defined. It is of interest to notice that the determinant criterion for the rank of M involves only addition, subtraction, and multiplication of its (polynomial) elements, and, hence, the rank may be computed within $F[\lambda]$! Likewise, the rank of M, when interpreted as the row rank of M, can be computed within $F[\lambda]$, since a set of rows of M is linearly independent over $F(\lambda)$ if and only if it is linearly independent over $F[\lambda]$.

Lemma 7.6. An $n \times n$ polynomial matrix M of rank r is equivalent to a diagonal matrix

$$D = \operatorname{diag}(h_1, h_2, \ldots, h_r, 0, \ldots, 0)$$

where each h_i is a monic polynomial and $h_1|h_2|\cdots|h_r$.

Proof. We may assume that $M = (f_j^i)$ has rank $r > 0$. Among the nonzero entries of M there is one of least degree which, by appropriate row and column interchanges, we may take as f_1^1. If $f_1^k \neq 0$, $k > 1$, then $f_1^k = qf_1^1 + r$, where $r = 0$ or is of lower degree than f_1^1. If q times the first row is subtracted from the kth row, the new matrix has r in the $(k, 1)$th position. If $r \neq 0$, it can be shifted to the (1,1)th position and we may repeat the operation with an f_1^1 of lower degree. In a finite number of such steps we obtain an equivalent matrix with $f_1^1 \neq 0$ and every remaining element of the first row and of the first column equal to 0.

Unless f_1^1 divides every element of this new matrix, its degree can be reduced still further. For if the (i, j)th entry in the new M is $f_j^i = qf_1^1 + r$, where r has lower degree than f_1^1, we may add the ith row to the first to place f_j^i in the $(1, j)$th position and then replace f_1^1 by r. Thus we may assume that M has the form

$$M' = \begin{pmatrix} f_1^1 & 0 & 0 & \cdots & 0 \\ 0 & f_2^2 & f_3^2 & \cdots & f_n^2 \\ 0 & f_2^3 & f_3^3 & \cdots & f_n^3 \\ \cdot & \cdot & \cdot & \cdots & \cdot \\ 0 & f_2^n & f_3^n & \cdots & f_n^n \end{pmatrix}$$

where f_1^1 is a monic polynomial that divides every f_j^i. If $r > 1$, the first row and column of M' are now ignored and the remaining submatrix reduced as above to obtain a matrix equivalent to M and of the form

$$M'' = \begin{bmatrix} f_1^1 & 0 & 0 & 0 & \cdots & 0 \\ 0 & f_2^2 & 0 & 0 & \cdots & 0 \\ 0 & 0 & f_3^3 & f_4^3 & \cdots & f_n^3 \\ 0 & 0 & f_3^4 & f_4^4 & \cdots & f_n^4 \\ \cdot & \cdot & \cdot & \cdot & \cdots & \cdot \\ 0 & 0 & f_3^n & f_4^n & \cdots & f_n^n \end{bmatrix}$$

where f_2^2 divides every element of M'' except possibly f_1^1. Since M'' is equivalent to M', every element of M'' is equal to a linear homogeneous function of the elements of M' and hence f_1^1 still divides every f_j^i.

By continuing this process we obtain a diagonal matrix with the required properties.

EXAMPLE 7.8

The equivalence of the characteristic matrix

$$\lambda I - A = \begin{bmatrix} \lambda + 5 & 3 & 2 & -4 \\ -2 & \lambda & -1 & 1 \\ -10 & -7 & \lambda - 4 & 9 \\ -2 & 0 & -1 & \lambda \end{bmatrix}$$

to the diagonal matrix

$$\mathrm{diag}(1, 1, \lambda + 1, \lambda^3 + 1)$$

can be shown by the following sequence of transformations, reading from left to right:

$$R_{13}^*(-2), R_{43}^*(\lambda), R_{14}(2), R_{24}(-1), R_{34}(\lambda - 4), R_{31}(2), R_{42}^*(\lambda^2 + 1), R_{13}(3),$$
$$R_{23}(\lambda), R_{41}^*(-3\lambda + 1), R_{24}, R_{13}^*, R_{12}, R_{23}, R_1^*(-1), R_2^*(-1)$$

The diagonal matrix of the preceding theorem is uniquely determined by the initial matrix M. To show this, we introduce a new notion, that of the kth *determinantal divisor* of M, $d_k(M)$, which is defined as the g.c.d. of the set of minors of order k of M if $1 \leq k \leq r$, the rank of M. In addition, it is convenient to define $d_0(M) = 1$. We know that a matrix N equivalent to M has the same rank as M and now prove a more general result.

Lemma 7.7. If N is equivalent to the polynomial matrix M of rank r, then $d_k(N) = d_k(M)$, $k = 1, 2, \ldots, r$.

Proof. It is sufficient to consider the case where N is obtained from M by a single row transformation. If $N = R_{ij}M$ or $R_i(c)M$, each kth-order minor of N is a constant multiple of the corresponding kth-order minor of M and the desired conclusion follows. If $N = R_{ij}(g)M$, then a square submatrix of N in which the ith row takes no part, or both of the ith and jth rows take part, has the same determinant as the corresponding submatrix of M. In the remaining case of a submatrix in which the jth row takes no part but the ith does, the determinant is a linear combination of minors from M of the same order, which implies that $d_k(M)|d_k(N)$. Since, conversely, M can be obtained from N by a type III transformation, $d_k(N)|d_k(M)$ and equality follows.

Recalling Lemma 7.6, we see that the determinantal divisors of D, and hence M, are

$$d_0 = 1,\ d_1 = h_1,\ d_2 = h_1h_2,\ \ldots,\ d_r = h_1h_2 \cdots h_r \tag{14}$$

so that h_i is characterized by the equation

$$h_i = \frac{d_i(M)}{d_{i-1}(M)} \qquad i = 1, 2, \ldots, r$$

The quotient $d_i(M)/d_{i-1}(M)$ is called the ith *invariant factor* of M. From Lemma 7.6 we may conclude that each invariant factor h_i of M is a polynomial, and $h_i|h_{i+1}$. Moreover, it is clear that these polynomials, which are uniquely defined by the determinantal divisors, in turn define the determinantal divisors. Those invariant factors of M which are different from 1 are called the *nontrivial invariant factors* of M.

Theorem 7.10. A polynomial matrix M of rank r is equivalent to a uniquely determined diagonal matrix

$$\operatorname{diag}(h_1, h_2, \ldots, h_r, 0, \ldots, 0)$$

such that each h_i is a monic polynomial which divides its successor; h_i is precisely the ith invariant factor of M. Two polynomial matrices are equivalent if and only if they have the same invariant factors.

Proof. The first assertion has already been shown. Next, Lemma 7.7 implies that if $N \overset{E}{\sim} M$, then their invariant factors agree. Conversely, if the invariant factors of M and N coincide, they are equivalent to one and the same diagonal matrix which displays these factors, and hence equivalent to each other. This completes the proof.

Thus, for the set of $n \times n$ polynomial matrices, the matrices described in the above theorem constitute a canonical set under equivalence. Moreover, the theorem contains a characterization of equivalence. In addition to this analogue of Theorem 3.8, there is one of Theorem 3.17 where the equivalence of two matrices over F is formulated as: $A \overset{E}{\sim} B$ if and only if there exist nonsingular matrices P and Q such that $B = PAQ$. The present version follows:

Theorem 7.11. If M and N are nth-order polynomial matrices, then $M \overset{E}{\sim} N$ if and only if there exist polynomial matrices P and Q having nonzero constant determinants such that $N = PMQ$.

Proof. The definition of elementary matrices and their utility in performing elementary transformations on a matrix carry over with no changes to the case in hand. Since the determinant of an elementary matrix is clearly a nonzero constant, the same is true for a product of such, so that $N \overset{E}{\sim} M$ implies $N = PMQ$, where P and Q have the required properties. For the converse we observe that if a polynomial matrix P has the property that $\det P$ is a nonzero constant, then $d_n(P) = 1$ and consequently each of its invariant factors is equal to 1. This means that $P \overset{E}{\sim} I$ (Theorem 7.10), or is equal to a product of elementary matrices. It follows that if $N = PMQ$, where both P and Q have nonzero constant determinants, N can be obtained from M by elementary transformations.

An example of this theorem is close at hand. Applying the row (column) transformations indicated in Example 7.8 to an identity matrix gives the following matrix $P(Q)$:

$$P = \begin{bmatrix} 0 & 0 & 0 & 1 \\ 2 & 0 & 1 & \lambda \\ 7 & 0 & 3 & 3\lambda + 2 \\ 2\lambda & 1 & \lambda & \lambda^2 - 1 \end{bmatrix} \qquad Q = \begin{bmatrix} 0 & 0 & 1 & -3\lambda + 1 \\ 0 & -1 & 0 & \lambda^2 + 1 \\ -1 & 0 & -2 & 7\lambda - 2 \\ 0 & 0 & 0 & 1 \end{bmatrix}$$

where we then have

$$P(\lambda I - A)Q = \operatorname{diag}(1, 1, \lambda + 1, \lambda^3 + 1)$$

We conclude this section with a brief discussion of a set of polynomials which can be defined in terms of the invariant factors of a polynomial matrix and which frequently have greater utility than the invariant factors. Let $h_1, h_2, \ldots, h_r$ be the invariant factors of M and $p_1, p_2, \ldots, p_s$ be the monic polynomials, irreducible over F, each of which divides at least one

h_i. Then every h_i can be written in the form

$$h_i = p_1^{e_{i1}} p_2^{e_{i2}} \cdots p_s^{e_{is}} \qquad i = 1, 2, \ldots, r$$

where each exponent e_{ij} is zero or a positive integer. The prime power factors (in any order)

$$p_1^{e_{11}}, \ldots, p_s^{e_{1s}}, \ldots, p_1^{e_{r1}}, \ldots, p_s^{e_{rs}}$$

whose exponents are positive are called the *elementary divisors* of M over F.

The following important distinction between the nature of invariant factors and that of elementary divisors should not escape the reader: The invariant factors of a fixed polynomial matrix M over $F[\lambda]$ are independent of the coefficient field F, while the elementary divisors are dependent upon F in such a way that, in general, they will change if F is enlarged. To verify the first statement, we recall that (i) the invariant factors are derived from g.c.d.'s of polynomials in $F[\lambda]$ and (ii) the g.c.d. of two polynomials in F is unchanged when F is enlarged (see Theorem 6.15). Concerning the second assertion, we need only recall that irreducibility is relative to the coefficient field at hand—an irreducible polynomial over $F[\lambda]$ may be reducible over $G[\lambda]$ if $G \supset F$.

It is clear that the invariant factors of M determine its elementary divisors over F uniquely; in addition, the elementary divisors in any order, together with the rank of M, determine its invariant factors uniquely. For this we recall the divisibility properties of the h_i's to conclude that h_r is the product of the highest powers of $p_1, p_2, \ldots, p_s$ which occur in the set of elementary divisors; h_{r-1} is similarly defined in terms of the remaining set of elementary divisors, etc. For example, if the elementary divisors of M, of rank 4, are

$$\lambda, \lambda^2, \lambda^2, (\lambda + 1)^2, (\lambda + 1)^4$$

then its invariant factors are

$$h_4 = \lambda^2(\lambda + 1)^4, h_3 = \lambda^2(\lambda + 1)^2, h_2 = \lambda, h_1 = 1$$

Theorem 7.12. Two polynomial matrices have the same invariant factors if and only if they have the same rank and the same elementary divisors.

The usefulness of elementary divisors is demonstrated by the following two theorems:

Theorem 7.13. If M is equivalent to $D = \text{diag}(f_1, f_2, \ldots, f_r, 0, \ldots, 0)$, where each f_i is a monic polynomial, then the prime power factors of the f_i's are the elementary divisors of M.

Proof. Let p denote any irreducible factor of any one of the f_i's, and arrange the f_i's according to ascending powers of p:

$$f_{i_1} = g_1 p^{k_1}, \ldots, f_{i_r} = g_r p^{k_r} \qquad \text{where } k_1 \leqq k_2 \leqq \cdots \leqq k_r$$

and each g is prime to p. Then the highest power of p which divides $d_i(M) = d_i(D)$ has exponent $k_1 + k_2 + \cdots + k_i$, and hence the highest power of p which divides $h_i = d_i/d_{i-1}$ is p^{k_i}, so that for each i such that $k_i > 0$, p^{k_i} is an elementary divisor of D and M. A repetition of this argument for each prime which divides some f_i shows that every prime power factor of every f_i is an elementary divisor of M and moreover that all elementary divisors of M are obtained in this way.

The following example illustrates how the preceding theorem can frequently ease the labor involved in finding invariant factors:

EXAMPLE 7.9

Assume that a polynomial matrix M has been reduced with elementary transformations to

$$\text{diag}(\lambda(\lambda + 1)^2, \lambda^2(\lambda + 1), \lambda^3(\lambda^2 + 1), \lambda^2 + 1, 0)$$

It follows that the elementary divisors of M over $R^{\#}$ are

$$\lambda, \lambda^2, \lambda^3, \lambda + 1, (\lambda + 1)^2, \lambda^2 + 1, \lambda^2 + 1$$

and hence that its invariant factors are

$$h_4 = \lambda^3(\lambda + 1)^2(\lambda^2 + 1),\ h_3 = \lambda^2(\lambda + 1)(\lambda^2 + 1),\ h_2 = \lambda,\ h_1 = 1$$

The reader is asked to supply a proof (using the preceding theorem) for the next result, which indicates an important property of elementary divisors that is not shared by invariant factors.

Theorem 7.14. The elementary divisors of a direct sum of polynomial matrices are the elementary divisors of all summands taken together.

PROBLEMS

1. Give a detailed proof of the results stated in (14) immediately following Lemma 7.7.

2. Supply a proof of Theorem 7.14.

3. Find the invariant factors and elementary divisors of each of the following polynomial matrices with entries in $R^{\#}[\lambda]$:

(*a*) $\text{diag}(\lambda^2, \lambda(\lambda + 1), \lambda^2 + 1, (\lambda + 1)^2)$

(*b*)
$$\begin{pmatrix} \lambda & \lambda - 2 & 2\lambda + 3 & \lambda^2 - \lambda + 1 \\ 2\lambda + 1 & \lambda - 1 & 3\lambda + 4 & \lambda^2 + \lambda + 2 \\ \lambda - 1 & \lambda + 2 & \lambda & \lambda^2 + 3\lambda - 1 \\ 2\lambda & 2\lambda & 3\lambda + 4 & 2\lambda^2 + 2\lambda + 1 \end{pmatrix}$$

(c) $$\begin{bmatrix} 2\lambda & 0 & 4\lambda & \lambda - 1 & 3\lambda - 3 \\ 3 & 6\lambda & 3\lambda + 6 & 0 & 1 - \lambda \\ 0 & \lambda & 0 & \lambda - 1 & 2\lambda - 2 \\ 1 & 2\lambda & \lambda + 2 & 0 & 0 \\ \lambda & 0 & 2\lambda & 0 & 0 \end{bmatrix}$$

(d) $$\begin{bmatrix} 2\lambda - 1 & \lambda & \lambda - 1 & 1 \\ 1 & \lambda^2 & 0 & 2\lambda - 2 \\ \lambda & 0 & 1 & 0 \\ 0 & 1 & \lambda & \lambda \end{bmatrix}$$

4. Find the invariant factors of the matrix

$$\begin{bmatrix} 0 & \lambda - 4 & 1 \\ \lambda & \lambda - 1 & 6\lambda - 7 \\ 0 & 2 & \lambda + 1 \end{bmatrix}$$

with entries in $J_5[\lambda]$.

5. Each of the following sets of polynomials in $R^{\#}[\lambda]$ is the set of elementary divisors of a polynomial matrix of rank 4. Determine the invariant factors in each case:

(a) $\lambda, \lambda - 1, \lambda - 2, \lambda - 3$
(b) $(\lambda + 1)^2, \lambda^2 + 1, (\lambda - 1)^2, (\lambda - 1)^3$
(c) $\lambda, \lambda^3, \lambda^4, (\lambda - 1)^2, (\lambda - 1)^3, \lambda + 1$

6. Compute the invariant factors and elementary divisors of the characteristic matrix of each matrix:

(a) $$\begin{bmatrix} 0 & 1 & 1 \\ 1 & 0 & 1 \\ 1 & 1 & 0 \end{bmatrix}$$

(b) $$\begin{bmatrix} -2 & -7 & 22 & 33 \\ 1 & 3 & -11 & -22 \\ 0 & 0 & -2 & -7 \\ 0 & 0 & 1 & 3 \end{bmatrix}$$

(c) $$\begin{bmatrix} 0 & 0 & 0 & 1 \\ 1 & 0 & 0 & 2 \\ 0 & 1 & 0 & 3 \\ 0 & 0 & 1 & 4 \end{bmatrix}$$

(d) $$\begin{bmatrix} 2 & 0 \\ 1 & 2 \end{bmatrix} \oplus \begin{bmatrix} 3 & 0 & 0 \\ 1 & 3 & 0 \\ 0 & 1 & 3 \end{bmatrix}$$

7.8. Canonical Representations of Linear Transformations. Having disposed of all necessary preliminaries, we are now in a position to derive the various well-known canonical representations of a linear transformation. Since our study of polynomial matrices (which was postponed until it became a necessity) has interrupted the discussion of **A**-bases, the reader is urged to review Sec. 7.6 at this time. The results of that section, when correlated with those of Sec. 7.7, form the basis of our next remarks.

Let **A** denote a linear transformation on the n-dimensional space V, $A = (\mathbf{A}; \alpha)$ a representation of **A**, and $h_1, h_2, \ldots, h_n$ the invariant factors of $\lambda I - A$. In addition, let m_i designate the degree of h_i, $i = 1, 2, \ldots, n$, and assume that

$$m_1 = m_2 = \cdots = m_{k-1} = 0 \qquad m_k > 0$$

in other words, that $\{h_k, h_{k+1}, \ldots, h_n\}$ is the set of nontrivial invariant factors of $\lambda I - A$. Interpreting $\lambda I - A$ as a complete relation matrix for the **A**-generating system $\{\alpha_1, \alpha_2, \ldots, \alpha_n\}$, the equivalence

$$\lambda I - A \overset{E}{\sim} H = \text{diag}(1, 1, \ldots, 1, h_k, \ldots, h_n)$$

implies that H is a complete relation matrix for some **A**-generating system $\{\beta_1, \beta_2, \ldots, \beta_n\}$. In addition, H being diagonal implies that $\{\beta_1, \beta_2, \ldots, \beta_n\}$ is an **A**-basis for V. Finally, since the ith diagonal term in H is the index polynomial of β_i, $i = 1, 2, \ldots, n$, we conclude that $\beta_1 = \beta_2 = \cdots = \beta_{k-1} = 0$ and consequently that $\{\beta_k, \beta_{k+1}, \ldots, \beta_n\}$ is an **A**-basis for V. It follows that V is the direct sum of the invariant spaces S_i, $i = k, k + 1, \ldots, n$, where S_i has the **A**-basis β_i and the ordinary basis indicated below:

$$(15)\qquad \begin{aligned} &V = S_k \oplus S_{k+1} \oplus \cdots \oplus S_n \\ &S_i = [\beta_i, \mathbf{A}\beta_i, \ldots, \mathbf{A}^{m_i-1}\beta_i] \qquad i = k + 1, k + 2, \ldots, n \end{aligned}$$

Our first canonical representation for **A** is simply the representation relative to the basis for V adapted to these spaces. In order to describe this result most effectively, it is desirable to introduce two definitions.

DEFINITION. *The* similarity factors *of an $n \times n$ matrix A are the invariant factors of $\lambda I - A$.*

Since $A \overset{S}{\sim} B$ implies that $\lambda I - A \overset{E}{\sim} \lambda I - B$ [indeed, if $B = PAP^{-1}$, then $\lambda I - B = \lambda I - PAP^{-1} = P(\lambda I - A)P^{-1} \overset{E}{\sim} \lambda I - A$ by Theorem 7.11], it follows from Theorem 7.10 that similar matrices have the same similarity factors. Consequently, it is possible to define the similarity factors of a linear transformation as those of any representation.

DEFINITION. *The* companion matrix *of a monic polynomial* $f(\lambda) = a_0 + a_1\lambda + \cdots + a_{m-1}\lambda^{m-1} + \lambda^m$ *is the matrix*

$$C(f) = \begin{bmatrix} 0 & 0 & \cdots & 0 & -a_0 \\ 1 & 0 & \cdots & 0 & -a_1 \\ 0 & 1 & \cdots & 0 & -a_2 \\ \cdot & \cdot & \cdots & \cdot & \cdot \\ 0 & 0 & \cdots & 1 & -a_{m-1} \end{bmatrix}$$

which has the negatives of the coefficients of f in the last column, 1's *in the diagonal just below the principal diagonal, and* 0's *elsewhere.*

The theorem below describes what we shall call the *first canonical representation of* **A**:

Theorem 7.15. A linear transformation **A** on V over F is represented by the direct sum of the companion matrices of its nontrivial similarity factors $h_k, h_{k+1}, \ldots, h_n$:

$$(\mathrm{R_I}) \qquad C(h_k) \oplus C(h_{k+1}) \oplus \cdots \oplus C(h_n)$$

Alternatively, a matrix A with nontrivial similarity factors $h_k, h_{k+1}, \ldots, h_n$ is similar to the matrix $(\mathrm{R_I})$.

Proof. We shall prove that $(\mathrm{R_I})$ represents **A** relative to the basis for V that is indicated in (15). To this end, let us denote an arbitrary β_i and its index polynomial h_i by simply β and $h = a_0 + a_1\lambda + \cdots + \lambda^m$ in order to simplify the notation. Then

$$\begin{aligned} \mathbf{A}(\mathbf{A}^j\beta) &= \mathbf{A}^{j+1}\beta \qquad 0 \leqq j < m - 1 \\ \mathbf{A}(\mathbf{A}^{m-1}\beta) &= \mathbf{A}^m\beta = (-a_0\mathbf{I} - a_1\mathbf{A} - \cdots - a_{m-1}\mathbf{A}^{m-1})\beta \\ &= -a_0\beta - a_1\mathbf{A}\beta - \cdots - a_{m-1}\mathbf{A}^{m-1}\beta \end{aligned}$$

since $h(\mathbf{A})\beta = a_0\beta + a_1\mathbf{A}\beta + \cdots + \mathbf{A}^m\beta = 0$. Thus in the space $S = [\beta, \mathbf{A}\beta, \ldots, \mathbf{A}^{m-1}\beta]$, **A** is represented by $C(h)$.

This theorem has many interesting consequences, which we shall enumerate after the next example, which illustrates the derivation of an **A**-basis and the accompanying canonical representation.

EXAMPLE 7.10

The matrix $\lambda I - A$ of Example 7.8 implicitly defines the matrix A, which we shall now assume defines a transformation $\mathbf{A}: X \to AX$ on $V = V_4(R^\#)$. Below are listed, in order of application, those row transformations applied to $\lambda I - A$ in reducing it to $\mathrm{diag}(1, 1, \lambda + 1, \lambda^3 + 1)$. Opposite each transformation appears the **A**-generating system accompanying that stage of the reduction; this **A**-generating system has been obtained from the one above it in accordance with the rules displayed near the end of Sec. 7.6.

	ϵ_1	ϵ_2	ϵ_3	ϵ_4
$R_{14}(2)$	ϵ_1	ϵ_2	ϵ_3	$\epsilon_4 - 2\epsilon_1$
$R_{24}(-1)$	ϵ_1	ϵ_2	ϵ_3	$\epsilon_4 - 2\epsilon_1 + \epsilon_2$
$R_{34}(\lambda - 4)$	ϵ_1	ϵ_2	ϵ_3	0
$R_{31}(2)$	$\epsilon_1 - 2\epsilon_3$	ϵ_2	ϵ_3	0
$R_{13}(3)$	$\epsilon_1 - 2\epsilon_3$	ϵ_2	$-3\epsilon_1 + 7\epsilon_3$	0
$R_{23}(\lambda)$	$\epsilon_1 - 2\epsilon_3$	ϵ_2	0	0
R_{24}	$\epsilon_1 - 2\epsilon_3$	0	0	ϵ_2
R_{12}	0	$\epsilon_1 - 2\epsilon_3$	0	ϵ_2
R_{23}	0	0	$\epsilon_1 - 2\epsilon_3$	ϵ_2

Thus, $\{\beta_1 = 0, \beta_2 = 0, \beta_3 = \epsilon_1 - 2\epsilon_3, \beta_4 = \epsilon_2\}$, or simply $\{\beta_3, \beta_4\}$, is an **A**-basis for V, and $V = S_3 \oplus S_4$, where S_3 has β_3 as an **A**-basis and ordinary basis, while S_4 has β_4 as an **A**-basis and $\{\beta_4, \mathbf{A}\beta_4, \mathbf{A}^2\beta_4\}$ as an ordinary basis. Hence

$$\beta_3 = \epsilon_1 - 2\epsilon_3\,,\ \beta_4 = \epsilon_2\,,\ \mathbf{A}\beta_4 = -3\epsilon_1 + 7\epsilon_3\,,\ \mathbf{A}^2\beta_4 = \epsilon_1 + \epsilon_2 - 2\epsilon_3 + \epsilon_4$$

is a basis for V relative to which **A** is represented by

$$C(\lambda + 1) \oplus C(\lambda^3 + 1) = \begin{pmatrix} -1 & 0 & 0 & 0 \\ 0 & 0 & 0 & -1 \\ 0 & 1 & 0 & 0 \\ 0 & 0 & 1 & 0 \end{pmatrix} = P^{-1} AP$$

where
$$P = \begin{pmatrix} 1 & 0 & -3 & 1 \\ 0 & 1 & 0 & 1 \\ -2 & 0 & 7 & -2 \\ 0 & 0 & 0 & 1 \end{pmatrix}$$

The corresponding representation of the transformation in Example 7.4, whose nontrivial similarity factors are

$$h_7 = \lambda - 2 \qquad h_8 = \lambda(\lambda - 2)^2\,(\lambda^2 - \lambda + 1)^2$$

is $C(h_7) \oplus C(h_8)$.

Theorem 7.16. Two $n \times n$ matrices over F are similar over F if and only if their respective similarity factors are the same.

Proof. We have already shown that the characteristic matrices of similar matrices are equivalent and hence have the same invariant factors. If, conversely, $\lambda I - A$ and $\lambda I - B$ have the same invariant factors, then A and B are both similar to the matrix ($\mathrm{R_I}$) constructed from the common nontrivial invariant factors and, therefore, similar to each other.

This theorem, with its serviceable characterization of similarity, climaxes our investigation of this relation. Moreover, we may conclude, once an agreement is made concerning the order in which the companion matrices of a set of polynomials will appear in a direct sum (*e.g*, in order of increasing size), that the matrices ($\mathrm{R_I}$) provide a canonical set under similarity. Finally, we infer that a linear transformation is determined to within similar transformations (see Sec. 7.4) by its similarity factors.

Theorem 7.17. In terms of the notation of this section, $h_i(\lambda)$ is both the minimum and characteristic function of $\mathbf{A}_i$, the transformation induced in S_i by $\mathbf{A}$, $i = k, k + 1, \ldots, n$.

Proof. We recall that $h_i(\lambda)$ is the index polynomial of β_i, hence, in particular, $h_i(\mathbf{A})\beta_i = h_i(\mathbf{A}_i)\beta_i = 0$. It follows that $h_i(\mathbf{A}_i) = \mathbf{0}$ on S_i since $h_i(\mathbf{A}_i)$ is seen to carry each basis vector $\mathbf{A}_i^j\beta_i$ for S_i into zero. Thus, if m_i denotes the minimum function of $\mathbf{A}_i$, it follows that $m_i | h_i$. On the other hand, $m_i(\mathbf{A}_i)\beta_i = 0$, and hence $h_i | m_i$. We may conclude, therefore, that $m_i = h_i$.

Next, since $C(h_i)$ represents $\mathbf{A}_i$, the characteristic function of $\mathbf{A}_i$ is $\det(\lambda I - C(h_i))$. It will follow that h_i is the characteristic function of $\mathbf{A}_i$ if we can prove that

$$\det(\lambda I - C(h)) = h(\lambda) \tag{16}$$

for any monic polynomial $h(\lambda) = \sum a_i\lambda^i$. But if in

$$\det(\lambda I - C(h)) = \det \begin{pmatrix} \lambda & 0 & \cdots & 0 & a_0 \\ -1 & \lambda & \cdots & 0 & a_1 \\ 0 & -1 & \cdots & 0 & a_2 \\ \cdot & \cdot & \cdots & \cdot & \cdot \\ 0 & 0 & \cdots & \lambda & a_{m-2} \\ 0 & 0 & \cdots & -1 & \lambda + a_{m-1} \end{pmatrix}$$

λ times each row is added to the preceding one, beginning with the last row, the desired conclusion is immediate.

Theorem 7.18. The last similarity factor, $h_n(\lambda)$, of $\mathbf{A}$ is the minimum function of $\mathbf{A}$. In matrical language, the last invariant factor of $\lambda I - A$ is the minimum function of A.

Proof. Since each similarity factor of $\mathbf{A}$ divides h_n, $h_n(\mathbf{A}) = \mathbf{0}$ on each space S_i, and consequently on V. If $m(\lambda)$ denotes the minimum function of A, we conclude that $m|h_n$. Conversely, since $m(\mathbf{A}) = \mathbf{0}$ on S_n, m is divisible by the minimum function, h_n, on S_n. Hence $m = h_n$.

Our next result follows from the preceding theorem together with the fact that the product of the similarity factors of $\mathbf{A}$ is equal to the characteristic function, $f(\lambda)$, of $\mathbf{A}$:

$$h_k h_{k+1} \cdots h_n = f \tag{17}$$

This formula is left as an exercise.

Theorem 7.19. A linear transformation is a root of its characteristic function. In matrical language, a square matrix is a root of its characteristic function (*Hamilton-Cayley theorem*).

Our next canonical representation, which is a refinement of the first, can be derived from our next result.

Lemma 7.8. Let $\mathbf{A}$ denote a linear transformation on a space S which has an $\mathbf{A}$-basis consisting of a single vector γ with index polynomial $h = fg$, where f and g are relatively prime monic polynomials. Then an $\mathbf{A}$-basis $\{\gamma_1, \gamma_2\}$ with complete relation matrix $\operatorname{diag}(f, g)$ can be found for S.

Proof. Adjoin 0 to the $\mathbf{A}$-basis γ to obtain an $\mathbf{A}$-basis $\{0, \gamma\}$ with complete relation matrix $\operatorname{diag}(1, fg)$. By assumption, there exist polynomials a, b such that $af + bg = 1$ (see Theorem 6.15). The sequence of elementary transformations $R_{21}(f)$, $R_{21}^*(-g)$, $R_{12}(-a)$, $R_{12}^*(b)$, $R_2(-1)$, R_{12} applied to the matrix $\operatorname{diag}(1, fg)$ produces the complete relation matrix $\operatorname{diag}(f, g)$ with accompanying $\mathbf{A}$-basis $\{b(\mathbf{A})g(\mathbf{A})\gamma, f(\mathbf{A})\gamma\}$.

Each of the spaces S_i of (15) has an $\mathbf{A}_i$-basis consisting of the single vector β_i, with index polynomial $h_i(\lambda)$. To relieve the notation we shall drop the subscript i for the moment. If h is decomposed into its prime power factors over F,

$$h = q_1 q_2 \cdots q_r$$

repeated use of the preceding lemma will determine a new $\mathbf{A}$-basis $\{\gamma_1, \gamma_2, \ldots, \gamma_r\}$ with complete relation matrix $\operatorname{diag}(q_1, q_2, \ldots, q_r)$. Recalling the proof of Theorem 7.15, it is clear that the representation of $\mathbf{A}$ relative to the ordinary basis for S determined by the $\mathbf{A}$-basis $\{\gamma_1, \gamma_2, \ldots, \gamma_r\}$ is the direct sum of the companion matrices of $q_1, q_2, \ldots, q_r$. Placing the subscript i back into the discussion, it is the adoption of the foregoing representation for each $\mathbf{A}_i$ that leads to the representation described in the next theorem. First, in order to avoid the introduction of further terminology, let us agree that by the *elementary divisors of a matrix A over F* shall be meant the elementary divisors over F of $\lambda I - A$ or, in other words, the collection of (highest) prime power factors of the similarity factors of A. The latter phrasing may then be applied to a linear transformation. Thus we may describe the *second canonical representation of* $\mathbf{A}$ as follows:

Theorem 7.20. In the first canonical representation of $\mathbf{A}$, the companion matrix of each similarity factor $h_i(\lambda)$ can be replaced by the direct sum of the companion matrices of the prime power factors over F of h_i. In other words, $\mathbf{A}$ is represented by the direct sum of the companion matrices of its elementary divisors, $q_1, q_2, \ldots, q_t$ over F:

$$(\mathrm{R}_{\mathrm{II}}) \qquad C(q_1) \oplus C(q_2) \oplus \cdots \oplus C(q_t)$$

Alternatively, a matrix A with elementary divisors $q_1, q_2, \ldots, q_t$ is similar to the matrix $(\mathrm{R}_{\mathrm{II}})$.

Upon reaching an agreement concerning the order of the direct summands, the matrices $(\mathrm{R}_{\mathrm{II}})$ provide us with a second canonical set for $n \times n$ matrices under similarity.

An important feature of Lemma 7.8 is that it describes the new $\mathbf{A}$-basis with relation matrix $\mathrm{diag}(f, g)$. Thus it provides a method for constructing a basis relative to which the representation $(\mathrm{R}_{\mathrm{II}})$ results. The next example illustrates this.

EXAMPLE 7.11

In Example 7.10 let $\mathbf{A}_4$ denote the transformation induced by $\mathbf{A}$ in S_4, which has β_4 as an $\mathbf{A}_4$-basis. Since the index polynomial of β_4 can be written as $(\lambda + 1)(\lambda^2 - \lambda + 1)$, it is possible to construct a new $\mathbf{A}_4$-basis $\{\gamma_1, \gamma_2\}$ with relation matrix $\mathrm{diag}(\lambda + 1, \lambda^2 - \lambda + 1)$. According to the proof of Lemma 7.8, we may choose

$$\gamma_1 = \tfrac{1}{3}(\mathbf{A}_4^2 - \mathbf{A}_4 + \mathbf{I})\beta_4 \qquad \gamma_2 = (\mathbf{A}_4 + \mathbf{I})\beta_4$$

where we have made use of the relation $(2 - \lambda)(\lambda + 1) + (\lambda^2 - \lambda + 1) = 3$. Thus $\{\gamma_1, \gamma_2, \mathbf{A}_4\gamma_2\}$ is a basis for S_4 relative to which $\mathbf{A}_4$ is represented by the matrix

$$C(\lambda + 1) \oplus C(\lambda^2 - \lambda + 1) = \begin{pmatrix} -1 & 0 & 0 \\ 0 & 0 & -1 \\ 0 & 1 & 1 \end{pmatrix} = Q^{-1}C(\lambda^3 + 1)Q$$

where
$$Q = \begin{bmatrix} \frac{1}{3} & 1 & 0 \\ -\frac{1}{3} & 1 & 1 \\ \frac{1}{3} & 0 & 1 \end{bmatrix}$$

expresses the new basis in terms of the former. Combining this result with Example 7.10, we conclude that relative to the basis $\{\beta_3, \gamma_1, \gamma_2, \mathbf{A}\gamma_2\}$ for V, $\mathbf{A}$ is represented by

$$C(\lambda + 1) \oplus C(\lambda + 1) \oplus C(\lambda^2 - \lambda + 1)$$
$$= \begin{bmatrix} -1 & 0 & 0 & 0 \\ 0 & -1 & 0 & 0 \\ 0 & 0 & 0 & -1 \\ 0 & 0 & 1 & 1 \end{bmatrix} = R^{-1}P^{-1}APR$$

where P is defined in Example 7.10 and $R = (1) \oplus Q$. This is the (R_{II}) representation of $\mathbf{A}$.

EXAMPLE 7.12

The (R_{II}) representation of the transformation in Example 7.4, which is

$$C(\lambda) \oplus C(\lambda - 2) \oplus C(\lambda - 2)^2 \oplus C(\lambda^2 - \lambda + 1)^2$$

accompanies the decomposition of the space into a direct sum of four spaces S_1, S_2, S_3, S_4 having

$$\gamma_1 = \epsilon_2 + \epsilon_3 + \epsilon_5 - \epsilon_7 \qquad \gamma_2 = 2\epsilon_1 + \epsilon_7$$
$$\gamma_3 = -\epsilon_2 - 2\epsilon_4 - \epsilon_5 - 2\epsilon_6 + \epsilon_8 \qquad \gamma_4 = \epsilon_3 - \epsilon_7$$

as their respective $\mathbf{A}$-bases.

To introduce the final canonical representation of $\mathbf{A}$, let us recall that the previous one accompanies a decomposition of V into invariant spaces S, each with an $\mathbf{A}$-basis consisting of a single vector γ with index polynomial of the form $q = p^e$, where $p = a_0 + a_1\lambda + \cdots + a_{j-1}\lambda^{j-1} + \lambda^j$ is irreducible over F. If $e > 1$, an ordinary basis may be found for S which determines a matrix that is nearer to diagonal form than $C(p^e)$. The basis that we have in mind is

$$(18) \quad \begin{aligned} &\delta = \gamma, \mathbf{A}\delta, \ldots, \mathbf{A}^{j-1}\delta, \delta_1 = p(\mathbf{A})\gamma, \mathbf{A}\delta_1, \ldots, \mathbf{A}^{j-1}\delta_1, \ldots, \\ &\qquad \delta_{e-1} = [p(\mathbf{A})]^{e-1}\gamma, \mathbf{A}\delta_{e-1}, \ldots, \mathbf{A}^{j-1}\delta_{e-1} \end{aligned}$$

The proof that this is a basis is immediate; it contains the required number of elements and if a linearly dependent set, the index polynomial of γ would be of degree less than je. Relative to this basis for S, the transformation induced in S by $\mathbf{A}$ is represented by the matrix

$$E(p^e) = \begin{bmatrix} C(p) & 0 & 0 & \cdots & 0 & 0 \\ N & C(p) & 0 & \cdots & 0 & 0 \\ 0 & N & C(p) & \cdots & 0 & 0 \\ \cdot & \cdot & \cdot & \cdots & \cdot & \cdot \\ 0 & 0 & 0 & \cdots & N & C(p) \end{bmatrix} \tag{19}$$

where N is a matrix with 1 in its upper right-hand corner and 0's elsewhere.

The only question that may arise concerning this assertion is in connection with the image under $\mathbf{A}$ of the last vector in each group. For this, let us examine $\mathbf{A}(\mathbf{A}^{j-1}\delta_i)$. We have [using the fact that $\mathbf{A}^j = -(a_0\mathbf{I} + a_1\mathbf{A} + \cdots + a_{j-1}\mathbf{A}^{j-1}) + p(\mathbf{A})$]

$$\begin{aligned} \mathbf{A}(\mathbf{A}^{j-1}\delta_i) &= \mathbf{A}^j[p(\mathbf{A})]^i\gamma \\ &= [-(a_0\mathbf{I} + a_1\mathbf{A} + \cdots + a_{j-1}\mathbf{A}^{j-1}) + p(\mathbf{A})][p(\mathbf{A})]^i\gamma \\ &= -a_0\delta_i - a_1\mathbf{A}\delta_i - \cdots - a_{j-1}\mathbf{A}^{j-1}\delta_i + \delta_{i+1} \qquad \text{if } i < e - 1 \\ &= -a_0\delta_i - a_1\mathbf{A}\delta_i - \cdots - a_{j-1}\mathbf{A}^{j-1}\delta_i \qquad \text{if } i = e - 1 \end{aligned}$$

It should be noted that $E(p^e)$ is uniquely defined by p^e in the sense that *$E(p^e)$ has p^e as its only elementary divisor.* This can be shown as follows: By the Laplace expansion theorem and equation (16)

$$\det(\lambda I - E(p^e)) = [\det(\lambda I - C(p))]^e = p^e$$

Next if we delete the first row and last column of $\lambda I - E(p^e)$, we obtain an upper triangular matrix with 1's along the principal diagonal. The determinant of this matrix is 1 so that the $(ej - 1)$st determinantal divisor is $d_{ej-1} = 1$. Hence $d_1 = d_2 = \cdots = d_{ej-1} = 1$, $d_{ej} = p^e$ and the only nontrivial invariant factor of $\lambda I - E(p^e)$ is $h_{ej} = p^e$, which proves the assertion.

This brings us to the *third canonical representation*, or the so-called *rational canonical representation* of a linear transformation, and in addition, a third canonical set for $n \times n$ matrices under similarity.

Theorem 7.21. In the second canonical representation of $\mathbf{A}$ (and A), the companion matrix $C(q)$ of an elementary divisor $q = p^e$, $e > 1$, may be replaced by a matrix of the type $E(p^e)$ in (19).

EXAMPLE 7.13

The (R_{II}) representation developed in Example 7.11 of the transformation defined in Example 7.10 is actually the rational canonical representation of that transformation, since each of its elementary divisors is simply a prime.

To find the rational canonical representation of the transformation **A** of Example 7.4 from its (R_{II}) representation in Example 7.12, we choose as new bases for S_3 and S_4, following (18),

$$\gamma_3\,,\ (\mathbf{A}-2\mathbf{I})\gamma_3 \qquad \text{and} \qquad \gamma_4\,,\ \mathbf{A}\gamma_4\,,\ (\mathbf{A}^2-\mathbf{A}+\mathbf{I})\gamma_4\,,\ \mathbf{A}(\mathbf{A}^2-\mathbf{A}+\mathbf{I})\gamma_4$$

respectively. The vectors γ_1 and γ_2, together with these six vectors, form a basis relative to which **A** is represented by

$$B = C(\lambda) \oplus C(\lambda-2) \oplus E(\lambda-2)^2 \oplus E(\lambda^2-\lambda+1)^2$$

Below, the vectors $\gamma_1, \gamma_2, \gamma_3, (\mathbf{A}-2\mathbf{I})\gamma_3, \gamma_4$, etc., appear as the columns of the matrix P; thus $P^{-1}AP = B$.

$$P = \begin{bmatrix} 0 & 2 & 0 & 0 & 0 & 0 & 0 & -1 \\ 1 & 0 & -1 & 2 & 0 & 1 & -1 & -1 \\ 1 & 0 & 0 & 0 & 1 & 0 & 0 & -2 \\ 0 & 0 & -2 & 0 & 0 & 0 & 1 & 1 \\ 1 & 0 & -1 & 1 & 0 & 1 & 0 & 0 \\ 0 & 0 & -2 & 0 & 0 & 1 & 1 & 1 \\ -1 & 1 & 0 & 0 & -1 & 0 & 0 & 2 \\ 0 & 0 & 1 & 1 & 0 & 0 & -1 & -1 \end{bmatrix}$$

$$P^{-1}AP = \begin{bmatrix} 0 & & & & & & & \\ & 2 & & & & & & \\ & & 2 & & & & & \\ & & 1 & 2 & & & & \\ & & & & 0 & -1 & & \\ & & & & 1 & 1 & & \\ & & & & & 1 & 0 & -1 \\ & & & & & & 1 & 1 \end{bmatrix}$$

where the entries that have not been filled in are all zero.

PROBLEMS

1. Determine the first canonical representation (with corresponding basis) of a transformation **A** on $V_4(R^\#)$ with the representation

$$A = \begin{pmatrix} 3 & 1 & -3 & 1 \\ -2 & -2 & 3 & 3 \\ 2 & 1 & -2 & 1 \\ -2 & -2 & 3 & 2 \end{pmatrix}$$

Follow the method of Example 7.10.

2. Verify equation (17).

3. In the proof of Theorem 7.17 it is shown directly that $h(\lambda)$ is the characteristic function of $C(h)$, and indirectly that $h(\lambda)$ is also the minimum function of $C(h)$. Give a direct proof of the latter result.

4. In the proof of Lemma 7.8 verify that the **A**-basis $\{0, \gamma\}$ becomes $\{b(\mathbf{A})g(\mathbf{A})\gamma, f(\mathbf{A})\gamma\}$ when the relation matrix diag$(1, fg)$ is changed to diag(f, g) using the transformations listed there.

5. Prove the following converse of Lemma 7.8: If S denotes a space with **A**-basis $\{\gamma_1, \gamma_2\}$ having the complete relation matrix diag(f, g), where f and g are relatively prime, then $\gamma = \gamma_1 + a(\mathbf{A})\gamma_2$ (where $af + bg = 1$) is an **A**-basis with index polynomial fg.

6. If S has γ as an **A**-basis with index polynomial fgh, where these polynomials are relatively prime in pairs, determine an **A**-basis $\{\gamma_1, \gamma_2, \gamma_3\}$ with relation matrix diag(f, g, h).

7. Determine a basis relative to which the transformation of Prob. 1 is represented by its second canonical representation.

8. Prove Theorem 7.20 by showing that the characteristic matrix of (R_{II}) has $q_1, q_2, \ldots, q_t$ as its elementary divisors.

9. Describe in detail the rational canonical representation of a linear transformation on a space over the real field $R^\#$. (HINT: An element of $R^\#[\lambda]$ can be written as a product of linear and irreducible quadratic factors.) Write out this canonical representation for the transformation whose nontrivial invariant factors are $(\lambda - 4)(\lambda^2 - 2\lambda + 3)^2$ and $(\lambda - 4)(\lambda^2 - 2\lambda + 3)^3$.

10. Determine a matrix P which transforms

$$A = \begin{pmatrix} 2 & 0 & -2 & -1 \\ -1 & -1 & 2 & 2 \\ 2 & 1 & -2 & -2 \\ -2 & -2 & 3 & 3 \end{pmatrix}$$

into its rational canonical form over $R^\#$.

7.9. The Classical Canonical Representation. If **A** is a linear transformation on a vector space V over a field F it may happen that an elementary divisor over F of **A** has the form $(\lambda - c)^e$. For example, this is true of *every* elementary divisor when F is the field of complex numbers C.

The matrix (19) in the rational canonical representation of **A** which is determined by such an elementary divisor has the form (assuming $e > 1$)

$$J_e(c) = \begin{bmatrix} c & & & & & \\ 1 & c & & & & \\ & 1 & c & & & \\ & & & \cdot & & \\ & & & & \cdot & \\ & & & & & \cdot \\ & & & & 1 & c \end{bmatrix} \tag{20}$$

with e c's along the principal diagonal, 1's on the next diagonal below, and 0's elsewhere. If $e = 1$, $J_1(c) = (c)$, the companion matrix of $\lambda - c$. We shall call a matrix of the type (20) a *Jordan matrix.* The next result is a corollary of Theorem 7.21.

Theorem 7.22. If every elementary divisor over F of a linear transformation **A** on V over F has the form $(\lambda - c)^e$, then **A** is represented by the direct sum of the Jordan matrices determined by its elementary divisors. In matrical language, if every elementary divisor over F of a square matrix A over F has the form $(\lambda - c)^e$, then A is similar to the direct sum of the Jordan matrices determined by its elementary divisors.

This is the so-called *classical canonical representation of* **A**, or the *classical* or *Jordan canonical form* of A under similarity. Of course, the similarity of a matrix of complex numbers to a classical canonical matrix holds without reservation. The same is true for a matrix A over any field F, provided one is willing to interpret A as a matrix over a suitable field including F. This remark is a consequence of the existence (which we shall not prove) of a field G containing F, such that every member of $F[\lambda]$, when regarded as a member of $G[\lambda]$, may be decomposed into linear factors. Indeed, the elementary divisors of A, as a matrix over G, have the form required for Theorem 7.22 and the assertion follows.

EXAMPLE 7.14

If the elementary divisors of **A** are $(\lambda - 4)$, $(\lambda - 3)^2$, $(\lambda - 2)^3$, then A is represented by the matrix

$$(4) \oplus \begin{bmatrix} 3 & 0 \\ 1 & 3 \end{bmatrix} \oplus \begin{bmatrix} 2 & 0 & 0 \\ 1 & 2 & 0 \\ 0 & 1 & 2 \end{bmatrix}$$

EXAMPLE 7.15

Let us find a matrix P that will transform the companion matrix $C = C(\lambda^3 - \lambda^2 - \lambda + 1)$ to its rational canonical form. The elementary divisors of $\lambda I - C$ are the prime power factors, namely, $(\lambda + 1)$ and $(\lambda - 1)^2$, of the only nontrivial invariant factor, $\lambda^3 - \lambda^2 - \lambda + 1$, of $\lambda I - C$. Thus $C \overset{S}{\sim} J_1(-1) \oplus J_2(1)$. To find P, we regard C as the representation of a transformation **C** relative to a basis for $V = V_4(R^{\#})$ determined by a **C**-basis consisting of the unit vector ϵ_1. Since $(3 - \lambda) \cdot (\lambda + 1) + (\lambda - 1)^2 = 4$, $\{\gamma_1, \gamma_2\}$, where $4\gamma_1 = (C - I)^2\epsilon_1$ and $4\gamma_2 = (C + I)\epsilon_1$, is another **C**-basis and $\{\gamma_1, \gamma_2, (C - I)\gamma_2\}$ is an ordinary basis of the type (18). Defining P as the matrix with these vectors as its columns, we have

$$4P = \begin{pmatrix} 1 & 1 & -1 \\ -2 & 1 & 0 \\ 1 & 0 & 1 \end{pmatrix} \quad \text{and} \quad P^{-1}CP = \begin{pmatrix} -1 & 0 & 0 \\ 0 & 1 & 0 \\ 0 & 1 & 1 \end{pmatrix}$$

If it is possible to represent a linear transformation **A** by a direct sum of Jordan matrices $J_e(c)$, the block pattern of this representation and the structure of **A** are known as soon as the various e's are known. A scheme that is in common usage to indicate these exponents which occur in the elementary divisors of **A** is the following: Write down within parentheses the e's associated with one characteristic value, then those associated with a second characteristic value, etc., and finally, unite all parentheses belonging to distinct characteristic values by brackets. Such an array is called the *Segre characteristic* of **A**. For example,

$$[(322)(21)(1)]$$

is the Segre characteristic of a transformation with three distinct characteristic roots a, b, and c and elementary divisors

$$(\lambda - a)^3, (\lambda - a)^2, (\lambda - a)^2, (\lambda - b)^2, (\lambda - b), (\lambda - c)$$

Thus **A** is represented by the 11×11 matrix

$$J_3(a) \oplus J_2(a) \oplus J_2(a) \oplus J_2(b) \oplus J_1(b) \oplus J_1(c)$$

which can be written directly from the characteristic.

PROBLEMS

1. Given the Segre characteristic $[(32)(411)(3)(2)]$ and the distinct characteristic roots a, b, c, d of a matrix A, determine its order, rank, the elementary divisors of $\lambda I - A$, the minimum function of A, and the classical canonical form of A.

2. Describe the Segre characteristic of a transformation which admits a diagonal representation.

3. Determine the form of all matrices which commute with a Jordan matrix. Use this result to determine the form of all matrices which commute with a matrix whose characteristic matrix has a single elementary divisor $(\lambda - c)^e$.

4. Determine the classical canonical form of the companion matrix $C = C(\lambda^4 - 4\lambda^2 + 4)$. If C represents a transformation **C** on $V_4(R^\#)$, determine a new basis relative to which **C** is represented by this canonical form.

5. Show that the Segre characteristic of a transformation **A** represented by the matrix

$$A = \begin{pmatrix} 2 & 6 & 0 & -27 & 9 & 0 \\ -2 & -10 & 1 & 50 & -18 & 0 \\ 0 & -4 & 2 & 18 & -6 & 0 \\ 0 & -4 & 0 & 20 & -6 & 0 \\ 2 & -4 & -1 & 22 & -4 & 0 \\ 2 & 1 & -1 & 0 & 1 & 2 \end{pmatrix}$$

is [(321)]. Write down the classical canonical representation of **A**, and find a corresponding basis for the space.

CHAPTER 8

UNITARY AND EUCLIDEAN VECTOR SPACES

8.1. Introduction. In our study of vector spaces in Chap. 2 one basic example was that of n-dimensional Euclidean space $V_n(R^{\#})$. Indeed Chap. 2 can be interpreted as a postulational development of a qualitative attribute of this familiar space, *viz.*, linearity. No mention has been made, for vector spaces in general, of another aspect of $V_n(R^{\#})$, the quantitative concepts of angle and length. This chapter is devoted to (i) an extension of these notions to a class of spaces (called unitary or Euclidean, according as the field of scalars is C or $R^{\#}$) which are important in physical applications and (ii) a study of linear transformations and certain functions and forms on these spaces.

We call the reader's attention to the fact that many of the definitions introduced in this chapter apply to infinite, as well as finite, dimensional spaces. Likewise, many theorems, although proved here only for finite dimensional spaces, have infinite dimensional analogues. Finally, let us remind the reader that the discussion presupposes no knowledge of Chap. 7 beyond Sec. 7.5.

A clue as to how quantitative concepts might be defined in vector spaces is suggested when it is observed that in $V_2(R^{\#})$ a single function on pairs of vectors to real numbers, called an *inner-product function* and symbolized by $(\ |\ )$, suffices for describing both lengths and angles. The function value at the pair $\{\xi, \eta\}$ where $\xi = (x^1, x^2)^T$ and $\eta = (y^1, y^2)^T$ is

$$(\xi|\eta) = x^1y^1 + x^2y^2$$

the so-called *inner product* of the two vectors ξ and η. The inner product of ξ with itself is

$$(\xi|\xi) = (x^1)^2 + (x^2)^2$$

the familiar formula for the square of the length of ξ. The length of ξ will be denoted by $||\xi||$; thus we write the preceding equation as

$$||\xi|| = [(x^1)^2 + (x^2)^2]^{\frac{1}{2}} = (\xi|\xi)^{\frac{1}{2}}$$

The notion of the distance between ξ and η is disposed of by defining it to be the length $||\xi - \eta||$.

Concerning angles, it is convenient (and adequate) to measure their cosines. If $\angle(\xi, \eta)$ denotes the angle between the nonzero vectors ξ and η, then applying the law of cosines to the triangle with sides ξ, η, $\eta - \xi$ (see Fig. 11) gives

$$||\eta - \xi||^2 = ||\xi||^2 + ||\eta||^2 - 2||\xi|| \cdot ||\eta|| \cos \angle(\xi, \eta)$$

A straightforward computation shows that $||\eta - \xi||^2 - ||\xi||^2 - ||\eta||^2 = -2(\xi|\eta)$ so that the above equation reduces to

$$\cos \angle(\xi, \eta) = \frac{(\xi|\eta)}{||\xi|| \cdot ||\eta||} \tag{1}$$

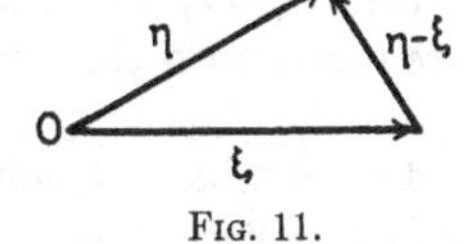

FIG. 11.

We may conclude from the foregoing that if the value $(\xi|\eta)$ of the inner-product function is known for all pairs of vectors, we can compute all lengths, distances, and angles. For lengths and distances are simply square roots of inner products of the form $(\xi|\xi)$, and, according to (1), angles can be expressed in terms of inner products and lengths, and hence in terms of inner products alone.

Unfortunately this function is inadequate for the same purposes in $V_2(C)$ if we are to preserve our intuitive notion of the positiveness of length. For example

$$||i\xi|| = (i\xi|i\xi)^{\frac{1}{2}} = -(\xi|\xi)^{\frac{1}{2}} = -||\xi|| \qquad i = \sqrt{-1}$$

so that if $\xi \neq 0$ not both of $||i\xi||$ and $||\xi||$ can be positive. An immediate remedy for this situation is to redefine $(\xi|\eta)$ as follows:

$$(\xi|\eta) = x^1\bar{y}^1 + x^2\bar{y}^2$$

where the bar denotes the complex conjugate. Then

$$(\xi|\xi) = x^1\bar{x}^1 + x^2\bar{x}^2$$

is never negative and is zero only when $\xi = 0$. However, this gain is at the expense of what appear to be very desirable properties of the earlier inner-product function. For example, whereas the earlier inner-product function is symmetric in its two arguments [that is, $(\xi|\eta) = (\eta|\xi)$ for all ξ, η], the new one is only Hermitian-symmetric [that is, $(\xi|\eta) = \overline{(\eta|\xi)}$ for all ξ, η]. This has the consequence that we cannot interchange arguments to deduce linearity in the second argument from linearity in the first [that is, $(a\xi + b\eta|\zeta) = a(\xi|\zeta) + b(\eta|\zeta)$], a property enjoyed by both functions.

However, if we were to develop the geometry of $V_2(C)$, we should find that not only is the new inner product satisfactory but that in addition all of its useful properties can be traced back to those we have already mentioned, *viz.*, Hermitian symmetry, linearity in the first argument, and positiveness. Thus a starting point for a postulational approach suggests itself: To imitate the quantitative aspects of Euclidean spaces in general vector spaces, we should assume the existence of a numerical function of ordered pairs of vectors possessing the properties of Hermitian symmetry, linearity, and positiveness. The development of this suggestion begins in the next section. We hasten to add that we shall confine our discussion of lengths, etc., to vector spaces over the real or complex field to avoid questions concerning the existence of square roots, the meaning of positiveness, and the like.

8.2. Unitary and Euclidean Vector Spaces. The following definitions are based upon the considerations of the previous section:

DEFINITION. *A* unitary vector space *is a vector space U over the complex field for which there is defined a* complex inner-product function $(\mid)$, *by which is meant a function on $U \times U$ to C possessing the following properties:*

I_1 . *Hermitian symmetry*: $(\xi|\eta) = \overline{(\eta|\xi)}$ *for all* ξ, η

I_2 . *Linearity*: $(a\xi + b\eta|\zeta) = a(\xi|\zeta) + b(\eta|\zeta)$

I_3 . *Positiveness*: $(\xi|\xi) \geqq 0$; $(\xi|\xi) = 0$ *if and only if* $\xi = 0$

A Euclidean vector space *is a vector space over the real field for which there is defined a real-valued inner-product function.*

Since an inner-product function on a Euclidean space E is real-valued, the conjugation in I_1 may be ignored to conclude that $(\mid)$ is symmetric and hence (using I_2) bilinear. Next, postulate I_3 implies that the quadratic function obtained by equating the arguments in $(\mid)$ is positive definite. If E is finite dimensional, we conclude from Theorem 5.1 (wherein we now refer both arguments to the same basis, $\{\alpha_1, \alpha_2, \ldots, \alpha_n\}$, for example) that an inner-product function is represented by a bilinear form X^TAY [where $A = ((\alpha_i|\alpha_j)) = A^T$] which reduces to a positive definite quadratic form when $Y = X$. A precise description of this form reads as follows: X^TAY is a real symmetric (I_1), bilinear (I_2 and I_1), positive definite (I_3) form. It is clear that, conversely, any such form defines an inner-product function, upon the selection of a basis.

A complex inner-product function is, in the terminology of Sec. 5.7, a Hermitian conjugate bilinear function which reduces to a positive definite Hermitian function when the arguments are equated. Also from Sec. 5.7 we recall that I_1 and I_2 imply the relation

$$(\zeta|a\xi + b\eta) = \bar{a}(\zeta|\xi) + \bar{b}(\zeta|\eta) \tag{2}$$

When U is of finite dimension, an inner-product function is represented by a Hermitian (I_1), conjugate bilinear [I_2 and (2)], positive definite (I_3) form $X^TH\bar{Y}$. Thus H is Hermitian ($H = H^*$) and positive definite. Conversely, a form $X^TH\bar{Y}$ constructed from such an H defines a complex inner-product function. The problem of finding a simple representation for such functions will be discussed later.

EXAMPLE 8.1

The space $V_n(R^*)$ becomes a Euclidean space upon defining the inner product of $\xi = (x^1, x^2, \ldots, x^n)^T$ and $\eta = (y^1, y^2, \ldots, y^n)^T$ as follows:

$$(\xi|\eta) = \sum x^i y^i$$

Again, $V_n(C)$ becomes a unitary vector space upon defining the inner product of ξ and η above as

$$(\xi|\eta) = \sum x^i \bar{y}^i$$

In the future, unless otherwise stated, we shall always assume that these inner products are used in $V_n(R^*)$ and $V_n(C)$, respectively.

EXAMPLE 8.2

The space $P(C)$ of all polynomials $\xi(t)$ in a real variable t with complex coefficients, $-1 \leqq t \leqq 1$, becomes a unitary space upon defining

$$(\xi|\eta) = \int_{-1}^{1} \xi(t)\overline{\eta(t)}\, dt$$

Many properties of unitary and Euclidean spaces can be proved simultaneously by examining only the unitary case and then specializing the field of scalars to the real field. Thus, until we reach a point in our discussion where properties peculiar to the complex field are used, we shall state definitions and theorems for unitary spaces and ask the reader to rephrase the statements for Euclidean spaces.

8.3. Schwarz's Inequality and Its Applications. This inequality is an immediate consequence of the following result, which, among other things, contains a criterion, in terms of inner products, for the linear dependence of a set of vectors:

Theorem 8.1. If $S = \{\alpha_1, \alpha_2, \ldots, \alpha_k\}$ is a finite set of vectors in a unitary space and G denotes the matrix

$$((\alpha_i|\alpha_j)) \qquad i, j = 1, 2, \ldots, k$$

then $\det G \geqq 0$, where the equality sign holds if and only if S is linearly dependent. If S is linearly independent, then G, which is Hermitian, is positive definite.

Proof. If S is linearly independent and $X = (x^1, x^2, \ldots, x^k)^T \neq 0$, then $\xi = \sum x^i \alpha_i \neq 0$ so that $0 < (\xi|\xi) = X^T G \bar{X}$. We conclude that G is positive definite, which in turn implies that $\det G > 0$. If S is linearly dependent, there exist complex numbers $c^1, c^2, \ldots, c^k$, not all zero, such that $\xi = \sum c^j \alpha_j = 0$ and consequently $(\alpha_i|\xi) = 0$, $i = 1, 2, \ldots, k$. It follows that the homogeneous system

$$(3) \quad x^1(\alpha_i|\alpha_1) + x^2(\alpha_i|\alpha_2) + \cdots + x^k(\alpha_i|\alpha_k) = 0 \qquad i = 1, 2, \ldots, k$$

has a nontrivial solution and hence that $\det G = 0$. Thus, in every case, $\det G \geqq 0$. It remains to prove that if $\det G = 0$, then S is a dependent set. With this assumption, (3) has a nontrivial solution $x^j = \bar{c}^j$, $j = 1, 2, \ldots, k$, and for $\xi = \sum c^j \alpha_j$ it follows that $(\alpha_i|\xi) = 0$, $i = 1, 2, \ldots, k$. Hence $\sum c^i(\alpha_i|\xi) = (\xi|\xi) = 0$, or $\xi = 0$. In other words, S is a dependent set.

The determinant of the matrix G in Theorem 8.1 is called the *Gramian* of the set S. Thus we may state that S is linearly dependent if and only if its Gramian vanishes.

In a unitary space, the *length*, $||\xi||$, of a nonzero vector ξ is defined as the positive square root $(\xi|\xi)^{\frac{1}{2}}$. The existence of this root is guaranteed by I_3. It should be kept in mind that length, according to this definition, is a relative notion, since it depends upon the inner-product function under consideration.

Upon specializing the previous theorem to the case $k = 2$ we obtain immediately the next theorem.

Theorem 8.2. (Schwarz's Inequality). For any pair of vectors $\{\alpha, \beta\}$ in a unitary space

$$(4) \qquad |(\alpha|\beta)| \leqq ||\alpha|| \cdot ||\beta||$$

where the equality sign holds if and only if $\{\alpha, \beta\}$ is a linearly dependent set.

Among the problems at the end of this section, several independent proofs of this theorem are outlined.

The next two theorems list minimum requirements for a length and distance function, respectively. The reader should convince himself that these are attributes of the familiar length and distance functions, respectively, in the case of ordinary Euclidean space.

Theorem 8.3. In a unitary space, length has the following properties:

L_1. $||a\xi|| = |a|\ ||\xi||$

L_2. $||\xi|| > 0$ unless $\xi = 0$

L_3. $||\xi + \eta|| \leqq ||\xi|| + ||\eta||$ (the triangular inequality)

Proof. Since $(a\xi|a\xi) = a\bar{a}(\xi|\xi)$, L_1 holds. Property L_2 is a corollary of property I_3 for an inner-product function. The proof of L_3 which follows uses Schwarz's inequality and the notation $\mathcal{R}[x]$ for the real part of the complex number x:

$$\begin{aligned} \|\xi + \eta\|^2 = (\xi + \eta|\xi + \eta) &= \|\xi\|^2 + (\xi|\eta) + (\eta|\xi) + \|\eta\|^2 \\ &= \|\xi\|^2 + 2\mathcal{R}[(\xi|\eta)] + \|\eta\|^2 \\ &\leqq \|\xi\|^2 + 2\|\xi\| \cdot \|\eta\| + \|\eta\|^2 \\ &= (\|\xi\| + \|\eta\|)^2 \end{aligned}$$

If we now define the *distance* between the vectors ξ and η of a unitary space as $\|\xi - \eta\|$, the following result holds:

Theorem 8.4. In a unitary space, distance has the following properties:
D_1 . $\|\xi - \eta\| > 0$ unless $\xi = \eta$
D_2 . $\|\xi - \eta\| = \|\eta - \xi\|$
D_3 . $\|\xi - \eta\| + \|\eta - \zeta\| \geqq \|\xi - \zeta\|$

Proof. Since $\|\xi - \xi\| = \|0\| = \|0\xi\| = 0\|\xi\| = 0$ by L_1 and $\|\xi - \eta\| > 0$ if $\xi \neq \eta$ by L_2, D_1 holds. The equation

$$\|\xi - \eta\| = \|(-1)(\eta - \xi)\| = |-1|\ \|\eta - \xi\| = \|\eta - \xi\|$$

proves D_2. Finally, D_3 follows from L_3 since

$$\|\xi - \eta\| + \|\eta - \zeta\| \geqq \|\xi - \eta + \eta - \zeta\| = \|\xi - \zeta\|$$

EXAMPLE 8.3

In the unitary space $V_n(C)$, Schwarz's inequality may be stated in the following form, which is known as *Cauchy's inequality*: For any two sets of complex numbers $\{x^1, x^2, \ldots, x^n\}$ and $\{y^1, y^2, \ldots, y^n\}$

$$\left|\sum_1^n x^i \bar{y}^i\right|^2 \leqq \sum_1^n |x^i|^2 \cdot \sum_1^n |y^i|^2$$

We have seen that if an inner-product function is defined in a vector space, a length function (or *norm*) possessing properties L_1, L_2, L_3 in Theorem 8.3 can be defined for vectors. However, if a norm, $\|\ \ \|$, satisfying L_1, L_2, L_3 is defined on a vector space, it may not be possible to find an inner-product function $(\ |\ )$, such that $(\xi|\xi)^{\frac{1}{2}} = \|\xi\|$, as the next example indicates.

EXAMPLE 8.4

Let $(\ |\ )$ denote the inner-product function for an n-dimensional Euclidean space E, and assume that a norm is defined in the usual way. If $\xi = \sum_1^n x^i \alpha_i = \alpha X$, where $\{\alpha_1, \alpha_2, \ldots, \alpha_n\}$ is a basis for E, then

$$\|\xi\|^2 = X^T A X$$

where A is the matrix whose (i, j)th entry is $(\alpha_i|\alpha_j)$. Since A is a positive definite symmetric matrix, the "unit sphere" in E, that is, the locus of the equation $||\xi|| = 1$, is an ellipsoid. On the other hand, the function

$$||\xi|| = \max (|x^1|, |x^2|, \ldots, |x^n|) \qquad \xi = \alpha X$$

is an acceptable norm in the space (in the sense that it satisfies L_1, L_2, L_3) for which the unit sphere is the "cube" whose vertices have α-coordinates $(\pm 1, \pm 1, \ldots, \pm 1)^T$. Hence, this norm cannot arise from an inner-product function.

PROBLEMS

1. Deduce Schwarz's inequality for the pair of vectors $\{\alpha, \beta\}$ from the inequality $0 \leqq ||a\alpha - b\beta||^2$, where $a = ||\beta||^2$ and $b = (\alpha|\beta)$.

2. Fill in the details of the following proof of Schwarz's inequality for vectors in a Euclidean space: The relation

$$0 \leqq ||x\alpha - y\beta||^2 = x^2||\alpha||^2 - 2xy(\alpha|\beta) + y^2||\beta||^2$$

shows that the quadratic form on the right is positive semidefinite and hence $(\alpha|\beta)^2 \leqq ||\alpha||^2 \cdot ||\beta||^2$.

3. Let α_1 and β_1 denote vectors of unit length in a Euclidean space. Expand $||\alpha_1 - \beta_1||^2$ and $||\alpha_1 + \beta_1||^2$ in turn to conclude that $(\alpha_1|\beta_1) \leqq 1$ and $(\alpha_1|\beta_1) \geqq -1$ and hence that $|(\alpha_1|\beta_1)| \leqq 1$. Deduce Schwarz's inequality for arbitrary vectors α, β by writing $\alpha = ||\alpha||\alpha_1$ and $\beta = ||\beta||\beta_1$.

4. Rewrite the formula $||\alpha + \beta||^2 + ||\alpha - \beta||^2 = 2(||\alpha||^2 + ||\beta||^2)$ as a theorem in plane geometry.

5. Let α denote a fixed vector of unit length in a unitary space. Show that the set S of all vectors β such that $(\beta|\alpha) = 1$ has only one unit vector in it and that σ in S implies $||\sigma|| \geqq 1$.

6. Let α and β denote fixed vectors in a unitary space U and T the set of all vectors ξ in U such that $||\xi - \alpha|| + ||\xi - \beta|| = ||\alpha - \beta||$. Show that for any τ in T, $\tau = a\alpha + b\beta$ where a, b are real and $a + b = 1$.

7. If $(\ |\ )$ denotes the inner-product function for a unitary space U, show that $\Re[(\ |\ )]$ is an inner-product function for the Euclidean space obtained from U by discarding all but the real scalars.

8. Given α, β such that $\mathscr{I}[(\alpha|\beta)] = 0$ (here $\mathscr{I}[x]$ is the imaginary part of the complex number x) in a unitary space, show that $(\alpha|\beta) = 0$ if and only if (i) $||\alpha + \beta|| = ||\alpha - \beta||$, or if and only if (ii) $||\alpha||^2 + ||\beta||^2 = ||\alpha \pm \beta||^2$.

9. Given α, β in a unitary space, show that $(\alpha|\beta) = 0$ if and only if (iii) $||\alpha + c\beta|| = ||\alpha - c\beta||$ for each scalar c, or if and only if (iv) $||\alpha + c\beta|| \geqq ||\alpha||$ for each scalar c.

10. The conditions of Probs. 7 and 8 are all equivalent for a Euclidean space. Consider now the Euclidean plane $V_2(R^{\#})$ with $||\xi||$ defined as in Example 8.4. Show that, with $\alpha = (1, 0)^T, \beta = (1, 2)^T, c = -\frac{1}{4}$, condition (i) is satisfied but not (iv). This gives another proof that $||\xi||$ cannot be derived from an inner product.

11. Theorem: If V is a vector space over the real field for which there is defined a norm satisfying L_1, L_2, L_3 and such that the unit sphere is an ellipsoid, then an inner product can be defined for V. Prove this for the two-dimensional case (see Example 8.4).

12. Let $[\alpha|\beta]$ be a "pseudo inner product" on a vector space V over C possessing the following properties: (i) it is real-valued and positive definite, (ii) it is real-linear in the first argument $\{[a\alpha + b\beta|\gamma] = a[\alpha|\gamma] + b[\beta|\gamma]$ for real $a, b\}$, (iii) $[\alpha|\beta] = [\beta|\alpha]$, (iv) $[\alpha|i\beta] = -[i\alpha|\beta]$. Prove that $(\alpha|\beta) = [\alpha|\beta] + i[\alpha|i\beta]$ defines an inner product in V.

13. The same as the preceding problem with (iv) replaced by $[i\alpha|i\beta] = [\alpha|\beta]$.

14. If $\alpha_i = (a_i^1, a_i^2, \ldots, a_i^n)^T, i = 1, 2, \ldots, n$, are n vectors in the unitary space $V_n(C)$, show that the Gramian of this set is equal to $|\det(a_j^i)|^2$.

8.4. Orthogonality. For a pair of nonzero vectors in a Euclidean space we may write Schwarz's inequality in the form $|(\xi|\eta)|/||\xi||\cdot||\eta|| \leqq 1$ to conclude that

$$-1 \leqq \frac{(\xi|\eta)}{||\xi|| \cdot ||\eta||} \leqq 1$$

Hence the middle term is the cosine of one and only one angle between 0 and π, which we can define as the angle between ξ and η. Our only interest in this proposal lies in the implied definition of a right angle, since we wish to formulate the concept of perpendicularity or orthogonality of two vectors.

DEFINITION. *In a unitary space a vector ξ is called* orthogonal *to a vector η (in symbols, $\xi \perp \eta$) if and only if $(\xi|\eta) = 0$. The vector ξ is called orthogonal to a subspace S (in symbols, $\xi \perp S$) if $\xi \perp \sigma$, all σ in S. A subspace T is called orthogonal to S (in symbols, $T \perp S$) if and only if $\tau \perp S$, all τ in T.*

Orthogonality is a symmetric relation since $(\xi|\eta) = 0$ implies $(\eta|\xi) = 0$ using I_1. Next, if $\xi \perp \eta$, then $a\xi \perp b\eta$ for all scalars a and b. Moreover, if $\xi \perp \eta$ and $\xi \perp \zeta$, then $\xi \perp (\eta + \zeta)$. The next theorem follows from these simple facts.

Theorem 8.5. If in a unitary space every member of the set $\{\xi_1, \xi_2, \ldots, \xi_r\}$ is orthogonal to every member of the set $\{\eta_1, \eta_2, \ldots, \eta_s\}$, then the space spanned by the ξ_i's is orthogonal to that spanned by the η_j's.

Many important theorems concerning unitary spaces stem from the following fundamental property:

Theorem 8.6. If S is a finite dimensional subspace of a unitary space U, each vector ξ in U can be expressed as a sum $\xi = \xi_S + \eta$, where ξ_S, the so-called (*orthogonal*) *projection of ξ on S*, is in S and η is orthogonal to S. Such a decomposition of ξ has the following properties:

(i) It is unique.
(ii) The Pythagorean theorem holds: $||\xi||^2 = ||\xi_S||^2 + ||\eta||^2$.
(iii) ξ_S is the best approximation, in the sense of length, to ξ in S.

Proof. The existence of a decomposition of the desired kind, which we shall establish first, is trivial if ξ is in S. For the case where ξ is not in S we present a proof by induction on the dimension of S. If $S = [\alpha]$ has dimension 1, then

$$\xi - \frac{(\xi|\alpha)}{(\alpha|\alpha)}\alpha \perp \alpha$$

as required. Assume the existence of an orthogonal projection on subspaces of U of dimension less than n, and consider the n-dimensional space $S = [\alpha_1, \alpha_2, \ldots, \alpha_n]$. By the induction assumption the projection of α_n on $T = [\alpha_1, \alpha_2, \ldots, \alpha_{n-1}]$ exists: $\alpha_n = \tau + \sigma$, where τ is in T and σ is orthogonal to T. Then

$$S = [\alpha_1, \alpha_2, \ldots, \alpha_{n-1}, \sigma] \qquad \text{where } \sigma \perp T = [\alpha_1, \alpha_2, \ldots, \alpha_{n-1}]$$

With S written in this form we project ξ on T to obtain $\xi = \xi_T + \zeta$ and, in turn, project ζ on $[\sigma]$ to get $\zeta = c\sigma + \eta$. Hence (see Fig. 12 for this construction) $\xi = (\xi_T + c\sigma) + \eta$, which is a decomposition of the required type. Clearly $\xi_T + c\sigma$ is in S. To prove that $\eta \perp S$, it is sufficient to verify that $\eta \perp \alpha_i$, $i = 1, 2, \ldots, n-1$, and $\eta \perp \sigma$. But $\eta \perp \sigma$ by construction, and since $\eta = (\xi - \xi_T) - c\sigma$, where $\xi - \xi_T = \zeta \perp \alpha_i$ and $\sigma \perp \alpha_i$, it follows that $\eta \perp \alpha_i$, $i = 1, 2, \ldots, n-1$. This completes the induction step and consequently the proof of the existence of the projection of a vector on a subspace.

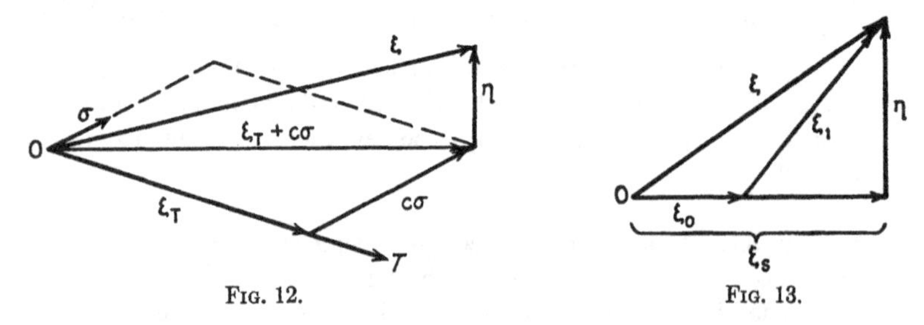

FIG. 12. FIG. 13.

If along with $\xi = \xi_S + \eta$ we have $\xi = \xi_S' + \eta'$ with ξ_S' in S and $\eta' \perp S$, then $\xi_S - \xi_S' = \eta' - \eta$, where the vector on the left is in S and that on the right is orthogonal to S. Only the zero vector can simultaneously satisfy these requirements and the desired uniqueness follows.

As for (ii), we have

$$\|\xi\|^2 = (\xi_S + \eta|\xi_S + \eta) = \|\xi_S\|^2 + (\xi_S|\eta) + (\eta|\xi_S) + \|\eta\|^2$$
$$= \|\xi_S\|^2 + \|\eta\|^2$$

Finally, to establish (iii), assume that $\xi = \xi_0 + \xi_1$, where ξ_0 is in S. We shall compare the lengths of ξ_1 and η (see Fig. 13). Since

$$\xi_1 = \xi - \xi_0 = (\xi_S + \eta) - \xi_0 = (\xi_S - \xi_0) + \eta$$

where $\eta \perp (\xi_S - \xi_0)$, the Pythagorean theorem may be applied to obtain

$$||\xi_1||^2 = ||\xi_S - \xi_0||^2 + ||\eta||^2$$

From this equation it is clear that if $\xi_0 \neq \xi_S$, $||\xi_1|| > ||\eta||$.

The reader will notice that the notion of the orthogonal projection of a vector ξ of a unitary space U on a subspace S of U is an immediate generalization of the familiar notion of projection in ordinary Euclidean space.

The set $S^\perp$ of all vectors in U orthogonal to S is easily seen to form a subspace which we call the *orthogonal complement* of S. The uniqueness of the representation of every vector in the form $\xi = \xi_S + \eta$ implies that

$$U = S \oplus S^\perp$$

There is associated with this decomposition of U the projection $\mathbf{E}$ on S along $S^\perp$ (see Sec. 6.6), which it is appropriate to label the *perpendicular projection* on S, under these circumstances. Observe that we need not specify the summand $S^\perp$ in describing $\mathbf{E}$ since it is uniquely determined by S. Later we shall derive a characterization of this type of projection.

PROBLEMS

1. In a unitary space let $\{\alpha, \beta, \gamma\}$ denote a linearly dependent set of vectors, where $\gamma \perp \alpha$ and $\gamma \perp \beta$. Show that $\{\alpha, \beta\}$ is a dependent set.

2. Find the projection of $\alpha = (0, 0, 2)$ on the subspace of $V_3(R^\#)$ spanned by $\beta = (1, 2, 1)$.

3. Use the method of proof of Theorem 8.6 to find the projection ξ_S of $\xi = (1, 2, 1, 1)$ on the subspace $S = [(0, 1, 1, 0), (0, 0, 1, 1)]$ of $V_4(R^\#)$. Determine $||\xi_S||$.

4. Find the orthogonal complement of the space S in Prob. 3.

5. If $S^\perp$ denotes the orthogonal complement of S in U, show that $(S^\perp)^\perp = S$.

6. Let S and T denote subspaces of a unitary space. Prove that $(S + T)^\perp = S^\perp \cap T^\perp$ and $(S \cap T)^\perp = S^\perp + T^\perp$.

8.5. Orthonormal Sets and Bases. In the Euclidean space $V_n(R^\#)$ the unit vectors ϵ_i, $i = 1, 2, \ldots, n$, have unit length and are mutually orthogonal, properties which often give them a preference in selecting a basis for $V_n(R^\#)$. It is one of our most important results that every finite dimensional unitary space has such a basis.

DEFINITION. *A set N of vectors in a unitary space is called an* orthogonal set *if each member is orthogonal to every other member. An orthogonal set where every vector has unit length is called an* orthonormal set.

Thus N is an orthonormal set of vectors if and only if for every pair $\{\xi, \eta\}$ of vectors in N, $(\xi|\eta) = 0$ or 1 according as $\xi \neq \eta$ or $\xi = \eta$. If N is a finite set $\{\xi_1, \xi_2, \ldots, \xi_n\}$, this criterion can be written as $(\xi_i|\xi_j) = \delta_{ij}$.

Every orthonormal set N is linearly independent since a relation $\sum a^i\xi_i = 0$ among the members of a finite subset $\{\xi_1, \xi_2, \ldots, \xi_k\}$ of N implies that

$$0 = (\sum_i a^i\xi_i|\xi_j) = \sum_i a^i(\xi_i|\xi_j) = \sum_i a^i\delta_{ij} = a^j$$

Consequently if U is a finite dimensional unitary space, the number of vectors in an orthonormal set does not exceed $d[U]$.

DEFINITION. *An orthonormal set N in a unitary space U is called* complete *if the zero vector alone is orthogonal to every member of N.*

An immediate criterion for the completeness of an orthonormal set N and one which motivates the usage of the phrase "maximal orthonormal set" in place of "complete orthonormal set," reads: N is complete if and only if it is contained in no larger orthonormal set. The reader should verify this as well as the following remark: In a finite dimensional unitary space the notions of an *orthonormal basis, i.e.*, a basis composed of an orthonormal set of vectors, and a complete orthonormal set are equivalent. Before presenting the theorem that asserts the existence of both, we mention the formula for the projection ξ_S of a vector on a space S spanned by a set $\{\beta_1, \beta_2, \ldots, \beta_m\}$ of orthonormal vectors:

$$\xi_S = \sum_i (\xi|\beta_i)\beta_i \tag{5}$$

To prove this, it is necessary only to observe that

$$(\xi - \xi_S|\beta_j) = (\xi|\beta_j) - \sum_i (\xi|\beta_i)(\beta_i|\beta_j) = (\xi|\beta_j) - (\xi|\beta_j) = 0$$

so that $(\xi - \xi_S) \perp S$. Since $(\xi|\beta_i)\beta_i$ is the projection of ξ on β_i (the reader should verify this in the case of ordinary Euclidean space), formula (5) expresses the projection of ξ on $S = [\beta_1, \beta_2, \ldots, \beta_m]$ as the sum of the projections of ξ on the various basis elements for S.

From (5) we also deduce that

$$||\xi_S||^2 = \sum_i (\xi|\beta_i)\overline{(\xi|\beta_i)} = \sum_i |(\xi|\beta_i)|^2$$

so that for any vector ξ and any orthonormal set $\{\beta_1, \beta_2, \ldots, \beta_m\}$ in U

$$\sum_i |(\xi|\beta_i)|^2 \leqq ||\xi||^2 \tag{6}$$

This is known as *Bessel's inequality.*

Theorem 8.7. A finite dimensional unitary space U has an orthonormal basis.

Proof (Gram-Schmidt construction). Let $\{\alpha_1, \alpha_2, \ldots, \alpha_n\}$ denote any basis of U. We shall construct an orthonormal basis $\{\beta_1, \beta_2, \ldots, \beta_n\}$ with the property that each β_j is a linear combination of $\alpha_1, \alpha_2, \ldots, \alpha_j$. For $j = 1$ define $\beta_1 = \alpha_1/||\alpha_1||$. Suppose that $\{\beta_1, \beta_2, \ldots, \beta_r\}$ have been found such that they form an orthonormal set and each β_j is a linear combination of $\alpha_1, \alpha_2, \ldots, \alpha_j$, $j = 1, 2, \ldots, r$. In order to determine β_{r+1} we subtract the projection $\sum_1^r(\alpha_{r+1}|\beta_i)\beta_i$ of α_{r+1} on $[\beta_1, \beta_2, \ldots, \beta_r]$ from α_{r+1} to obtain the nonzero vector

$$\eta = \alpha_{r+1} - \sum_1^r (\alpha_{r+1}|\beta_i)\beta_i$$

orthogonal to β_j, $j = 1, 2, \ldots, r$. Moreover, this vector is a linear combination of $\alpha_1, \alpha_2, \ldots, \alpha_{r+1}$. If we define $\beta_{r+1} = \eta/||\eta||$, $\{\beta_1, \beta_2, \ldots, \beta_{r+1}\}$ is an orthonormal set with the desired property and the induction step is established.

In a numerical example the computations are usually simplified if the above construction is modified to first finding merely an orthogonal basis and then normalizing each vector. The following example illustrates this procedure:

EXAMPLE 8.5

In $E = V_4(R^\#)$, let $S = [\alpha_1, \alpha_2]$ and $T = [\alpha_3, \alpha_4]$, where

$\alpha_1 = (1, 1, 0, 0)$, $\alpha_2 = (0, 0, 1, -1)$, $\alpha_3 = (1, 1, 1, 0)$, $\alpha_4 = (0, 0, 0, 1)$

We shall find an orthonormal basis for E such that certain subsets of this basis will form bases for $S \cap T$, S, $S + T$, and $S^\perp$. Clearly $d[S] = d[T] = 2$. Since $\alpha_1 + \alpha_2 - \alpha_3 + \alpha_4 = 0$ and $\{\alpha_1, \alpha_2, \alpha_3\}$ is a linearly independent set, $d[S + T] = 3$. Hence $d[S \cap T] = 1$. Rewriting $\alpha_1 + \alpha_2 - \alpha_3 + \alpha_4 = 0$ as $\alpha_1 + \alpha_2 = \alpha_3 - \alpha_4$ determines a vector $\beta_1 = (1, 1, 1, -1)$ in both S and T which we may take as a generator for $S \cap T$.

Since β_1 is in S and not a multiple of α_1, $\{\beta_1, \alpha_1\}$ is a basis for S; we propose to replace α_1 by a vector in S orthogonal to β_1. The projection of α_1 on β_1 has the form $c\beta_1$, so that there exists a vector of the form $(\alpha_1 - c\beta_1) \perp \beta_1$. This condition implies that $c = \frac{1}{2}$. Thus $\alpha_1 - \frac{1}{2}\beta_1$, or (multiplying by 2 to avoid fractions)

$$\beta_2 = 2\alpha_1 - \beta_1 = (1, 1, -1, 1)$$

is orthogonal to β_1 and $\{\beta_1, \beta_2\}$ is a basis for S.

Next we look for a vector β_3 in $S + T$ that is orthogonal to S. We may begin with α_3, which is in $S + T$ but not in S. If we subtract from α_3 its projection on S, we shall get a vector orthogonal to S. Thus we shall choose the scalars c and d so that $(\alpha_3 - c\beta_1 - d\beta_2|\beta_i) = 0$, $i = 1, 2$. We find that $c = \frac{3}{4}$ and $d = \frac{1}{4}$; consequently

$$\alpha_3 - \tfrac{3}{4}\beta_1 - \tfrac{1}{4}\beta_2 = (0, 0, \tfrac{1}{2}, \tfrac{1}{2})$$

is a vector in $S + T$ that is orthogonal to S. Multiplying by 2, we set $\beta_3 = (0, 0, 1, 1)$.

Finally we seek a vector $\beta_4 = (x^1, x^2, x^3, x^4)$ orthogonal to $S + T$. The constraints on these components corresponding to the conditions $(\beta_i|\beta_4) = 0$, $i = 1, 2, 3$ are

$$x^1 + x^2 + x^3 - x^4 = 0 \qquad x^1 + x^2 - x^3 + x^4 = 0 \qquad x^3 + x^4 = 0$$

or

$$x^1 = -x^2 \qquad x^3 = x^4 = 0$$

Setting $x^1 = 1$, we have $\beta_4 = (1, -1, 0, 0)$. Then $\{\beta_1, \beta_2, \beta_3, \beta_4\}$ is an orthogonal basis for E, and $\{\beta_3, \beta_4\}$ is a basis for $S^\perp$. If desired, this basis can be normalized by dividing each vector by the square root of the sum of the squares of the components.

Now that the existence of an orthonormal basis is settled, we return to (5) to conclude that if $\{\beta_1, \beta_2, \ldots, \beta_n\}$ is an orthonormal basis for U then

$$\xi = \sum_{i=1}^{n} (\xi|\beta_i)\beta_i$$

or *the coordinates of* ξ *relative to an orthonormal basis are the inner products of* ξ *with the various basis elements.* Moreover, (6) now reads

$$||\xi||^2 = \sum_{i=1}^{n} |(\xi|\beta_i)|^2 \tag{7}$$

which is known as *Parseval's equation.* This equation yields another criterion, as the reader may verify, for the completeness of an orthonormal set: Completeness is equivalent to the existence of Parseval's equation for every vector in the space. This criterion can be extended to the case of infinite dimensional spaces.

We close this section with two further illustrations of the merits of an orthonormal basis. We have mentioned (see Sec. 8.2) that the inner-product function defined in a unitary space U is represented by a form $X^T H \overline{Y}$ once a basis is chosen. If $\{\alpha_1, \alpha_2, \ldots, \alpha_n\}$ is the basis and $\xi = \alpha X$, $\eta = \alpha Y$, then

$$(\xi|\eta) = X^T H \overline{Y} \qquad \text{where } H = (h_{ij}),\ h_{ij} = (\alpha_i|\alpha_j)$$

If the basis is orthonormal, it is observed that $h_{ij} = \delta_{ij}$ or $H = I$ and the inner-product function is represented by $X^T \overline{Y}$. In other words, relative to an orthonormal basis, the inner-product function assumes its simplest form:

$$\text{If } \xi = \sum x^i \alpha_i \text{ and } \eta = \sum y^i \alpha_i\text{, then } (\xi|\eta) = \sum x^i \overline{y}^i$$

Finally, in order to indicate the abundance of orthonormal bases in a unitary space, we remark that the superdiagonal representation of a linear transformation, whose existence is assured by Theorem 7.9, can always be realized relative to an orthonormal basis in the case of a unitary space. The proof of this is postponed until later (Theorem 8.14) so that the matrical interpretation can be stated in its most useful form.

PROBLEMS

1. Verify the following statement made in the text: An orthonormal set N is complete if and only if it is contained in no larger orthonormal set.

2. Show that in a finite dimensional unitary space, N is a complete orthonormal set if and only if it is an orthonormal basis.

3. Show that Parseval's equation (7) holds for an orthonormal set N if and only if N is complete.

4. Given an orthonormal set of vectors $\{\xi_1, \xi_2, \ldots, \xi_k\}$ in a unitary space U together with any set of k complex numbers $c_1, c_2, \ldots, c_k$, show that there exists a vector α in U such that $(\alpha|\xi_i) = c_i$, $i = 1, 2, \ldots, k$.

5. Let $\{\xi_1, \xi_2, \ldots, \xi_k\}$ denote an orthonormal set of vectors in a unitary space U, and let M denote the set of all vectors α in U such that $||\alpha||^2 = \sum(\alpha|\xi_i)^2$. Prove that M is a subspace of U.

6. Find an orthonormal basis for the subspaces $S = [(1, 2, 0, 0), (0, 1, 1, 0), (0, 1, 1, 1)]$ of $V_4(R^\#)$. Extend this basis to an orthonormal basis for $V_4(R^\#)$. Find the coordinates of $(1, 3, 1, 0)$ relative to this basis.

7. In the proof of Theorem 8.7, show that the members of the orthonormal basis obtained from the basis $\{\alpha_1, \alpha_2, \ldots, \alpha_n\}$ can be expressed as follows:

$$\beta_k = (g_{k-1}g_k)^{-\frac{1}{2}} \det \begin{bmatrix} (\alpha_1|\alpha_1) & (\alpha_1|\alpha_2) & \cdots & (\alpha_1|\alpha_{k-1}) & \alpha_1 \\ (\alpha_2|\alpha_1) & (\alpha_2|\alpha_2) & \cdots & (\alpha_2|\alpha_{k-1}) & \alpha_2 \\ \cdot & \cdot & \cdots & \cdot & \cdot \\ (\alpha_k|\alpha_1) & (\alpha_k|\alpha_2) & \cdots & (\alpha_k|\alpha_{k-1}) & \alpha_k \end{bmatrix}$$

where $g_0 = 1$ and g_k denotes the Gramian of $\{\alpha_1, \alpha_2, \ldots, \alpha_k\}$, $k = 1, 2, \ldots, n$.

8. Use the formula of the preceding problem to derive an orthonormal basis for the subspace $[1, t, t^2]$ of the unitary space defined in Example 8.2.

9. Show that the Gram-Schmidt construction may be applied to deduce an orthonormal basis for a unitary space having a countably infinite basis. If this construction is applied to the basis $\{1, t, t^2, \ldots, t^n, \ldots\}$ for the space of Example 8.2, the $(n + 1)$th term of the resulting sequence is the product of $(n + \frac{1}{2})^{\frac{1}{2}}$ by the polynomial of degree n which is known as the Legendre polynomial of degree n.

8.6. Unitary Transformations. In view of our interest in transformations on vector spaces, it is natural to inquire for those transformations on a unitary space U which preserve the inner-product function $(\ |\ )$ defined on U, since such transformations will a fortiori preserve length and orthogonality.

DEFINITION. *A transformation* **T** *on a unitary space* U *is called* unitary *if it preserves the inner-product function in* U:

$$(\mathbf{T}\xi|\mathbf{T}\eta) = (\xi|\eta) \qquad \textit{all } \xi, \eta \textit{ in } U$$

The next theorem provides some insight into the structure of such transformations.

Theorem 8.8. The following three conditions on a transformation **T** on a unitary space U are equivalent to each other:

(i) **T** is unitary.

(ii) **T** is linear and length-preserving (where length-preserving means $||\mathbf{T}\xi|| = ||\xi||$, all ξ in U).

(iii) **T** preserves the origin ($\mathbf{T}0 = 0$) and distances ($||\mathbf{T}\xi - \mathbf{T}\eta|| = ||\xi - \eta||$, all ξ, η in U).

Proof. If **T** is unitary, it obviously preserves lengths, which are square roots of inner products. To show that **T** is linear, we must prove that

$$\mathbf{T}(a\xi) = a\mathbf{T}\xi \qquad \mathbf{T}(\xi + \eta) = \mathbf{T}\xi + \mathbf{T}\eta$$

for all a, ξ, η. To verify the first equation, it is sufficient to show that $||\mathbf{T}(a\xi) - a\mathbf{T}\xi||^2 = 0$. This is done below:

$$\begin{aligned} ||\mathbf{T}(a\xi) - a\mathbf{T}\xi||^2 &= ||\mathbf{T}(a\xi)||^2 - \bar{a}(\mathbf{T}a\xi|\mathbf{T}\xi) - a(\mathbf{T}\xi|\mathbf{T}a\xi) + a\bar{a}||\mathbf{T}\xi||^2 \\ &= ||a\xi||^2 - \bar{a}(a\xi|\xi) - a(\xi|a\xi) + a\bar{a}||\xi||^2 \\ &= a\bar{a}(\xi|\xi) - a\bar{a}(\xi|\xi) - a\bar{a}(\xi|\xi) + a\bar{a}(\xi|\xi) = 0 \end{aligned}$$

The additivity of **T** is shown similarly. Thus we conclude that (i) implies (ii).

If (ii) holds for **T**, then clearly $\mathbf{T}0 = 0$ and $||\mathbf{T}\xi - \mathbf{T}\eta|| = ||\mathbf{T}(\xi - \eta)|| = ||\xi - \eta||$, which is (iii).

Finally, if (iii) holds, we deduce (i) from the identity

$$2\mathcal{R}[(\xi|\eta)] = ||\xi||^2 + ||\eta||^2 - ||\xi - \eta||^2 \tag{8}$$

First we remark that (iii) implies that **T** preserves lengths, since $||\mathbf{T}\xi|| = ||\mathbf{T}\xi - \mathbf{T}0|| = ||\xi - 0|| = ||\xi||$. Thus, if ξ and η are replaced by $\mathbf{T}\xi$ and $\mathbf{T}\eta$ in the right side of (8), it is unchanged, member for member. It follows that $\mathcal{R}[(\mathbf{T}\xi|\mathbf{T}\eta)] = \mathcal{R}[(\xi|\eta)]$ and, therefore, that $\mathcal{I}[(\xi|\eta)] = \mathcal{R}[-i(\xi|\eta)]$ is preserved by **T**. Hence $(\mathbf{T}\xi|\mathbf{T}\eta) = (\xi|\eta)$ as desired.

We call to the reader's attention that the above theorem shows that a unitary transformation is necessarily linear.

By virtue of (iii) above, a unitary transformation may be called a *rotation*. A transformation on a unitary space which merely preserves distance is named a *rigid motion*. Thus a rotation is a rigid motion. The same is true of a translation (see Prob. 10, Sec. 6.3) and, consequently, of the product of a rotation and a translation. That every rigid motion is of this form is proved next.

Theorem 8.9. A rotation followed by a translation is a rigid motion; conversely, every rigid motion is of this form.

Proof. It remains to prove the second assertion. Let $\mathbf{M}$ denote a rigid motion, and set $\mathbf{M}0 = \alpha$. We define the transformation $\mathbf{R}$ on the space U by the equation

$$\mathbf{R}\xi = \mathbf{M}\xi - \alpha \qquad \text{all } \xi \text{ in } U$$

Then $\mathbf{M} = \mathbf{T}_\alpha\mathbf{R}$, where $\mathbf{R}$ is a rotation and $\mathbf{T}_\alpha$ is the translation with defining equation

$$\mathbf{T}_\alpha\eta = \eta + \alpha \qquad \text{all } \eta \text{ in } U$$

In addition to preserving lengths, distance, and orthogonality, the unitary transformations possess additional important properties, which we shall now discuss.

Theorem 8.10. A unitary transformation is nonsingular. The set of all unitary transformations on a unitary space U forms a multiplicative group, the so-called *unitary group*. The unitary group is a subgroup of the full linear group, $L_n(C)$.

Proof. If $\mathbf{T}$ is unitary, $N(\mathbf{T}) = O$, which implies that $\mathbf{T}$ is nonsingular (Theorem 6.8). The verification of the group properties is left as an exercise.

Theorem 8.11. The image of an orthonormal basis of a unitary space U under a unitary transformation is an orthonormal basis. Conversely, if the linear transformation $\mathbf{T}$ on U is such that the image of some one orthonormal basis is again an orthonormal basis, then $\mathbf{T}$ is unitary.

Proof. Since a unitary transformation preserves length and orthogonality, it is clear that it carries an orthonormal basis into another. For the converse, suppose that $\{\alpha_1, \alpha_2, \ldots, \alpha_n\}$ is the orthonormal basis which is carried into another by $\mathbf{T}$; then $(\mathbf{T}\xi|\mathbf{T}\eta) = (\xi|\eta)$ for ξ, η in the set $\{\alpha_1, \alpha_2, \ldots, \alpha_n\}$. Using the linearity of $\mathbf{T}$ gives the same conclusion for all ξ, η in V.

The advantages (*e.g.*, simplicity of the inner-product function) that are offered by orthonormal bases for unitary spaces could have led us to inquire for those linear transformations which preserve the characteristic properties of such bases. The above theorem states that we would have been led to precisely those transformations that we have labeled unitary.

Another important consequence of this theorem can be stated as soon as some facts are known about the representations of unitary transformations. The representations relative to orthonormal bases are easily determined.

Theorem 8.12. Relative to an orthonormal basis for a unitary space U, a matrix $A = (a_j^i)$ represents a unitary transformation $\mathbf{T}$ if and only if its column vectors A_j, interpreted as elements in $V_n(C)$, form an orthonormal set: $A_i^{\mathrm{T}}\overline{A}_j = \sum_k a_i^k \overline{a}_j^k = \delta_{ij}$.

Proof. If $\mathbf{T}$ is represented by A relative to the orthonormal basis $\{\alpha_1, \alpha_2, \ldots, \alpha_n\}$, then

$$(9) \qquad \mathbf{T}\alpha_j = a_j^1\alpha_1 + a_j^2\alpha_2 + \cdots + a_j^n\alpha_n \qquad j = 1, 2, \ldots, n$$

Since $\{\mathbf{T}\alpha_1, \mathbf{T}\alpha_2, \ldots, \mathbf{T}\alpha_n\}$ is an orthonormal basis (Theorem 8.11), we have $(\mathbf{T}\alpha_i|\mathbf{T}\alpha_j) = \delta_{ij}$, which, when expanded, gives the desired conclusion.

For the converse, we begin with the matrix A and the orthonormal α-basis and consider the transformation $\mathbf{T}$ defined by (9). The assumptions concerning A imply that $(\mathbf{T}\alpha_i|\mathbf{T}\alpha_j) = \delta_{ij}$. Hence $\{\mathbf{T}\alpha_1, \mathbf{T}\alpha_2, \ldots, \mathbf{T}\alpha_n\}$ is an orthonormal basis, and consequently $\mathbf{T}$ is unitary by Theorem 8.11. This completes the proof.

The conditions on the matrix A of the previous theorem can be collected in the matrix equation $A^{\mathrm{T}}\overline{A} = I$, or its equivalent $A^*A = I$, where, as usual, $A^* = \overline{A}^{\mathrm{T}}$. It follows that A^* is a left inverse of A and, hence, that A is nonsingular (which, of course, is implied by Theorem 8.10) with inverse $A^{-1} = A^*$. Therefore, $AA^* = I$, or the rows of A form an orthonormal set. Since our argument is reversible, it is clear that, conversely, the orthogonality of the rows of A implies the orthogonality of its columns.

DEFINITION. *A matrix A over the complex field is called* unitary *if AA^* $(= A^*A) = I$.*

The proof of the next theorem is left as an exercise.

Theorem 8.13. If A is a unitary matrix, then A^{T}, $\overline{A}$, A^*, and A^{-1} are all unitary; moreover, $|\det A| = 1$. The set of all $n \times n$ unitary matrices forms a multiplicative group.

We can now state the implication of Theorem 8.11 that was promised earlier. It is this: In the light of Sec. 6.5, the study of the simplifications that are possible relative to orthonormal bases, in the equations of figures defined by Hermitian forms, is the study of Hermitian forms under the unitary group. In turn, according to Theorem 8.12, this amounts to the reduction of Hermitian forms $X^T H \bar{X}$ by means of *unitary substitutions, i.e.*, substitutions $X = PY$ where P is unitary. This topic is discussed in Sec. 8.9.

We conclude this section with the proof of a refinement for unitary spaces of Theorem 7.9 which guarantees the existence of a superdiagonal representation for a linear transformation on a vector space over the complex field.

Theorem 8.14. For a linear transformation $\mathbf{T}$ on an n-dimensional unitary space U, there exists an orthonormal basis $\{\beta_1, \beta_2, \ldots, \beta_n\}$ such that $(\mathbf{T}; \beta)$ is a superdiagonal representation. In matrical language, for any $n \times n$ matrix A over C, there exists a unitary matrix P such that PAP^{-1} is superdiagonal.

Proof. Assume that $\mathbf{T}$ is represented by a superdiagonal matrix relative to the α-basis. If this is not orthonormal, the Gram-Schmidt process may be used to construct an orthonormal basis $\{\beta_1, \beta_2, \ldots, \beta_n\}$, where β_j is a linear combination of $\alpha_1, \alpha_2, \ldots, \alpha_j$, $j = 1, 2, \ldots, n$ (consult the proof of Theorem 8.7 for this). It follows that the representation $(\mathbf{T}; \beta)$ is still superdiagonal.

Concerning the final assertion of the theorem, let A define the transformation $\mathbf{T}$ relative to the orthonormal α-basis. Now $\mathbf{T}$ has a superdiagonal representation relative to some orthonormal β-basis. If $\alpha = \beta P$, then P is unitary and $(\mathbf{T}; \beta) = PAP^{-1}$.

The next example illustrates this theorem.

EXAMPLE 8.6

In Example 7.7 a superdiagonal representation, which we now label B, was found for a transformation $\mathbf{A}$, initially defined by a representation A relative to the ϵ-basis for $V_4(C)$:

$$A = \begin{bmatrix} 8 & 9 & -9 & 0 \\ 0 & 2 & 0 & 2 \\ 4 & 6 & -4 & 0 \\ 0 & 0 & 0 & 3 \end{bmatrix} \qquad B = \begin{bmatrix} 2 & 0 & -6 & 2 \\ 0 & 2 & -3 & 0 \\ 0 & 0 & 2 & -2 \\ 0 & 0 & 0 & 3 \end{bmatrix} = (PQ)^{-1}A(PQ)$$

where

$$PQ = \begin{bmatrix} 0 & 3 & 0 & 0 \\ 1 & -2 & 0 & 0 \\ 1 & 0 & 1 & 0 \\ 0 & 0 & 0 & 1 \end{bmatrix}$$

The basis accompanying B we now label

$$\beta_1 = (0, 1, 1, 0)^T \qquad \beta_2 = (3, -2, 0, 0)^T$$
$$\beta_3 = (0, 0, 1, 0)^T \qquad \beta_4 = (0, 0, 0, 1)^T$$

The following orthonormal basis has been constructed from the β-basis by the Gram-Schmidt process:

$$\gamma_1 = \frac{\beta_1}{\sqrt{2}}, \gamma_2 = \frac{(\beta_1 + \beta_2)}{\sqrt{11}}, \gamma_3 = \frac{(-13\beta_1 - 2\beta_2 + 22\beta_3)}{3\sqrt{22}}, \gamma_4 = \beta_4$$

Defining the matrix R by the equation $\gamma = \beta R$, we find that **A** is represented by the matrix C below, relative to the orthonormal γ-basis:

$$C = R^{-1}BR = \begin{bmatrix} 2 & 0 & -2\sqrt{11} & \sqrt{2} \\ 0 & 2 & -11\sqrt{2} & -2/\sqrt{11} \\ 0 & 0 & 2 & -3\sqrt{2}/\sqrt{11} \\ 0 & 0 & 0 & 3 \end{bmatrix}$$

where
$$PQR = \begin{bmatrix} 0 & 3/\sqrt{11} & -2/\sqrt{22} & 0 \\ 1/\sqrt{2} & -1/\sqrt{11} & -3/\sqrt{22} & 0 \\ 1/\sqrt{2} & 1/\sqrt{11} & 3/\sqrt{22} & 0 \\ 0 & 0 & 0 & 1 \end{bmatrix}$$

Finally, let us mention that our results include a unitary matrix which transforms the original matrix A to superdiagonal form. Indeed, the above matrix PQR determines the final orthonormal γ-basis in terms of the initial orthonormal ϵ-basis ($\gamma = \epsilon PQR$) and, hence, is unitary; moreover, we have shown above that $(PQR)^{-1}A(PQR) = C$.

PROBLEMS

1. Complete the proof that (i) implies (ii) in Theorem 8.8 by demonstrating the additivity of **T**.

2. Verify the identity $4(\xi|\eta) = ||\xi + \eta||^2 - ||\xi - \eta||^2 + i||\xi + i\eta||^2 - i||\xi - i\eta||^2$ and use it to prove directly that (ii) implies (i) in Theorem 8.8.

3. Complete the proof of Theorem 8.10.

4. Supply a proof of Theorem 8.13.

5. Show that the matrix

$$A = \begin{bmatrix} -\sin^2\varphi + i\cos^2\varphi, & (1+i)\sin\varphi\cos\varphi \\ (1+i)\sin\varphi\cos\varphi, & -\cos^2\varphi + i\sin^2\varphi \end{bmatrix}$$

is unitary for all φ. If A defines the transformation **T** on $V_2(C)$ relative to the ϵ-basis, show that the image under **T** of the orthonormal basis $\{(0, i)^T, (-1, 0)^T\}$ is again an orthonormal basis. Determine the characteristic values of A and the characteristic vector of unit length accompanying each characteristic value. Show that these vectors form an orthonormal basis for $V_2(C)$ and represent **T** relative to this basis.

6. In the Euclidean space $V_3(R^{\#})$, a transformation **T** is defined by the matrix

$$\begin{bmatrix} 11 & 1 & -[illegible] \\ 4 & -7 & 14 \\ 4 & -7 & 14 \end{bmatrix}$$

relative to the ϵ-basis. Represent **T** by a superdiagonal matrix relative to some orthonormal basis.

7. Show that the $n \times n$ matrix whose (r, s)th entry is $e^{ir(s-1)\theta}/\sqrt{n}$, where $\theta = 2\pi/n$, is unitary.

8. Prove that if A is a real symmetric matrix and B is a real skew-symmetric matrix of the same order, then the matrix $(I + B + iA)(I - B - iA)^{-1}$ is unitary.

8.7. Normal Transformations. The problems that arise in connection with unitary transformations are the derivation of simple representations and the derivation of a canonical set of Hermitian forms under the unitary group. The latter is discussed in Sec. 8.9. A direct attack upon the former problem would yield the existence of a diagonal representation relative to an orthonormal basis. The question would then arise as to a description of *all* such transformations. These are the so-called normal transformations. Their theory is known and, accordingly, an alternative plan of exposition is possible. Namely, we can begin with a definition of normal transformations, discuss their properties and representations, and thereby obtain the desired results concerning unitary transformations as by-products.

The adoption of the above plan necessitates the verification of the following existence theorem: For each linear transformation **T** on a unitary space U, there can be found a linear transformation $\mathbf{T}^*$ on U, the so-called *adjoint* of **T**, such that

$$(\mathbf{T}\xi|\eta) = (\xi|\mathbf{T}^*\eta) \qquad \text{for all } \xi, \eta \text{ in } U$$

The first part of this section is devoted to a proof† of this assertion in the case where U is finite dimensional. The first step is the proof of a lemma which the reader will observe is valid under less restrictive assumptions than those chosen.

Lemma 8.1. Let f and g denote two linear functions on a vector space V over F, with values in the field F, such that $f(\xi) = 0$ whenever $g(\xi) = 0$. Then $f = cg$ for some scalar c.

†The author is indebted to A. Wilansky for suggesting this proof.

Proof. If $f \neq cg$, then there exist vectors ξ and η such that $f(\xi)g(\eta) \neq f(\eta)g(\xi)$. Consequently, the linear system

$$af(\xi) + bf(\eta) = 1$$

$$ag(\xi) + bg(\eta) = 0$$

has a solution (a, b), such that $f(a\xi + b\eta) = af(\xi) + bf(\eta) = 1$, while $g(a\xi + b\eta) = 0$, contrary to hypothesis.

The existence of the adjoint transformation is an immediate consequence of the next result.

Lemma 8.2. If g denotes a complex-valued linear function defined on a unitary space U, then there exists a uniquely determined vector ζ in U such that

$$g(\xi) = (\xi|\zeta) \qquad \text{for all } \xi \text{ in } U$$

Proof. If $g = 0$, then we may choose $\zeta = 0$. Otherwise, let M denote the subspace of all vectors ξ for which $g(\xi) = 0$, and let $M^\perp$ denote the orthogonal complement of M. Then $M^\perp$ contains a nonzero vector ζ_0 which we may assume has unit length. Since $g(\xi) = 0$ implies $(\xi|\zeta_0) = 0$, we conclude from Lemma 8.1 that $(\xi|\zeta_0) = cg(\xi)$. We evaluate c by choosing $\xi = \zeta_0$; this gives $c = 1/g(\zeta_0)$ and

$$g(\xi) = g(\zeta_0)(\xi|\zeta_0) = (\xi|\zeta_0\overline{g(\zeta_0)}) = (\xi|\zeta)$$

To prove the uniqueness of the solution, suppose that $(\xi|\zeta_1) = (\xi|\zeta_2)$ for all ξ; then $(\xi|\zeta_1 - \zeta_2) = 0$ for all ξ and, in particular, for $\xi = \zeta_1 - \zeta_2$. Thus $||\zeta_1 - \zeta_2||^2 = 0$, and $\zeta_1 = \zeta_2$. This completes the proof.

Since, for a given linear transformation $\mathbf{T}$ on U, the inner product $(\mathbf{T}\xi|\eta)$ is a linear function of ξ, assuming that η is fixed, the above lemma implies the existence of a unique vector ζ such that $(\mathbf{T}\xi|\eta) = (\xi|\zeta)$ for all ξ. If we allow η to range over U, this procedure makes correspond to each η a ζ depending, of course, upon η; we write $\zeta = \mathbf{T}^*\eta$. The defining property of $\mathbf{T}^*$ is therefore:

$$(\mathbf{T}\xi|\eta) = (\xi|\mathbf{T}^*\eta)$$

This establishes the existence of the adjoint of $\mathbf{T}$ on U. Direct computations, which are left as exercises, show that $\mathbf{T}^*$ is linear and, moreover, that $(\mathbf{T}^*)^* = \mathbf{T}$. Additional algebraic properties of the adjoint are collected below.

Lemma 8.3. The following computational rules are valid for the adjoint:

(i) $\mathbf{0}^* = \mathbf{0}$

(ii) $\mathbf{I}^* = \mathbf{I}$

(iii) $(\mathbf{S} + \mathbf{T})^* = \mathbf{S}^* + \mathbf{T}^*$

(iv) $(a\mathbf{T})^* = \bar{a}\mathbf{T}^*$

(v) $(\mathbf{ST})^* = \mathbf{T}^*\mathbf{S}^*$

(vi) $(\mathbf{T}^{-1})^* = (\mathbf{T}^*)^{-1}$, if $\mathbf{T}^{-1}$ exists.

(vii) If $p(\lambda) = \sum a_i\lambda^i$ is a polynomial in λ with complex coefficients and $\bar{p}(\lambda) = \sum \bar{a}_i\lambda^i$ is the so-called complex conjugate of p, then

$$[p(\mathbf{T})]^* = \bar{p}(\mathbf{T}^*)$$

Proof. The proofs of all these relations are elementary. To indicate the procedure, we carry out the computations for (v) and (vii). The former is a consequence of the relation

$$(\mathbf{ST}\xi|\eta) = (\mathbf{T}\xi|\mathbf{S}^*\eta) = (\xi|\mathbf{T}^*\mathbf{S}^*\eta)$$

The latter follows immediately from (ii) to (v).

To exhibit the relationship between a transformation $\mathbf{T}$ and its adjoint $\mathbf{T}^*$ in the light of their representations, we introduce the concept of the *dual basis* to a basis $\{\alpha_1, \alpha_2, \ldots, \alpha_n\}$ for the space U as a set of n vectors $\{\beta_1, \beta_2, \ldots, \beta_n\}$ such that $(\alpha_i|\beta_j) = \delta_{ij}$, $i, j = 1, 2, \ldots, n$. If such a set of vectors exists, it is clearly linearly independent and, therefore, a basis for U. The next lemma settles the existence question.

Lemma 8.4. If $\{\alpha_1, \alpha_2, \ldots, \alpha_n\}$ is any basis for the unitary space U and $\{c_1, c_2, \ldots, c_n\}$ is any set of n complex numbers, then there exists a unique vector β such that $(\alpha_i|\beta) = c_i$, $i = 1, 2, \ldots, n$.

Proof. If $\beta = \sum b^i\alpha_i$, then

$$(\alpha_i|\beta) = (\alpha_i|\sum_j b^j\alpha_j) = \sum_j \bar{b}^j(\alpha_i|\alpha_j)$$

so that it is sufficient to show that the system of linear equations $\sum_j(\alpha_i|\alpha_j)x^j = c_i$, $i = 1, 2, \ldots, n$, has a unique solution. But this is an immediate consequence of Theorem 8.1 and completes the proof.

It follows from this lemma that for each $j = 1, 2, \ldots, n$, a unique vector β_j in U can be found for which $(\alpha_i|\beta_j) = \delta_{ij}$, $i = 1, 2, \ldots, n$; thus $\{\beta_1, \beta_2, \ldots, \beta_n\}$ is a dual basis to $\{\alpha_1, \alpha_2, \ldots, \alpha_n\}$. In passing, we notice that a new characterization of orthonormal bases presents itself: The orthonormal bases for U are precisely the self-dual bases. We can now prove the assertion which relates the representations of a linear transformation and those of its adjoint.

Theorem 8.15. If the matrix A represents the linear transformation $\mathbf{T}$ on U relative to the α-basis, then the associated adjoint matrix† $A^* = \overline{A}^T$ represents the adjoint $\mathbf{T}^*$ relative to the dual basis.

Proof. Let $A = (a_j^i) = (\mathbf{T}; \alpha)$ and $B = (b_j^i) = (\mathbf{T}^*; \beta)$, where the β-basis is the dual of the α-basis. Upon comparing the end terms in the chain of equations

$$a_j^i = (\sum_k a_j^k \alpha_k | \beta_i) = (\mathbf{T}\alpha_j | \beta_i) = (\alpha_j | \mathbf{T}^*\beta_i) = (\alpha_j | \sum_k b_i^k \beta_k) = \overline{b}_i^j$$

we conclude that $A = B^*$, or $A^* = B$ as required.

We now turn to the classification of several types of linear transformations in terms of the concept of adjoint.

Theorem 8.16. A linear transformation $\mathbf{T}$ on a finite dimensional unitary space U is unitary if and only if $\mathbf{T}^* = \mathbf{T}^{-1}$.

Proof. Let $\mathbf{T}$ denote a unitary transformation. The computation

$$0 = (\mathbf{T}\xi | \mathbf{T}\eta) - (\xi | \eta) = (\mathbf{T}^*\mathbf{T}\xi | \eta) - (\xi | \eta) = ((\mathbf{T}^*\mathbf{T} - \mathbf{I})\xi | \eta)$$

valid for all ξ, η implies that $\mathbf{T}^*\mathbf{T} = \mathbf{I}$, or $\mathbf{T}^* = \mathbf{T}^{-1}$. Conversely, if the equation $\mathbf{T}^* = \mathbf{T}^{-1}$ is valid for a linear transformation $\mathbf{T}$, then, for all ξ,

$$||\mathbf{T}\xi||^2 = (\mathbf{T}\xi | \mathbf{T}\xi) = (\xi | \mathbf{T}^*\mathbf{T}\xi) = (\xi | \xi) = ||\xi||^2$$

Recalling Theorem 8.8, we conclude that $\mathbf{T}$ is unitary.

Theorems 8.15 and 8.16 provide us with a new proof of the fact that, relative to an orthonormal basis, a unitary transformation is represented by a matrix A such that $A^{-1} = A^*$, that is, a unitary matrix.

DEFINITION. *A linear transformation* $\mathbf{T}$ *on a unitary space* U *is called* Hermitian *or* self-adjoint *if* $\mathbf{T} = \mathbf{T}^*$.

It is left as an exercise to show that, relative to an orthonormal basis, a Hermitian transformation is represented by a Hermitian matrix and that, conversely, the linear transformation defined by a Hermitian matrix relative to an orthonormal basis is Hermitian.

DEFINITION. *A linear transformation* $\mathbf{T}$ *on a unitary space is called* normal *if it commutes with its adjoint*: $\mathbf{T}\mathbf{T}^* = \mathbf{T}^*\mathbf{T}$. *A matrix of complex numbers is called* normal *if it commutes with its adjoint.*

†For the remainder of this chapter we shall refer to A^* as simply the adjoint of A. This will cause no confusion with an earlier use of this word.

Just as unitary and Hermitian transformations are represented by unitary and Hermitian matrices, respectively, relative to orthonormal bases, so are normal transformations represented by normal matrices relative to orthonormal bases. Conversely, a normal matrix defines a normal transformation relative to an orthonormal basis.

It is obvious that both unitary and Hermitian transformations are normal, using Theorem 8.16 and the definition of a Hermitian transformation, respectively. Consequently a discussion of the representations of normal transformations will include that of both unitary and Hermitian transformations. For this, several preliminary results are needed.

Theorem 8.17. If $\mathbf{T}$ is a normal transformation and $p(\lambda)$ denotes any polynomial with complex coefficients, then $p(\mathbf{T})$ is normal and $p(\mathbf{T})\xi = 0$ if and only if $\bar{p}(\mathbf{T}^*)\xi = 0$, where $\bar{p}(\lambda)$ is the complex conjugate of $p(\lambda)$. In particular, ξ is a characteristic vector of $\mathbf{T}^*$ if and only if it is a characteristic vector of $\mathbf{T}$; if $\mathbf{T}\xi = c\xi$, then $\mathbf{T}^*\xi = \bar{c}\xi$.

Proof. A direct computation demonstrates the first assertion. Next, if $p(\mathbf{T})\xi = 0$, we have [using (vii) of Lemma 8.3]

$$0 = (p(\mathbf{T})\xi|p(\mathbf{T})\xi) = (\xi|\bar{p}(\mathbf{T}^*)p(\mathbf{T})\xi) = (\xi|p(\mathbf{T})\bar{p}(\mathbf{T}^*)\xi)$$
$$= (\bar{p}(\mathbf{T}^*)\xi|\bar{p}(\mathbf{T}^*)\xi)$$

or $\bar{p}(\mathbf{T}^*)\xi = 0$. Since our steps are reversible, the converse is also true. This result, when applied to the polynomial $\lambda - c$, yields the final assertion.

From the last sentence of the above theorem we obtain immediately the following result:

COROLLARY 1. Each characteristic value of a Hermitian transformation is real. Each characteristic value of a unitary transformation has absolute value 1.

Defining a *characteristic value of a matrix* A over F as a root in F of the polynomial $\det(\lambda I - A)$, and recalling that Hermitian and unitary matrices represent Hermitian and unitary transformations, respectively, we can restate the above corollary as follows:

COROLLARY 2. Every characteristic value of a Hermitian matrix is real. Every characteristic value of a unitary matrix has absolute value 1.

Theorem 8.18. If $\mathbf{T}$ is a normal transformation on a unitary space, then characteristic vectors associated with distinct characteristic values are orthogonal.

Proof. If $\mathbf{T}\xi_i = c_i\xi_i$, $i = 1, 2$, then

$$c_1(\xi_1|\xi_2) = (\mathbf{T}\xi_1|\xi_2) = (\xi_1|\mathbf{T}^*\xi_2) = (\xi_1|\bar{c}_2\xi_2) = c_2(\xi_1|\xi_2)$$

where the previous theorem has been applied to rewrite $\mathbf{T}^*\xi_2$ as $\bar{c}_2\xi_2$. It follows that $(\xi_1|\xi_2) = 0$, as required, provided $c_1 \neq c_2$.

Theorem 8.19. If $\mathbf{T}$ is a normal transformation on a unitary space, then its minimum function is equal to a product of distinct irreducible (therefore linear) factors.

Proof. We shall show that the criterion of Theorem 7.7 is satisfied for each characteristic value c of $\mathbf{T}$. For this it is sufficient to show that a member ξ of $N(\lambda - c)^2$ is a member of $N(\lambda - c)$. Defining $\eta = (\mathbf{T} - c\mathbf{I})\xi$, this implication can be restated as: If $(\mathbf{T} - c\mathbf{I})\eta = 0$, then $\eta = 0$. Thus, by the definition of η, $\mathbf{T}\xi = c\xi + \eta$, and, by assumption, $\mathbf{T}\eta = c\eta$. We have, therefore,

$$c(\xi|\eta) + (\eta|\eta) = (c\xi + \eta|\eta) = (\mathbf{T}\xi|\eta) = (\xi|\mathbf{T}^*\eta) = (\xi|\bar{c}\eta) = c(\xi|\eta)$$

Comparing the end members, we find that $(\eta|\eta) = 0$, or $\eta = 0$ as desired.

Using the preceding two theorems, it is possible to show that if $\mathbf{T}$ is a normal transformation on the unitary space U, there exists an orthonormal basis relative to which $\mathbf{T}$ is represented by a diagonal matrix. More generally, it is possible to prove the following characterization of normal transformations:

Theorem 8.20. The normal transformations on a finite dimensional unitary space U are precisely those linear transformations on U which admit a diagonal representation relative to some orthonormal basis for U.

Proof. Let $\{c_1, c_2, \ldots, c_r\}$ denote the set of distinct characteristic values of the normal transformation $\mathbf{T}$ [therefore the minimum function of $\mathbf{T}$ is $\prod_1^r(\lambda - c_i)$], and set $V_i = N(\lambda - c_i)$, $i = 1, 2, \ldots, r$. Since every nonzero vector in V_i is a characteristic vector of $\mathbf{T}$, Theorem 8.18 implies that $V_i \perp V_j$, $i \neq j$. Thus, if an orthonormal basis is chosen for each V_i, the composite set of vectors is an orthonormal basis for V relative to which $\mathbf{T}$ is represented by

$$D = \text{diag}(c_1, \ldots, c_1, c_2, \ldots, c_2, \ldots, c_r, \ldots, c_r)$$

where c_i appears a number of times equal to $d[V_i]$, $i = 1, 2, \ldots, r$.

To prove the converse of this result, assume that a linear transformation $\mathbf{T}$ is represented by a diagonal matrix, for example D above, relative to an orthonormal basis for V. Relative to the same basis, $\mathbf{T}^*$ is represented by

$D^* = \text{diag}(\bar{c}_1, \ldots, \bar{c}_r)$ (see Theorem 8.15). Since $DD^* = D^*D$, it follows that $\mathbf{TT}^* = \mathbf{T}^*\mathbf{T}$, or $\mathbf{T}$ is normal.

Next we state an immediate corollary of this result.

Theorem 8.21. Each unitary matrix and each Hermitian matrix can be transformed to diagonal form by a unitary matrix. In other words, if M denotes a unitary or Hermitian matrix, there exists a unitary matrix P, such that PMP^{-1} is diagonal.

We remark that Theorem 8.20 follows from the existence of a superdiagonal representation relative to an orthonormal basis (Theorem 8.14); this method of proof is outlined in Prob. 13 at the end of this section. The proof that we have presented is in keeping with our emphasis on linear transformations rather than matrices as the fundamental concept for study; in addition, our proof provides a practical method for computing a diagonal representation. In brief, if the representation A relative to the ϵ-basis of a normal transformation $\mathbf{T}$ is known, the first step is to determine the characteristic values of $\mathbf{T}$. If the characteristic value c is of (algebraic) multiplicity a, then the homogeneous system $(cI - A)X = 0$ has rank $n - a$ and a solution space of dimension a (that is, the geometric multiplicity agrees with its algebraic multiplicity; see Theorem 7.8) for which an orthonormal basis can be found. Repeating this step for each characteristic value yields a basis of the desired type. The next example illustrates this for the case of a unitary transformation.

EXAMPLE 8.7

The characteristic values of the unitary transformation $\mathbf{T}$ represented by the (unitary) matrix

$$A = \begin{bmatrix} \dfrac{i + \omega + 2\omega^2}{4} & \dfrac{\omega - \omega^2}{2\sqrt{2}} & \dfrac{1 + i\omega^2}{2\sqrt{2}} & \dfrac{-1 - i\omega}{4} \\[2ex] \dfrac{\omega - \omega^2}{2\sqrt{2}} & \dfrac{i + 2\omega + \omega^2}{4} & \dfrac{-1 - i\omega^2}{4} & \dfrac{-1 - i\omega}{2\sqrt{2}} \\[2ex] \dfrac{-1 - i\omega^2}{2\sqrt{2}} & \dfrac{1 + i\omega^2}{4} & \dfrac{3i + \omega^2}{4} & 0 \\[2ex] \dfrac{1 + i\omega}{4} & \dfrac{1 + i\omega}{2\sqrt{2}} & 0 & \dfrac{3i + \omega}{4} \end{bmatrix}$$

are i, i, ω, ω^2, where $\omega = (-1 + \sqrt{3}i)/2$ is a primitive cube root of unity. Accompanying the twofold characteristic value i is the two-dimensional solution space of $(iI - A)X = 0$, for which $\{\rho_1, \rho_2\}$ (see below) is an orthonormal basis. For the characteristic values ω and ω^2, the vectors ρ_3 and ρ_4, respectively, are characteristic vectors of unit length. If P is the (unitary) matrix whose columns are

$$\rho_1 = \begin{bmatrix} 0 \\ \frac{1}{2} \\ -\frac{i}{2} \\ -\frac{i}{\sqrt{2}} \end{bmatrix} \quad \rho_2 = \begin{bmatrix} \frac{1}{2} \\ 0 \\ \frac{i}{\sqrt{2}} \\ -\frac{i}{2} \end{bmatrix} \quad \rho_3 = \begin{bmatrix} -\frac{i}{2} \\ -\frac{i}{\sqrt{2}} \\ 0 \\ \frac{1}{2} \end{bmatrix} \quad \rho_4 = \begin{bmatrix} -\frac{i}{\sqrt{2}} \\ \frac{i}{2} \\ -\frac{1}{2} \\ 0 \end{bmatrix}$$

then (see Example 7.5)

$$P^{-1}AP = P^*AP = \text{diag}(i, i, \omega, \omega^2)$$

which represents $\mathbf{T}$ relative to the orthonormal ρ-basis.

We conclude our discussion of normal transformations with a characterization of unitary and Hermitian transformations, among the normal transformations, in terms of characteristic values. The proof is left as an exercise.

Theorem 8.22. A normal transformation $\mathbf{T}$ is Hermitian if and only if all of its characteristic values are real; $\mathbf{T}$ is unitary if and only if all of its characteristic values have absolute value 1.

PROBLEMS

1. Let $\mathbf{T}$ denote a linear transformation on a unitary space U with orthonormal basis $\{\beta_1, \beta_2, \ldots, \beta_n\}$. Define the transformation $\mathbf{T}'$ by the equation

$$\mathbf{T}'\eta = \sum_{i=1}^{n} (\eta|\mathbf{T}\beta_i)\beta_i$$

Verify that $\mathbf{T}'$ is linear and that $(\xi|\mathbf{T}'\eta) = (\mathbf{T}\xi|\eta)$, in other words, that $\mathbf{T}' = \mathbf{T}^*$. The reader will observe that, in contrast to this definition of the adjoint of $\mathbf{T}$, the definition in the text makes no use of a basis for the space.

2. Using the definition given in the text, verify the properties of the adjoint mentioned just before Lemma 8.3. Complete the proof of Lemma 8.3, and, in addition, supply a proof of the following statement: If $f(\lambda)$ is the characteristic function of $\mathbf{T}$, then $\bar{f}(\lambda)$ is the characteristic function of $\mathbf{T}^*$.

3. In the proof of Theorem 8.16 the following result is used: If $(\mathbf{A}\xi|\eta) = 0$ for all ξ, η, then $\mathbf{A} = \mathbf{0}$. Prove this. Prove, more generally, that if $\mathbf{A}$ is a linear transformation on a unitary space, such that $(\mathbf{A}\xi|\xi) = 0$ for all ξ, then $\mathbf{A} = \mathbf{0}$. HINT: Verify and then apply the identity

$$a\bar{b}(\mathbf{A}\xi|\eta) + \bar{a}b(\mathbf{A}\eta|\xi) = (\mathbf{A}(a\xi + b\eta)|a\xi + b\eta) - |a|^2(\mathbf{A}\xi|\xi) - |b|^2(\mathbf{A}\eta|\eta)$$

by first choosing $a = b = 1$ and then $a = i$, $b = 1$.

4. Verify that, relative to an orthonormal basis, a Hermitian transformation is represented by a Hermitian matrix and, conversely, that the linear transformation defined by a Hermitian matrix, relative to an orthonormal basis, is Hermitian.

5. An *involution* is a linear transformation $\mathbf{T}$ such that $\mathbf{T}^2 = \mathbf{I}$. Show that a linear transformation on a unitary space, which satisfies any two of the three conditions of being involutory, Hermitian, or unitary, also satisfies the third.

6. Using Prob. 3, show that a linear transformation $\mathbf{A}$ on a unitary space U is Hermitian if and only if $(\mathbf{A}\xi|\xi)$ is real for all ξ in U. Show that the corresponding statement for a Euclidean space is false.

7. Prove: If $\mathbf{A}$ and $\mathbf{B}$ are Hermitian transformations, then $\mathbf{AB}$ and $\mathbf{BA}$ are Hermitian if and only if $\mathbf{AB} = \mathbf{BA}$.

8. Prove: If $\mathbf{A}$ is Hermitian, then, for an arbitrary $\mathbf{B}$, $\mathbf{BAB}^*$ is Hermitian; if $\mathbf{B}$ has an inverse and $\mathbf{BAB}^*$ is Hermitian, then so is $\mathbf{A}$.

9. Let $\mathbf{T}$ denote a linear transformation on the finite dimensional unitary space U, and S a subspace of U invariant under $\mathbf{T}$. Show that $S^\perp$ is invariant under $\mathbf{T}$ if and only if $\mathbf{T}$ is normal.

10. Verify the Corollary to Theorem 8.17.

11. Prove Theorem 8.19 using the criterion of Theorem 7.8. HINT: Use the results in Prob. 9 to conclude that if c is a characteristic value of a normal transformation $\mathbf{T}$, then the orthogonal complement of $N(\lambda - c)$ is also an invariant space of $\mathbf{T}$.

12. In the proof of Theorem 8.20, verify that the relation $DD^* = D^*D$ implies that $\mathbf{TT}^* = \mathbf{T}^*\mathbf{T}$.

13. (An alternate proof of Theorem 8.20.) Let $\mathbf{T}$ denote a normal transformation on U, $\{\alpha_1, \alpha_2, \ldots, \alpha_n\}$ an orthonormal basis for U, and $A = (\mathbf{T}; \alpha)$. According to Theorem 8.14, we may assume that the α-basis has been chosen so that A is superdiagonal. Prove that actually A is diagonal by comparing the diagonal terms of the equal matrices AA^* and A^*A.

14. Prove Theorem 8.22.

15. Show that the matrix $\begin{bmatrix} 0 & 1 \\ 0 & 0 \end{bmatrix}$ is not normal. More generally, prove that a triangular matrix is normal if and only if it is diagonal.

16. Is the matrix below (i) normal, (ii) unitary, (iii) Hermitian?

$$\begin{bmatrix} 1 & 0 & 1 & i \\ \frac{1}{2} & 1 & 1 & i \\ 0 & \frac{1}{2} & -\frac{3}{4} & -\frac{i}{4} \\ -\frac{i}{2} & -\frac{i}{2} & -\frac{5i}{4} & \frac{7}{4} \end{bmatrix}$$

17. Let A denote a skew-Hermitian matrix (that is, $A = -A^*$). Show that the characteristic values of A are pure imaginaries and that A can be transformed to a diagonal matrix.

8.8. Normal, Orthogonal, and Symmetric Transformations on Euclidean Spaces. Early in the chapter we warned the reader that we would essentially ignore the Euclidean spaces, so long as the definitions and theorems for unitary spaces carried over unchanged or had apparent analogues in the real case. The point has now been reached in our development where we must take cognizance of the fact that not every polynomial over the real field, $R^\#$, has a root in $R^\#$ and in particular, therefore, that not every polynomial can be decomposed into a product of linear factors over $R^\#$. Since it is in Sec. 8.6 that the terminology and results for Euclidean spaces begin to diverge from the corresponding items for unitary spaces, we shall reexamine our exposition, beginning with Sec. 8.6, from the viewpoint of Euclidean spaces.

In Euclidean spaces, unitary transformations are called *orthogonal* and Hermitian ones *symmetric*. The characterizations of unitary transformations presented in Theorems 8.8 and 8.11 are valid for orthogonal transformations; we remark that in the proof of Theorem 8.8, the left member of equation (8) will be replaced by $2(\xi|\eta)$. The set of all orthogonal transformations on an n-dimensional Euclidean space forms a multiplicative group, called the *orthogonal group*; it is a subgroup of the full linear group $L_n(R^\#)$. In the real case, Theorem 8.12 states that, relative to an orthonormal basis, a matrix A represents an orthogonal transformation if and only if its column vectors, interpreted as elements of $V_n(R^\#)$, form an orthonormal set, or in other words, that $A^TA = I$. In turn, this means that $A^{-1} = A^T$; such a matrix is called *orthogonal*. It follows immediately from the definition that if A is orthogonal then $\det A = \pm 1$. The remark after Theorem 8.13 has the following analogue: The study of the simplifications that are possible in quadratic forms X^TAX by means of *orthogonal substitutions*, *i.e.*, substitutions $X = PY$, where P is orthogonal, can be interpreted as the study of the equivalence of quadratic forms under the orthogonal group on $V_n(R^\#)$.

Theorem 8.14 is an example (our first, in fact) of a result for unitary spaces that has no analogue in the Euclidean case. To support this statement, we need remark only that the existence of a superdiagonal representation for a linear transformation **T** implies that the characteristic function of **T** is a product of linear factors.

Turning next to Sec. 8.7, it is seen immediately that the definition of the adjoint of a linear transformation is meaningful and that the basic properties summarized in Lemma 8.3 are valid if the conjugation signs are ignored. It is now the transpose, A^T (to which $A^* = \bar{A}^T$ reduces over $R^\#$), that represents the adjoint transformation **T*** when A represents **T** (Theorem

8.15). In terms of the adjoint transformation, an orthogonal transformation $\mathbf{T}$ is characterized by the equation $\mathbf{T}^* = \mathbf{T}^{-1}$, while the corresponding characterization of a symmetric transformation is given by the equation $\mathbf{T}^* = \mathbf{T}$. Parallel to the result that relative to an orthonormal basis an orthogonal transformation is represented by an orthogonal matrix, we find that, relative to such a basis, a symmetric transformation is represented by a symmetric matrix.

Transferring, unchanged, the definition of a normal transformation to Euclidean spaces, it is clear that Theorem 8.17 is valid. However, we remind the reader that characteristic vectors need not exist in the real case; a rotation through $\pi/2$ of every vector of the Euclidean plane illustrates this. Consequently, the corollaries to this theorem now tell us (i) nothing about characteristic values of a symmetric transformation (or matrix) and (ii) that the only admissible characteristic values of an orthogonal transformation (or matrix) are ± 1. To obtain more information about such transformations, and normal transformations in general, within the confines of the real field,† we must dig deeper. The first result is the analogue of Theorem 8.19.

Theorem 8.23. The minimum function of a normal transformation $\mathbf{T}$ on a finite dimensional Euclidean space is equal to a product of distinct irreducible (therefore linear or quadratic) factors.

Proof. Since every irreducible member of $R^{\#}[\lambda]$ is linear or quadratic, the minimum function, $m(\lambda)$, of $\mathbf{T}$ can be decomposed into a product of such factors. The proof of Theorem 8.19 shows immediately that no linear factor that occurs is repeated.

To prove the corresponding result for a quadratic factor $q(\lambda) = \lambda^2 + a\lambda + b$, we observe that the argument in the second half of the proof of Theorem 7.7 applies to any factor of $m(\lambda)$. Thus we need merely show that a member ξ of $N(q^2)$ is a member of $N(q)$. For this we introduce the transformation $\mathbf{A} = q(\mathbf{T}) - b\mathbf{I}$ and the vector $\eta = q(\mathbf{T})\xi$. Then $(\mathbf{A} + b\mathbf{I})\eta = q(\mathbf{T})\eta = 0$ by the definitions of η and ξ. According to Theorem 8.17, the normality of $\mathbf{T}$ implies the normality of $\mathbf{A}$ and, in turn, that $(\mathbf{A}^* + b\mathbf{I})\eta = 0$, or $\mathbf{A}^*\eta = -b\eta$. This, together with the fact that $\mathbf{A}\xi = \eta - b\eta$, permits us to conclude that $(\eta|\eta) = 0$ or, $\eta = 0$ which completes the proof; the appropriate computation follows:

$$(\eta|\eta) - b(\xi|\eta) = (\eta - b\xi|\eta) = (\mathbf{A}\xi|\eta) = (\xi|\mathbf{A}^*\eta) = (\xi|-b\eta) = -b(\xi|\eta)$$

The analogue of Theorem 8.18 now follows, as we proceed to show:

†The author became interested in the question of how neatly a discussion of the transformations under consideration could be carried out entirely within the real field He believes that the method devised may hold some interest.

Theorem 8.24. If $\mathbf{T}$ is a normal transformation on a finite dimensional Euclidean space, the null spaces determined by different irreducible factors of the minimum function of $\mathbf{T}$ are orthogonal.

Proof. Let p and q denote two distinct irreducible factors of the minimum function of $\mathbf{T}$, and let α, β denote members of $N(p)$ and $N(q)$, respectively. Since p and q are relatively prime, there exist polynomials r and s such that $rp + sq = 1$. Replacing λ by $\mathbf{T}$ in this equation gives a representation of the identity transformation which, when applied to α and β in turn, gives

$$s(\mathbf{T})q(\mathbf{T})\alpha = \alpha \qquad \text{and} \qquad r(\mathbf{T})p(\mathbf{T})\beta = \beta$$

since $p(\mathbf{T})\alpha = q(\mathbf{T})\beta = 0$ by assumption. Consequently, using Lemma 8.3 and Theorem 8.17,

$$\begin{aligned}(\alpha|\beta) = (s(\mathbf{T})q(\mathbf{T})\alpha|r(\mathbf{T})p(\mathbf{T})\beta) &= (s(\mathbf{T})\alpha|q(\mathbf{T}^*)r(\mathbf{T})p(\mathbf{T})\beta) \\ &= (s(\mathbf{T})\alpha|r(\mathbf{T})p(\mathbf{T})q(\mathbf{T}^*)\beta) \\ &= 0\end{aligned}$$

This completes the proof.

The two preceding theorems imply that, accompanying a normal transformation $\mathbf{T}$ with minimum function $m(\lambda)$ on a finite dimensional Euclidean space E, there is an orthonormal basis for E relative to which $\mathbf{T}$ is represented by a direct sum of matrices A_i (see Theorem 7.3) which represent $\mathbf{T}$ on the null spaces determined by the irreducible factors of $m(\lambda)$. Of course, an A_i accompanying a linear factor is diagonal. An A_i accompanying a quadratic factor can be chosen as a direct sum of 2×2 matrices. The final result in this connection is stated as Theorem 8.27. Although a proof could be supplied now (see Prob. 5), one obtains greater insight into the geometric interpretation of the result if the structure of orthogonal and symmetric transformations is known. The analysis of these special types of normal transformations is easy with our present knowledge of normal transformations in general. For example, the structure of symmetric transformations is settled with the aid of the next lemma.

Lemma 8.5. If $\mathbf{T}$ is a symmetric transformation ($\mathbf{T} = \mathbf{T}^*$) on an n-dimensional Euclidean space, then its minimum function is equal to a product of distinct linear factors over R^*.

Proof. Let $q = \lambda^2 - 2a\lambda + b$ denote a quadratic factor of the minimum function of $\mathbf{T}$. Then there exists a vector α, which we may assume has unit length, such that $(\mathbf{T}^2 - 2a\mathbf{T} + b\mathbf{I})\alpha = 0$. The computation below shows that the discriminant of q is nonnegative, or q is reducible over $R^{\#}$:

$$
\begin{aligned}
0 \leqq (a\alpha - \mathbf{T}\alpha | a\alpha - \mathbf{T}\alpha) &= a^2 - 2a(\alpha|\mathbf{T}\alpha) + (\mathbf{T}\alpha|\mathbf{T}\alpha) \\
&= a^2 - 2a(\alpha|\mathbf{T}\alpha) + (\alpha|\mathbf{T}^2\alpha) \\
&= a^2 - 2a(\alpha|\mathbf{T}\alpha) + (\alpha|2a\mathbf{T}\alpha - b\alpha) \\
&= a^2 - 2a(\alpha|\mathbf{T}\alpha) + 2a(\alpha|\mathbf{T}\alpha) - b = a^2 - b
\end{aligned}
$$

In the light of Theorems 8.23, 8.24, and 7.5 this lemma immediately yields the following theorem:

Theorem 8.25. Let $\mathbf{T}$ denote a symmetric transformation on an n-dimensional Euclidean space E. Then there exists an orthonormal basis for E relative to which $\mathbf{T}$ is represented by a diagonal matrix. In particular, $\mathbf{T}$ has n characteristic values, counting each a number of times equal to its algebraic multiplicity. In matrical language, a real symmetric matrix can be transformed to diagonal form by an orthogonal matrix.

Since the commonplace applications of this theorem are in connection with quadratic forms, an example is postponed until the next section.

Finally, let us analyze the orthogonal transformations! For this the analogue of Lemma 8.5 is needed.

Lemma 8.6. A linear factor of the minimum function $m(\lambda)$ of an orthogonal transformation $\mathbf{T}$ on an n-dimensional Euclidean space E has the form $\lambda \pm 1$. An irreducible quadratic factor of $m(\lambda)$ has the form

$$q(\lambda) = \lambda^2 - 2a\lambda + 1 \qquad \text{where } a = (\alpha|\mathbf{T}\alpha)$$

for any vector α of unit length in $N(q)$, the null space of $q(\mathbf{T})$.

Proof. The first assertion is a consequence of an earlier remark that the only possible characteristic values of an orthogonal transformation are ± 1. Next, if $q = \lambda^2 - 2a\lambda + b$ is an irreducible quadratic factor of $m(\lambda)$, consider the null space $N(q)$, that is, the set of all vectors α in E such that $q(\mathbf{T})\alpha = 0$. Using Theorem 8.17, $N(q)$ can also be described as the set of all vectors α such that $q(\mathbf{T}^*)\alpha = 0$ or, since $\mathbf{T}^* = \mathbf{T}^{-1}$, as the set of all vectors α such that $(\mathbf{I} - 2a\mathbf{T} + b\mathbf{T}^2)\alpha = 0$. Thus

$$N(\lambda^2 - 2a\lambda + b) = N(b\lambda^2 - 2a\lambda + 1) = N\left(\lambda^2 - \frac{2a\lambda}{b} + \frac{1}{b}\right)$$

But since q is the minimum function of the transformation induced by $\mathbf{T}$ in $N(q)$ (see Theorem 7.3), it follows from Theorem 7.2 that $\lambda^2 - \dfrac{2a\lambda}{b} + \dfrac{1}{b}$ divides q, or $b = 1$.

Finally, to evaluate a, let α denote a unit vector in $N(q)$. The computation below shows that $a = (\alpha|\mathbf{T}\alpha)$, which thereby completes the proof:

$$(\alpha|\mathbf{T}\alpha) = (\mathbf{T}\alpha|\mathbf{T}^2\alpha) = (\mathbf{T}\alpha|2a\mathbf{T}\alpha - \alpha) = 2a(\mathbf{T}\alpha|\mathbf{T}\alpha) - (\mathbf{T}\alpha|\alpha)$$
$$= 2a - (\alpha|\mathbf{T}\alpha)$$

It is now possible to describe the structure of an orthogonal transformation.

Theorem 8.26. Let $\mathbf{T}$ denote an orthogonal transformation on an n-dimensional Euclidean space E. An orthonormal basis can be found for E relative to which $\mathbf{T}$ is represented by a matrix of the form

$$\text{(10)} \qquad \text{diag}(1, R_{\theta_1}, \ldots, R_{\theta_1}, \ldots, R_{\theta_m}, \ldots, R_{\theta_m}, -1)$$

where

$$\text{(11)} \quad R_{\theta_i} = \begin{bmatrix} \cos\theta_i & -\sin\theta_i \\ \sin\theta_i & \cos\theta_i \end{bmatrix} \qquad 0 \leqq \theta_i \leqq \pi,\ i = 1, 2, \ldots, m$$

and where not all of the three indicated types of diagonal blocks need appear. In other words, $\mathbf{T}$ is a direct sum of rotations in mutually orthogonal planes (*i.e.*, two-dimensional spaces), together with a possible reflection (the effect of the -1) in an $(n - 1)$-dimensional space. In matrical language, an orthogonal matrix can be transformed to the form (10) with a second orthogonal matrix. The final -1 appears if and only if $\det A$ (which is necessarily ± 1) is equal to -1.

Proof. Referring to the paragraph following Theorem 8.24, we already know the following facts about $\mathbf{T}$ since it is normal: If the minimum function of $\mathbf{T}$ is

$$m(\lambda) = q_1 q_2 \cdots q_k$$

then E decomposes into a direct sum of null spaces

$$E = N(q_1) \oplus N(q_2) \oplus \cdots \oplus N(q_k)$$

where the minimum function of $\mathbf{T}_i$, the transformation induced in $N(q_i)$ by $\mathbf{T}$, is q_i and $N(q_i) \perp N(q_j)$, $i \neq j$. Now let us examine each $N(q_i)$ in the light of Lemma 8.6.

If $q_i = \lambda - 1$, then $\mathbf{T}_i = \mathbf{I}$ and is represented by an identity matrix I relative to any basis, and in particular relative to an orthonormal basis. If $d[N(\lambda - 1)] > 1$, we may write I in the form

$$I = 1 \oplus R_0 \oplus R_0 \oplus \cdots \oplus R_0$$

using (11); the initial 1 appears if and only if the order of I is odd. If

$q_i = \lambda + 1$, a similar discussion demonstrates the existence of a representation for $\mathbf{T}_i$ of the form

$$-I = R_\pi \oplus R_\pi \oplus \cdots \oplus R_\pi \oplus (-1)$$

where the final (-1) appears if the order of $-I$ is odd.

If $q_i = \lambda^2 - 2a\lambda + 1$, we assert that an orthonormal basis can be found for $N(q_i)$ relative to which $\mathbf{T}_i$ is represented by a direct sum of rotation matrices (11), where θ is the uniquely determined angle between 0 and π, such that $\cos\theta = a$. (This definition is meaningful since, with q_i irreducible, $a^2 < 1$.) We begin with any unit vector α in $N(q_i)$; then β, which is defined by the equation

$$\sqrt{1 - a^2}\,\beta = -a\alpha + \mathbf{T}\alpha$$

is a unit vector orthogonal to α, since $a = (\alpha|\mathbf{T}\alpha)$. The space $S_\alpha = [\alpha, \beta]$ is invariant under $\mathbf{T}_i$, and if it does not exhaust $N(q_i)$, its orthogonal complement $S_\alpha^\perp$ in $N(q_i)$ is also invariant under $\mathbf{T}_i$, as the reader can easily verify. We then repeat the preceding step.

Continuing in this way yields the desired orthonormal basis for $N(q_i)$. To show this, it is sufficient to examine the effect of $\mathbf{T}_i$ on S_α, since it behaves in exactly the same way on any other two-dimensional space of the same type. A direct computation shows that

$$\begin{aligned} \mathbf{T}_i\alpha &= \cos\theta\cdot\alpha + \sin\theta\cdot\beta \\ \mathbf{T}_i\beta &= -\sin\theta\cdot\alpha + \cos\theta\cdot\beta \end{aligned} \qquad (12)$$

so that, relative to the orthonormal basis whose construction we have indicated, $\mathbf{T}_i$ is represented by a direct sum of the form $R_\theta \oplus R_\theta \oplus \cdots \oplus R_\theta$. The proof of the theorem is completed with the observation that a transformation $\mathbf{T}_i$, defined by (12), relative to an orthonormal basis $\{\alpha, \beta\}$, is a plane rotation of each vector through a counterclockwise angle of θ (see Example 3.3, reinterpreting the coordinate transformation there as a vector transformation).

In view of the above canonical representation for an orthogonal transformation, we label such a transformation a *symmetry* or a *rotation* of the Euclidean space, according as the final -1 does or does not appear in its representation (10). The next example will clarify the use of the former term.

EXAMPLE 8.8

In $V_3(R^\#)$ the possible canonical representations of an orthogonal transformation $\mathbf{T}$ are

$$(1) \oplus R_\theta \quad \text{and} \quad R_\theta \oplus (-1) \qquad 0 \leqq \theta \leqq \pi \qquad (13)$$

relative to an orthonormal basis for the space. The reader should write out the expanded version of each of these types for $\theta = 0$, $\theta = \pi$, and $0 < \theta < \pi$ to obtain the six possible canonical forms for **T**. Each instance of the second type in (13) involves a reflection in a plane; this effects a "symmetry" of the space in the sense explained in Prob. 7, Sec. 6.3.

EXAMPLE 8.9

Determine an orthonormal basis for $V_3(R^{\#})$ relative to which the orthogonal transformation **T**, defined by the matrix

$$A = \begin{bmatrix} 0 & 0 & 1 \\ 1 & 0 & 0 \\ 0 & 1 & 0 \end{bmatrix}$$

is represented by a matrix of the type in (11).

The characteristic and minimum function of **T** is $\lambda^3 - 1 = (\lambda - 1)(\lambda^2 + \lambda + 1)$. The unit vector $\alpha_1 = (1/\sqrt{3}, 1/\sqrt{3}, 1/\sqrt{3})^T$ is a characteristic vector accompanying 1. Next, $\alpha_2 = (-1/\sqrt{6}, 2/\sqrt{6}, -1/\sqrt{6})^T$ is a unit vector in $N(\lambda^2 + \lambda + 1)$. As in the proof of Theorem 8.26, we extend this vector to a basis for this null space by adjoining a vector α_3, defined according to the prescription for β in Theorem 8.26:

$$\sqrt{\frac{3}{4}}\alpha_3 = \frac{1}{2}\alpha_2 + A\alpha_2 \qquad \alpha_3 = \left(-\frac{1}{\sqrt{2}}, 0, \frac{1}{\sqrt{2}}\right)^T$$

Thus $\{\alpha_1, \alpha_2, \alpha_3\}$ is an orthonormal basis relative to which **T** is represented by the matrix

$$P^{-1}AP = \begin{bmatrix} 1 & 0 & 0 \\ 0 & -\frac{1}{2} & -\frac{\sqrt{3}}{2} \\ 0 & \frac{\sqrt{3}}{2} & -\frac{1}{2} \end{bmatrix} \qquad \text{where } P = \begin{bmatrix} \frac{1}{\sqrt{3}} & -\frac{1}{\sqrt{6}} & -\frac{1}{\sqrt{2}} \\ \frac{1}{\sqrt{3}} & \frac{2}{\sqrt{6}} & 0 \\ \frac{1}{\sqrt{3}} & -\frac{1}{\sqrt{6}} & \frac{1}{\sqrt{2}} \end{bmatrix}$$

Of course, we could have written down this canonical representation as soon as the minimum function of A was determined.

We turn finally to the derivation of a simple representation for a normal transformation **T** on a Euclidean space. The reader should observe that, in effect, we prove that **T** is the product of a symmetric transformation and an orthogonal transformation, which means that **T** is a rotation followed by a dilatation.

Theorem 8.27. Let $\mathbf{T}$ be a normal transformation on an n-dimensional Euclidean space E. Then there exists an orthonormal basis for E relative to which $\mathbf{T}$ is represented by a matrix of the form

$$\text{diag}(C, R_1, \ldots, R_1, \ldots, R_m, \ldots, R_m)$$

where C is a real diagonal matrix and each R_i is a 2×2 matrix of the form

$$R_i = \begin{pmatrix} c_i & -d_i \\ d_i & c_i \end{pmatrix}$$

The matrix C exhibits the (real) characteristic values of $\mathbf{T}$ (if any exist), each repeated a number of times equal to its algebraic multiplicity. One or more matrices R_i accompanies an irreducible quadratic factor q_i of the minimum function of $\mathbf{T}$; the numbers c_i and d_i are the real and imaginary parts of one of the pair of conjugate imaginary roots of q_i.

Proof. The first paragraph (with the reference to Lemma 8.6 replaced by one to Theorem 8.23) of the proof of Theorem 8.26 is applicable to the present proof and we shall continue from that point. Analogous to the case of an orthogonal transformation, the linearity of a factor $q_i(\lambda)$ of $m(\lambda)$ implies that $\mathbf{T}_i$ is represented by a scalar matrix relative to any basis for $N(q_i)$.

If $q_i(\lambda) = \lambda^2 - 2a\lambda + b$ is an irreducible quadratic, we define the scalar transformation $\mathbf{S}_i$ on $N(q_i)$ by the equation

$$\mathbf{S}_i\,\alpha = \sqrt{b}\,\alpha \qquad \text{for all } \alpha \text{ in } N(q_i)$$

Then it is seen immediately that $\mathbf{S}_i^{-1}\mathbf{T}_i = \mathbf{R}_i$, let us say, is a normal transformation on $N(q_i)$ with irreducible minimum function

$$q_i'(\lambda) = \lambda^2 - 2a'\lambda + 1 \qquad \text{where } a' = \frac{a}{\sqrt{b}}$$

Actually $\mathbf{R}_i$ is orthogonal. To show this we form

$$\begin{aligned}(\mathbf{R}_i + \mathbf{R}_i^*)(\mathbf{R}_i - \mathbf{R}_i^*) = \mathbf{R}_i^2 - \mathbf{R}_i^{*2} &= 2a'\mathbf{R}_i - \mathbf{I} - (2a'\mathbf{R}_i^* - \mathbf{I}) \\ &= 2a'(\mathbf{R}_i - \mathbf{R}_i^*)\end{aligned}$$

where we have used the fact that $q'(\lambda)$ is the minimum function of both $\mathbf{R}_i$ and $\mathbf{R}_i^*$. We leave it as an exercise (see Prob. 6) to show that the factor $\mathbf{R} - \mathbf{R}_i^*$ may be canceled from the initial and final members of the above chain of equalities, to conclude that

$$\mathbf{R}_i + \mathbf{R}_i^* = 2a'\mathbf{I}$$

Multiplying this equation by $\mathbf{R}_i$ and then substituting for $\mathbf{R}_i^2$ gives the desired formula, *viz.*, $\mathbf{R}_i\mathbf{R}_i^* = \mathbf{I}$.

We infer from Theorem 8.26 that there exists an orthonormal basis for

$N(q_i)$ relative to which $\mathbf{R}_i$ is represented by a direct sum of matrices (11) where $\cos\theta_i = a'$. Relative to the same basis, $\mathbf{T}_i = \mathbf{S}_i\mathbf{R}_i$ is represented by a direct sum of the matrices

$$\begin{bmatrix} \sqrt{b}\cos\theta_i & -\sqrt{b}\sin\theta_i \\ \sqrt{b}\sin\theta_i & \sqrt{b}\cos\theta_i \end{bmatrix} = \begin{bmatrix} c_i & -d_i \\ d_i & c_i \end{bmatrix}$$

where c_i and d_i are the numbers described in the theorem. This completes the proof.

PROBLEMS

1. Let (l_i, m_i, n_i), $i = 1, 2, 3$, denote the direction cosines of three mutually perpendicular lines in three-dimensional Euclidean space. Show that the matrix whose rows are these triplets is orthogonal.

2. Verify that the matrix

$$\frac{1}{k}\begin{bmatrix} 1 + l^2 - m^2 - n^2 & 2(lm + n) & 2(nl - m) \\ 2(lm - n) & 1 - l^2 + m^2 - n^2 & 2(mn + l) \\ 2(nl + m) & 2(mn - l) & 1 - l^2 - m^2 + n^2 \end{bmatrix}$$

where k, l, m, and n are real numbers, such that $l^2 + m^2 + n^2 = k - 1$, is orthogonal. Find its characteristic values.

3. In the proof of Lemma 8.6 fill in the details related to the proof that an irreducible quadratic factor has the form $\lambda^2 - 2a\lambda + 1$.

4. In the proof of Theorem 8.26, show that (i) β is a unit vector orthogonal to α, (ii) $S_\alpha = [\alpha, \beta]$ is invariant under $\mathbf{T}_i$, (iii) $S_\alpha^\perp$ is invariant under $\mathbf{T}_i$. Finally, verify equation (12).

5. Mould a proof for Theorem 8.27 along the lines of that for Theorem 8.26. The result in Prob. 9 of the previous section will be needed.

6. Verify the assertion made in the proof of Theorem 8.27 that $\mathbf{R}_i - \mathbf{R}_i^*$ may be canceled as stated by showing that this transformation is nonsingular. HINT: Since the minimum function of $\mathbf{R}_i$ is $q_i^{\prime}$, $\mathbf{R}_i$ is not symmetric by Lemma 8.5 and consequently, $\mathbf{R}_i - \mathbf{R}_i^* \neq \mathbf{0}$.

7. Determine a basis for $V_4(R^\#)$ relative to which the orthogonal transformation defined by the matrix

$$A = \begin{bmatrix} 0 & 0 & 0 & 1 \\ 1 & 0 & 0 & 0 \\ 0 & 1 & 0 & 0 \\ 0 & 0 & 1 & 0 \end{bmatrix}$$

is represented by a matrix of the type (10).

8.9. Hermitian Forms under the Unitary Group. Possibly it will not be out of place to recall the background pertinent to the forthcoming discussion. The earlier study (Sec. 5.7) of unrestricted nonsingular substitutions in Hermitian forms was described in Sec. 6.5 as the study of Hermitian forms under the full linear group $L_n(C)$. In alibi terminology the principal result may be stated thus: For the set of Hermitian forms in $x^1, x^2, \ldots, x^n$, the forms

$$\sum_1^p y^i\bar{y}^i - \sum_{p+1}^r y^i\bar{y}^i \qquad p \leqq r \leqq n$$

constitute a canonical set under $L_n(C)$. This implies that a suitable nonsingular linear transformation will reduce the locus

$$X^{\mathrm{T}}H\bar{X} = 1 \tag{14}$$

in $V_n(C)$ to one with an equation of the form

$$\sum_1^p y^i\bar{y}^i - \sum_{p+1}^r y^i\bar{y}^i = 1$$

From a remark in Sec. 8.6 it follows that reductions in (14) within the confines of orthonormal bases for the space amount to the reductions that are possible with unitary transformations. This matter will be settled (reverting to the alias viewpoint) by the following discussion of Hermitian forms under unitary substitutions.

Making the substitution $X = P^{\mathrm{T}}Y$† in the form $X^{\mathrm{T}}H\bar{X}$ replaces the Hermitian matrix H by the Hermitian matrix

$$PHP^* \qquad \text{where } P^* = \bar{P}^{\mathrm{T}} \tag{15}$$

If P is unitary, then $P^* = P^{-1}$, so that the matrix in (15) becomes PHP^{-1} or a matrix similar to H! We apply Theorem 8.21 to conclude that $X^{\mathrm{T}}H\bar{X}$ can be reduced to a diagonal form $\sum_1^n c_i y^i\bar{y}^i$ where the c_i's are the characteristic values of H. Since (Cor. 2, Theorem 8.17) these are real numbers, we may assume that the y^i's are so ordered (a unitary substitution will do this!) that

$$c_1 \geqq c_2 \geqq \cdots \geqq c_n$$

As characteristic values of H, the c_i's are uniquely determined by the given form. This fact, coupled with our agreement concerning their order, establishes the next result.

Theorem 8.28. A Hermitian form $X^{\mathrm{T}}H\bar{X}$ can be reduced to a uniquely determined diagonal form

$$c_1y^1\bar{y}^1 + c_2y^2\bar{y}^2 + \cdots + c_ny^n\bar{y}^n \qquad c_1 \geqq c_2 \geqq \cdots \geqq c_n \tag{16}$$

by a unitary substitution. The coefficients c_i are the (real) characteristic values of H.

†Writing the matrix of a substitution as a transpose is merely for the purpose of gaining uniformity in our notation.

From the alibi viewpoint this result means that, in the set of all Hermitian forms in $x^1, x^2, \ldots, x^n$, the forms of the type (16) constitute a canonical set under the unitary group. Of course, a form X^THX may be regarded as the representation h_α of a Hermitian function h on an n-dimensional unitary space U, indeed as that representation corresponding to the α-basis for U. We may and shall assume the α-basis is orthonormal. Then, so states the preceding theorem, together with Theorem 8.11, h is represented by a form of the type (16) relative to a suitable orthonormal basis for U. The forms (16) are designated as the canonical representations of Hermitian functions under the unitary group, and the accompanying orthonormal basis, in each case, is called a basis of *principal axes* for the function.

It is appropriate that we recast the foregoing result for matrices. If A is any square matrix of complex numbers, then PAP^* is both conjunctive and similar to A when P is unitary; thus *two of the fundamental matrix equivalence relations that have arisen in our studies coalesce* (see the second and fourth entries of the table in Sec. 7.4).

DEFINITION. *If A is a square matrix of complex numbers, a matrix of the form PAP^*, where P is unitary, is called* unitary similar *to A.*

Unitary similarity is an equivalence relation over the set of all $n \times n$ complex matrices. As such, it induces an equivalence relation over the set of $n \times n$ Hermitian matrices, which, since the property of being Hermitian is a unitary similarity invariant, is still unitary similarity. Then Theorem 8.28 yields the following result:

Theorem 8.29. A Hermitian matrix H is unitary similar to a diagonal matrix (whose diagonal entries are necessarily the characteristic values of H). Two Hermitian matrices are unitary similar if and only if they have the same characteristic values.

It is clear that each equivalence class of Hermitian matrices contains exactly one diagonal matrix $\operatorname{diag}(c_1, c_2, \ldots, c_n)$, such that $c_1 \geqq c_2 \geqq \cdots \geqq c_n$. These matrices provide us with a canonical set of Hermitian matrices under unitary similarity. As a working method for finding the canonical form of a Hermitian matrix H (and consequently, of a Hermitian form with matrix H) under unitary similarity, we may use that of Example 8.7; this entails interpreting H as the representation of a Hermitian transformation. This method is further illustrated by Examples 8.10 and 8.11 in the next section.

With one minor modification, the above discussion of unitary similarity over the set of Hermitian matrices is applicable to the more extensive set of $n \times n$ normal matrices of complex numbers. Indeed, since the property of being normal is invariant under unitary similarity, unitary similarity restricted to the set of $n \times n$ normal matrices is the same relation. Then Theorem 8.20 yields Theorem 8.29 restated for normal matrices. It is possible to describe a canonical set of normal matrices under unitary similarity as soon as a rule is devised for ordering a set of n complex numbers. One such is to order first according to decreasing absolute values and then order those numbers of equal absolute value by decreasing amplitude.

Turning next to the real case, the entire section up to this point can be reread with the conjugation bar omitted, "unitary matrix" replaced by "orthogonal matrix," etc. We shall state the principal results for reference.

Theorem 8.30. A real quadratic form $X^T AX$ can be reduced to a uniquely determined diagonal form

$$c_1(y^1)^2 + c_2(y^2)^2 + \cdots + c_n(y^n)^2 \qquad c_1 \geqq c_2 \geqq \cdots \geqq c_n \tag{17}$$

by an orthogonal substitution. The coefficients are the characteristic values of the symmetric matrix A.

This result relies upon Theorem 8.25. The forms of the type (17) constitute a canonical set under the orthogonal group.

The unitary similarity relation translates into *orthogonal similarity* in the real case; thus A and B are orthogonal similar if and only if there exists an orthogonal matrix P such that $B = PAP^T$. This relation amounts to a merger of the equivalence relations for real matrices which we have called congruence and similarity. The matrical version of Theorem 8.30 is stated next.

Theorem 8.31. A real symmetric matrix A is orthogonal similar to a diagonal matrix whose diagonal entries are the characteristic values of A. Two real symmetric matrices are orthogonal similar if and only if they have the same characteristic values.

The real symmetric matrices of the form $\text{diag}(c_1, c_2, \ldots, c_n)$ where $c_1 \geqq c_2 \geqq \cdots \geqq c_n$ constitute a canonical set under orthogonal similarity.

Analogous to the complex case, we may consider the orthogonal similarity relation over the set of normal real matrices. For this we use Theorem 8.27. As soon as a rule is established for distinguishing in each equivalence class a single matrix of the type displayed in that theorem, one obtains a canonical set of normal real matrices under orthogonal similarity.

PROBLEMS

1. Verify the assertion made prior to Theorem 8.28 that a renumbering of the set of variables in a form can be effected by a unitary (indeed, an orthogonal) substitution. The matrix of such a substitution is called a *permutation matrix*, and is characterized by the property that each row and each column contains a single 1, all other entries being 0. Show that the $n \times n$ permutation matrices and multiplication form a subgroup of the orthogonal group isomorphic to the symmetric group S_n.

2. Show that the property of being Hermitian (normal) is a unitary similarity invariant.

3. For the Hermitian matrix

$$A = \begin{pmatrix} 1 & i & 0 \\ -i & 1 & -i \\ 0 & i & 1 \end{pmatrix}$$

determine its canonical form D under unitary similarity as well as the unitary matrix P such that $PAP^* = D$.

4. Prove that if H is a fixed positive definite Hermitian matrix and X is a variable positive semidefinite Hermitian matrix, then the minimum value of $\det(H + X)$ is $\det A$ and is attained when and only when $X = 0$.

8.10. Reduction of Quadric Surfaces to Principal Axes. The term "principal axes" introduced in the previous section has its origin in the geometric problem of determining a orthonormal basis for a Euclidean space, relative to which the equation of a locus defined by a quadratic function is of the form

$$\sum c_i(y^i)^2 = a \tag{18}$$

where on the left appears the canonical representation of the function. For simplicity, let us consider the locus of the equation

$$X^{\mathrm{T}}AX = a \tag{19}$$

in $V_n(R^\#)$. The notation tacitly assumes that vectors X are referred to the orthonormal basis of unit vectors ϵ_i. We shall assume that A has rank n with characteristic values $c_1 \geqq c_2 \geqq \cdots \geqq c_n$ and shall label the locus, in accordance with an earlier definition, a central quadric surface. We know that there exists an orthogonal substitution $X = P^{\mathrm{T}}Y$ which reduces the above form to $\sum_1^n c_i(y^i)^2$. Recalling the structure of an orthogonal transformation, it follows that the given basis of unit vectors can be rotated (indeed by a sequence of plane rotations) into an orthonormal basis, so-called "principal axes" of the figure, relative to which the surface is represented by equation (18) or, after an obvious manipulation,

$$\pm \frac{(y^1)^2}{a_1^2} \pm \frac{(y^2)^2}{a_2^2} \pm \cdots \pm \frac{(y^n)^2}{a_n^2} = 0 \text{ or } 1 \tag{20}$$

Now let us specialize our discussion to the case $n = 3$. Assuming that the locus exists and is not a point, only the following cases exist, if we ignore the ordering of the a_i's:

(i) $$\frac{(y^1)^2}{a_1^2} + \frac{(y^2)^2}{a_2^2} + \frac{(y^3)^2}{a_3^2} = 1 \qquad \text{(ellipsoid)}$$

(ii) $$\frac{(y^1)^2}{a_1^2} + \frac{(y^2)^2}{a_2^2} - \frac{(y^3)^2}{a_3^2} = 1 \qquad \text{(hyperboloid of one sheet)}$$

(iii) $$\frac{(y^1)^2}{a_1^2} - \frac{(y^2)^2}{a_2^2} - \frac{(y^3)^2}{a_3^2} = 1 \qquad \text{(hyperboloid of two sheets)}$$

(iv) $$\frac{(y^1)^2}{a_1^2} + \frac{(y^2)^2}{a_2^2} \pm \frac{(y^3)^2}{a_3^2} = 0 \qquad \text{(cone)}$$

Observe that one can distinguish among these cases by the signs of the characteristic values of A. Of course the classification can be extended to the case of quadric surfaces (19) with A singular; this will yield various types of cylinders.

In the following two examples, the reduction of quadric surfaces to principal axes is discussed:

EXAMPLE 8.10

Consider the quadric surface defined by the equation

(21) $$2(x^1)^2 + (x^2)^2 - 4x^1x^2 - 4x^2x^3 = 4$$

relative to the basis of unit vectors for $V_3(R^\#)$. The matrix

$$A = \begin{bmatrix} 2 & -2 & 0 \\ -2 & 1 & -2 \\ 0 & -2 & 0 \end{bmatrix}$$

of the quadratic form in (21) is nonsingular with characteristic values 4, 1, -2. Let us determine the principal axes and corresponding equation of the figure (a hyperboloid of one sheet, according to the signs of the characteristic values). Our method stems from the interpretation of A as the representation of a symmetric transformation **A** relative to the orthonormal ϵ-basis (see Example 8.7 and the paragraph preceding it). The null spaces $N(\lambda - 4)$, $N(\lambda - 1)$, and $N(\lambda + 2)$, accompanying the factors of the minimum function $(\lambda - 4)(\lambda - 1)(\lambda + 2)$ of **A**, determine an orthonormal basis of characteristic vectors for $V_3(R^\#)$, relative to which **A** is represented by

$$\text{diag}(4, 1, -2) = P^{-1}AP = P^{\text{T}}AP$$

The columns of P are characteristic vectors ρ_1, ρ_2, ρ_3 of unit length corresponding to 4, 1, -2, respectively, in other words, unit vectors which generate the three null spaces. Since P is orthogonal, $\det P = 1$, or -1; the latter case can always be avoided by reversing the direction of one of the ρ_i's if necessary. With $\det P = 1$, we can then *rotate* the ϵ-basis into the ρ-basis.

To determine the ρ_i's, we solve in turn the systems of linear equations

$$(4I - A)X = 0 \qquad (I - A)X = 0 \qquad (-2I - A)X = 0$$

We obtain

$$3\rho_1 = \begin{pmatrix} 2 \\ -2 \\ 1 \end{pmatrix} \quad 3\rho_2 = \begin{pmatrix} 2 \\ 1 \\ 2 \end{pmatrix} \quad 3\rho_3 = \begin{pmatrix} 1 \\ 2 \\ 2 \end{pmatrix} \quad \text{or} \quad 3P = \begin{pmatrix} 2 & 2 & 1 \\ -2 & 1 & 2 \\ 1 & 2 & 2 \end{pmatrix}$$

Then the substitution $X = PY$ reduces (21), or $X^T AX = 4$, to

$$Y^T(P^T AP)Y = 4(y^1)^2 + (y^2)^2 - 2(y^3)^2 = 4$$

or

$$\frac{(y^1)^2}{1} + \frac{(y^2)^2}{4} - \frac{(y^3)^2}{2} = 1$$

EXAMPLE 8.11

Consider the quadric surface defined by the equation

$$X^T AX = 2(x^1)^2 + 2(x^2)^2 + 2(x^3)^2 - 2x^1x^2 + 2x^1x^3 - 2x^2x^3 = 4$$

relative to the basis of unit vectors for $V_3(R^\#)$. Let us determine principal axes and the corresponding equation of the figure. The characteristic function of the matrix A, implicitly defined above, is $f(\lambda) = (\lambda - 1)^2(\lambda - 4)$, and, consequently, the minimum function of A is $(\lambda - 1)(\lambda - 4)$. The presence of $(\lambda - 1)^2$ in $f(\lambda)$ implies that $d[N(\lambda - 1)] = 2$. In contrast to the preceding example, where three one-dimensional null spaces occurred, with corresponding generators of unit length determined uniquely apart from sign, we can now choose an orthonormal basis $\{\rho_1, \rho_2\}$ for $N(\lambda - 2)$ in infinitely many ways. Any such basis, when extended to a basis for $V_3(R^\#)$ by adjoining a generator ρ_3 of unit length for $N(\lambda - 4)$, is a set of principal axes for the figure. The vector ρ_3 may always be chosen so that the ϵ-basis can be rotated into the ρ-basis.

To determine ρ_1 and ρ_2, we first solve the system $(I - A)X = 0$ of rank 1 and then replace the basis of the solution space by an orthonormal basis. Of course, ρ_3 is obtained from $(4I - A)X = 0$. The results appear in the column of the matrix

$$P = \begin{bmatrix} \frac{1}{\sqrt{2}} & \frac{1}{\sqrt{6}} & -\frac{1}{\sqrt{3}} \\ \frac{1}{\sqrt{2}} & -\frac{1}{\sqrt{6}} & \frac{1}{\sqrt{3}} \\ 0 & -\frac{2}{\sqrt{6}} & -\frac{1}{\sqrt{3}} \end{bmatrix}$$

which defines the substitution $X = PY$, which, in turn, reduces the equation of the surface to

$$(y^1)^2 + (y^2)^2 + 4(y^3)^2 = 4 \quad \text{or} \quad \frac{(y^1)^2}{4} + \frac{(y^2)^2}{4} + \frac{(y^3)^2}{1} = 1$$

The geometric explanation of the arbitrariness in the choice of $\{\rho_1, \rho_2\}$ is now clear—the figure is an ellipsoid of revolution.

It is clear that the computational procedure used above can be described in simple language, *i.e.*, without reference to symmetric transformations, etc.

PROBLEMS

1. Determine an orthogonal substitution that reduces the quadratic form $x^1x^2 + x^2x^3 + x^3x^1$ to a diagonal form. What is this diagonal form?

2. Determine a matrix orthogonal similar to

$$A = \begin{bmatrix} 2 & -1 & 1 \\ -1 & 2 & -1 \\ 1 & -1 & 2 \end{bmatrix}$$

3. Determine a basis of principal axes for the quadratic form $4(x^1)^2 + (x^2)^2 - 8(x^3)^2 + 4x^1x^2 - 4x^1x^3 + 8x^2x^3$.

4. Determine a set of principal axes for the quadric surface $2(x^1)^2 + 2(x^2)^2 - (x^3)^2 + 8x^1x^2 - 4x^1x^3 - 4x^2x^3 = 2$.

8.11 Maximum Properties of Characteristic Values. In $V_3(R^\#)$ it is possible to describe the characteristic values of a (real) symmetric matrix A as solutions of a sequence of maximum problems associated with the quadric surface $X^TAX = 1$. When the problem is formulated appropriately, it admits an extension to the n-dimensional case. The outcome is an interesting and useful characterization of the characteristic values of a symmetric matrix.

To simplify both the writing and reading of the state of affairs in three dimensions, we shall consider the case of a positive definite matrix A, so that the associated quadric surface is an ellipsoid E. When referred to a set of principal axes, the equation of E has the form

$$Y^{\mathrm{T}}DY = \frac{(y^1)^2}{a_1^2} + \frac{(y^2)^2}{a_2^2} + \frac{(y^3)^2}{a_3^2} = 1$$

where $c_i = 1/a_i^2$, $i = 1, 2, 3$, are the characteristic values of A. We may and shall assume that $c_1 \geqq c_2 \geqq c_3 > 0$ and, consequently, that $a_1^2 \leqq a_2^2 \leqq a_3^2$. Then it is clear that a_1 is equal to the shortest† distance from the origin O to E (this is the shortest semiaxis of E), while a_2 is equal to the shortest distance from O to E in the space (a plane) orthogonal to the first principal axis, etc. Since $c_i = 1/a_i^2$, the characteristic values of A therefore appear as the solutions of certain maximum problems.

The above observation concerning a_1 can be stated as follows: The square of the shortest semiaxis (that is, a_1^2) is the minimum value in the minimum problem

$$Y^{\mathrm{T}}Y = \min \qquad \text{for } Y^{\mathrm{T}}DY = 1 \tag{22}$$

Then 1 is the maximum value in the maximum problem

$$Y^{\mathrm{T}}DY = \max \qquad \text{for } Y^{\mathrm{T}}Y = a_1^2$$

Normalizing the side condition in this problem, we conclude that $1/a_1^2 = c_1$, the greatest characteristic value of A, is the maximum value in the maximum problem

$$Y^{\mathrm{T}}DY = \max \qquad \text{for } Y^{\mathrm{T}}Y = 1 \tag{23}$$

In turn, the maximum value in (23) may be determined as that in

$$\rho(Y) = \frac{Y^{\mathrm{T}}DY}{Y^{\mathrm{T}}Y} = \max \qquad \text{for } Y \neq 0 \tag{24}$$

where this equation defines the function ρ [on $V_3(R^\#)$ to $R^\#$] which we call a *Rayleigh quotient.*

From a study of the Rayleigh quotient in the general case will emerge the characterization of the characteristic values of a symmetric matrix that we have in mind. So we begin again! Let A denote an $n \times n$ symmetric matrix, and define the Rayleigh quotient ρ for A as follows:

$$\rho(X) = \frac{X^{\mathrm{T}}AX}{X^{\mathrm{T}}X} \qquad X \text{ in } V_n(R^\#) \text{ and } X \neq 0 \tag{25}$$

†The usage throughout this section of the words "maximum" and "minimum" (rather than "least upper bound" and "greatest lower bound," respectively) is justified, since in every case it is a question of a continuous function defined on a bounded closed set in a finite dimensional Euclidean space. Under these circumstances, the least upper bound of the function is attained at some point of the set; similarly for the greatest lower bound.

To study ρ, we introduce an auxiliary function u of X in $V_n(R^{\#})$ and a real variable ρ, defined by the equation

$$u(X, \rho) = X^T AX - \rho X^T X \tag{26}$$

Then the equation

$$u(X, \rho) = 0 \qquad \text{where } X \neq 0$$

defines ρ as the Rayleigh quotient (25). The justification for introducing u into the discussion is found in the next result.

Lemma 8.7. If $\{X_1, \rho_1\}$ is a solution of the problem of maximizing ρ on the set where $X \neq 0$ [thus $\rho(X_1) = \rho_1 = \max$, $X_1 \neq 0$], then X_1 maximizes $u(X, \rho_1)$ on the set where $X \neq 0$. Conversely, if there exists an $X_0 \neq 0$ maximizing $u(X, \rho_0)$ for an arbitrary but fixed number ρ_0, then the maximum value is 0 and $\rho_0 = \rho_1$.

Proof. The assumptions for the first part of the lemma imply that (i) $u(X_1, \rho_1) = 0$ and $X_1 \neq 0$ and (ii) if $u(X, \rho) = 0$ for $X \neq 0$, then $\rho_1 \geqq \rho$. We have to prove [since $u(X_1, \rho_1) = 0$] that $u(X, \rho_1) \leqq 0$ for all $X \neq 0$. Assume, contrariwise, that

$$u(X_2, \rho_1) = X_2^T AX_2 - \rho_1 X_2^T X_2 > 0 \qquad X_2 \neq 0$$

Since $X_2^T X_2 > 0$, it is clear that there exists a $\rho_2 > \rho_1$ with $u(X_2, \rho_2) = 0$. But this contradicts (ii).

For the converse, we first show that if $u(X, \rho_0)$ can be maximized on the set where $X \neq 0$, then its maximum value is 0; this is a consequence of the homogeneity property $u(cX, \rho_0) = c^2 u(X, \rho_0)$ of u. Thus, if X_0 maximizes $u(X, \rho_0)$, then $u(X_0, \rho_0) = 0$, which implies that ρ_0 is a value of ρ in (25): $\rho_0 = \rho(X_0)$. Since ρ_1 is the maximum value of ρ, it follows that $\rho_0 \leqq \rho_1$. To prove equality, assume that $\rho_0 < \rho_1$. Then since

$$u(X_1, \rho_1) = X_1^T AX_1 - \rho_1 X_1^T X_1 = 0$$

it follows that $u(X_1, \rho_0) > 0$, which is contrary to the assumption that X_0 maximizes $u(X, \rho_0)$ with maximum value 0. This completes the proof.

The importance of this lemma lies in the fact that it converts the problem of maximizing ρ in (25) by a choice of $X \neq 0$ to that of maximizing u in (26) for a fixed but unknown ρ. The steps involved in the conversion are of the type that one takes in establishing the rule known as *Lagrange's method of undetermined multipliers* for investigating the stationary points of a function whose variables are constrained by a side condition. In effect, we have validated this rule for the case at hand. By restricting our attention to this particular problem we have been able to obtain more precise information (in terms of maximum rather than stationary values) than the Lagrange method gives in general.

Our principal concern is the determination of the ρ-value, namely, the maximum value ρ_1 of ρ, for which it is possible to maximize u. To do this, we recall that at a critical point X of $u(X, \rho_1)$, we have

$$\frac{\partial u}{\partial x^i} = 0 \qquad i = 1, 2, \ldots, n \tag{27}$$

This system of equations can be collapsed to the single matrix equation $AX - \rho_1 X = 0$ to conclude that a solution X_1 is a characteristic vector accompanying ρ_1, which is a characteristic value of A. On the other hand, for any characteristic value ρ_0 of A and accompanying characteristic vector X_0, we have $u(X_0, \rho_0) = X_0^T(AX_0 - \rho_0 X_0) = 0$. This proves the following result:

Theorem 8.32. The maximum value of the Rayleigh quotient

$$\rho(X) = \frac{X^T A X}{X^T X}$$

for X in $V_n(R^*)$ and different from 0, is the maximum characteristic value c_1 of the symmetric matrix A, and a vector X_1 for which this maximum is attained is a characteristic vector associated with c_1.

This result can be paraphrased as follows: If c_1 denotes the largest characteristic value of A and X is any nonzero vector, then

$$\frac{X^T A X}{X^T X} \leqq c_1$$

where the equality sign holds for each characteristic vector associated with c_1.

The remaining characteristic values and associated vectors of A may be obtained by using the suggestion, made earlier, in conjunction with the ellipsoid. If the characteristic values of A are $c_1, c_2, \ldots, c_n$ (each counted a number of times equal to its algebraic multiplicity), we may assume that $c_1 \geqq c_2 \geqq \cdots \geqq c_n$ and then describe c_2 as follows: It is the maximum of the Rayleigh quotient ρ when X is restricted to the orthogonal complement in $V_n(R^*)$ of the space generated by X_1. In general, the kth characteristic value c_k is obtained after $c_1, c_2, \ldots, c_{k-1}$ (together with associated characteristic vectors $X_1, X_2, \ldots, X_{k-1}$) have been found, as the maximum of ρ under the side condition of orthogonality to the space $[X_1, X_2, \ldots, X_{k-1}]$.

One ulterior purpose of the foregoing exposition has been to draw the reader's attention to the intimate relationship among the diverse notions of characteristic values of a symmetric matrix, semiaxes of a quadric surface, and Lagrange's method of undetermined multipliers.

Another purpose has been to develop the background that is necessary to justify the following method for deriving a set of principal axes for a real quadratic function q when some approximation method must be used to estimate maximum values of a function. Let X^TAX be the value of q at the vector ξ in $V_n(R^\#)$ with ϵ-coordinates X, so that $q(\xi) = X^TAX$. If α_1 maximizes q on the unit sphere, that is, α_1 is a vector of unit length such that $q(\alpha_1) = \max$, then $q(\alpha_1) = \rho_1$, the largest characteristic value of A, and α_1 is an accompanying unit characteristic vector. If we choose α_1 as the first member of a new orthonormal basis $\{\alpha_1, \alpha_2, \ldots, \alpha_n\}$ for the space, then, of course, $q(\xi) = Y^TBY$ where Y designates the α-coordinates of ξ. Actually, $q(\xi)$ has the form

$$q(\xi) = \rho_1(y^1)^2 + q_1(\xi)$$

where $q_1(\xi)$ is a quadratic form in $y^2, y^3, \ldots, y^n$. (The reader can verify this statement directly by showing that since q assumes its maximum value at the vector with α-coordinates $(1, 0, \ldots, 0)^T$, Y^TBY contains no cross product terms involving y^1.) Thus, $q_1(\xi)$ is a quadratic function on the $(n-1)$-dimensional orthogonal complement of α_1, and we may reapply the same procedure of choosing a new orthonormal basis $\{\beta_2, \beta_3, \ldots, \beta_n\}$ such that $q_1(\beta_2) = \max$; this splits another diagonal term off the form. Continuing in this manner, one finally obtains a basis of principal axes $\{\alpha_1, \beta_2, \gamma_3, \ldots\}$ for q, together with a diagonal representation of the type described in equation (17).

PROBLEMS

1. Imitate the discussion at the beginning of this section through equation (24) for the ellipse whose equation is $x^2 + 4y^2 = 1$.

2. Do the same, with appropriate modifications, for the hyperbola whose equation is $x^2 - 4y^2 = 1$.

3. Determine a set of principal axes for the quadratic form $(x^1)^2 + 2x^1x^2 + x^1x^3$, using the method outlined after Theorem 8.32.

8.12. The Reduction of Pairs of Hermitian Forms. In the analysis of various physical problems, there arises the question of the simultaneous simplification of a pair of Hermitian, or quadratic, forms by a single nonsingular substitution. We shall discuss one such problem, *viz.*, when one form is positive definite, not only to supply an answer but, in addition, to show how neatly it fits into our study of unitary spaces.

Let

$$f = X^TH\overline{X} \qquad \text{and} \qquad g = X^TK\overline{X}$$

be the given forms, where H is positive definite and both H and K are of order n. We now introduce the vector space $V_n(C)$ into the discussion and mold it into a unitary space to fit our needs by defining an inner-product function as follows:

$$(X|Y) = X^T H \overline{Y}$$

(It is left to the reader to validate the alleged accomplishment.) As such, we have introduced a notion of length, orthogonality, etc., into $V_n(C)$ and, consequently, have defined the unitary group for the space. Now let us construct an orthonormal basis for the space. This may be done by the Gram-Schmidt process, starting with the basis of unit vectors, which has tacitly been assumed to be the basis for $V_n(C)$ up to this point. A nonsingular substitution

$$X = P^T Y \tag{28}$$

is thereby defined which reduces f to $Y^T \overline{Y}$ (that is, $PHP^* = I$). Of course, P is not unitary in general. Under the same substitution, g is reduced to

$$Y^T(PKP^*)\overline{Y} \tag{29}$$

According to Theorem 8.28, there exists a unitary substitution

$$Y = Q^T Z \qquad Q^T \text{ unitary} \tag{30}$$

which reduces (29) to the diagonal form

$$Z^T D \overline{Z} \qquad D = \operatorname{diag}(d_1, d_2, \ldots, d_n)$$

where the d_i's are the characteristic values of PKP^* and hence the roots of the polynomial equation

$$\begin{aligned} 0 = \det(\lambda I - PKP^*) &= \det(\lambda PHP^* - PKP^*) \\ &= \det P(\lambda H - K)P^* \\ &= \det(\lambda H - K) \end{aligned}$$

Here we have used the fact that $PHP^* = I$.

Since the substitution (30) is unitary, it does not disturb $Y^T \overline{Y}$, that is, $Y^T \overline{Y} = Z^T \overline{Z}$. We conclude, therefore, that the substitution

$$X = P^T Q^T Z = (QP)^T Z$$

simultaneously reduces the original forms f and g to

$$Z^T \overline{Z} = z^1 \bar{z}^1 + z^2 \bar{z}^2 + \cdots + z^n \bar{z}^n$$

and

$$Z^T D \overline{Z} = d_1 z^1 \bar{z}^1 + d_2 z^2 \bar{z}^2 + \cdots + d_n z^n \bar{z}^n$$

This proves our final result.

Theorem 8.33. Let $X^T H\overline{X}$ and $X^T K\overline{X}$ be Hermitian forms in n variables, with H positive definite. Then there exists a nonsingular substitution $X = RZ$ which reduces the forms to $Z^T\overline{Z}$ and $Z^T D\overline{Z}$, respectively, where $D = \text{diag}(d_1, d_2, \ldots, d_n)$. For any such substitution, the numbers $d_1, d_2, \ldots, d_n$ are the roots of the polynomial equation $\det(\lambda H - K) = 0$.

PROBLEMS

1. Illustrate the exposition in this section with the forms whose matrices are

$$\begin{pmatrix} 1 & i \\ -1 & 2 \end{pmatrix} \quad \text{and} \quad \begin{pmatrix} 3 & 1 \\ 1 & 3 \end{pmatrix}$$

2. Show that two Hermitian forms with matrices H and K, respectively, can be reduced simultaneously to their respective canonical forms (*i.e.*, exhibiting the characteristic values of H and K) by a *unitary* substitution, if and only if H and K commute: $HK = KH$.

REFERENCES

ARTIN, E.: "Galois Theory," 2d ed., University of Notre Dame, 1944.

BIRKHOFF, G., and S. MACLANE: "A Survey of Modern Algebra," The Macmillan Company, New York, 1941.

HALMOS, P. R.: "Finite Dimensional Vector Spaces," Princeton University Press, Princeton, N. J., 1942.

HASSE, H.: "Höhere Algebra," Vols. I and II, Walter De Gruyter & Company, Berlin, 1933.

JULIA, G.: "Introduction mathématique aux théories quantiques," Vol. I, Gauthier-Villars & Cie, Paris, 1936.

MACDUFFEE, C. C.: "The Theory of Matrices," Chelsea Publishing Co., New York, 1946.

MACDUFFEE, C. C.: Vectors and Matrices, *Carus Math. Mon.* 7, Mathematical Association of America, 1943.

SCHREIER, O., and E. SPERNER: "Einführung in die analytische Geometrie und Algebra," Vols. I and II, B. G. Teubner, Leipzig, 1931.

TURNBULL, H. W., and A. C. AITKEN: "An Introduction to the Theory of Canonical Matrices," Blackie & Son, Ltd., Glasgow, 1948.

INDEX OF SPECIAL SYMBOLS

Symbol	*Page*
(N)	3
$X = (x^1, x^2, \ldots, x^n)^{\mathrm{T}}$	3
(N_e)	9
(H)	17
$(\mathrm{H}^{\mathrm{T}})$	20
$F, +, \cdot$	24
$R^{\#}$	24
R	24
C	25
J	25
J_p	26
$\epsilon_1, \epsilon_2, \ldots, \epsilon_n$	32
$V_n(F)$	33
$P_n(C)$	34
O	36
$[\alpha_1, \alpha_2, \ldots, \alpha_n]$	37
$S + T$	37
$S \cap T$	37
$d[V]$	39
$S \oplus T$	51
$\xi \equiv \eta(\mathrm{mod}\ S)$	52
V/S	52
$V \simeq V'$	53
$\delta_{ij} = \delta_j^i$	57
I_n	57
A^{T}	57
R_{ij}	59
$R_i(c)$	59
$R_{ij}(c)$	59
$r(A)$	62
$A \overset{E}{\sim} B$	65
J_r	65
α-basis	67
α-coordinates	67
$n(A)$	85
$\det A$	87
$\mathrm{adj}\, A$	95
$A \overset{C}{\sim} B$	112
$D_k(A)$	113
A^*	130
$\mathbf{A}$	134
$\mathbf{A}\alpha$	135
$(\mathbf{A}; \alpha)$	137
$\mathbf{I}$	138
$L_n(F)$	144

Symbol	*Page*	*Symbol*	*Page*
$R(\mathbf{A})$	151	$d_k(\dot{M})$	193
$N(\mathbf{A})$	151	h_i	194
$r(\mathbf{A})$	151	$C(f)$	200
$n(\mathbf{A})$	151	$E(p^e)$	206
$\mathbf{0}$	156	$J_e(c)$	209
$f \mid g$	160	$(\mid)$	212
$\mathbf{A}_1 \oplus \mathbf{A}_2 \oplus \cdots \oplus \mathbf{A}_r$	168	$\|\|\xi\|\|$	212
$A_1 \oplus A_2 \oplus \cdots \oplus A_r$	168	$\xi \perp \eta$	219
$N(f)$	169	$S^{\perp}$	221
$A \overset{S}{\sim} B$	175	$\mathbf{T}^*$	231

INDEX

(See also Index of Special Symbols)

A

Abelian group, 139
Adjoint matrix, 95
Adjoint transformation, 231–234, 240
Affine transformation, 145
Alias, 138
Alibi, 138
Alternating group, 147
Associated adjoint matrix, 130, 234
Associates, 160
Augmented matrix, 56

B

α-basis, 67
A-basis, 189–191
Basis, adapted to subspaces, 168
  of solution set of (H), 19
  of vector space, 41–42
Bessel's inequality, 222
Bilinear forms, 107–110
  canonical set under equivalence, 109
  matrix of, 108
  rank of, 108
  reduction of, 109–110
Bilinear function, 107–109
  rank of, 108
  representation of, 107–109
   canonical, 109
Binary composition, xv
Binary relation, 47

C

Canonical form, 49–50
  (*See also* Hermitian matrix; etc.)
Canonical representation, of bilinear function, 109
  of Hermitian function, 131, 250
  of linear transformation, 200–209
  of real quadratic function, 119, 251
Canonical set, 49–50
  (*See also* Congruence; Conjunctivity; Bilinear form; etc.)
Cauchy's inequality, 217
Cayley's theorem, 146
Central quadric surface, 124, 252–255
Characteristic function, 177, 203
Characteristic matrix, 190
Characteristic value, 182–184, 235
  algebraic multiplicity of, 183–184
  geometric multiplicity of, 183–184
Characteristic vector, 182
Classical canonical form, 209
Classical canonical representation, 209
Coefficient matrix, 56
Cofactor, 93
Column rank, 61
Column space, 61
Commutative group, 139
Commutative ring, 155
Companion matrix, 200
Complete orthonormal set, 222
Complete relation matrix, 189
Congruence, 112–113
  canonical set under, 121
Congruence modulo S, 52
Congruence transformation, 113
  C_{I}, 113, 116
  C_{III}, 113, 116
Congruent matrices, 112–113
Conjunctive matrices, 130
Conjunctivity, 130
  canonical set under, 131–132
Consistency condition, 6
Consistent system of equations, 6
α-coordinates, 67
Coordinates, 41
Coset, 141
Cramer's rule, 96
Cyclic group, 143

D

Determinant function, 87–94
  development of, 94
  evaluation of, 100–103

Determinant function, of product, 92
of transposed matrix, 92
Determinantal divisor, 193–194
Diagonal matrix, 56
Diagonal representation, 180, 236
Dimension, 39–40
n-dimensional Euclidean space, 33
Direct sum, of linear transformations, 168
of matrices, 168
of subspaces, 51–52
Division algorithm, 159
Dual basis, 233

E

Echelon matrix, 60
Echelon system of equations, 8
Elementary column transformation, 65, 81–82
Elementary divisor, of linear transformation, 204
of matrix, 204
of polynomial matrix, 196
Elementary matrix, 78–80
Elementary row transformation, 59, 77–78
Elementary transformation, 65
Ellipsoid of inertia, 127
Equivalence, 65, 82
canonical set under, 65, 194–195
under a group, 148–151
Equivalence relation, 48–50, 178–179
canonical form under, 49–50
canonical set under, 49–50
(*See also* Congruence; Conjunctivity; etc.)
Equivalent bilinear forms, 108
Equivalent Hermitian forms, 130
Equivalent matrices, 65–67, 82–84
Equivalent quadratic forms, 112, 149
Equivalent systems of equations, 7
Euclidean vector space, 214

F

Field, 23–27, 140
addition in, 23
multiplication in, 23
subfield of, 24
unit element of, 23
zero element of, 23
Finite field, 25
Finite group, 143
Form, 106
(*See also* Bilinear forms; Linear form; etc.)
Full linear group, 144
Functions, xiv–xv
equality of, 5
scalar multiple of, 5
sum of, 5

G

Generating set, 37
A-generating system, 189–191
relation for, 189
relation matrix for, 189
Gram-Schmidt construction, 222
Gramian, 216
Greatest common divisor, 160
Group, 139–148
of correspondences, 146
order of, 144
of symmetries, 145

H

Hamilton-Cayley theorem, 203
Hermitian conjugate bilinear form, 129
Hermitian conjugate bilinear function, 129
Hermitian forms, 130
canonical set, under equivalence, 132
under unitary group, 250
theorems about, 132, 249
Hermitian function, 130
representation of, 130
canonical, 131, 250
signature of, 132
Hermitian matrix, 129–132
canonical form, under conjunctivity, 131–132
under unitary similarity, 250
theorems about, 131–132, 250
Hermitian transformation, 234–235, 238
Homogeneous system of equations, 4
solution of, nontrivial, 15
trivial, 15
Homomorphic image, 53
Homomorphism function, 53

I

Ideal, 161
generated by $\{f_1, f_2, \ldots, f_r\}$, 161
Idempotent transformation, 153
Identity matrix, 57

Identity transformation, 138
Improper subspace, 38
Inconsistent system of equations, 2
Index, 123
Index polynomial, 188
Induced transformation, 166
Inner product, 212
Inner-product function, 212–215
Intersection of subspaces, 37
Invariant factor, 194
Invariant function, 63
Invariant space, 166–168
ℛ-invariants, 49
  complete set of, 49
Inverse function, 44
Inverse matrix, 79–80, 83–84, 96
Inverse transformation, 137
Involution, 239
Irreducible polynomial, 162
Isomorphic fields, 45
Isomorphic spaces, 45–46
Isomorphic systems, 45–46
Isomorphism function, 45–46

J

Jacobian, 97
Jordan matrix, 209

K

Kronecker delta, 57

L

Least common multiple, 162
Linear equation, 1
  solution of, 1
Linear form, 4, 8, 106
  length of, 8
  (*See also* Set of linear forms)
Linear function, 45, 134–136
  representation of, 105–107
Linear transformation, 136
  canonical representations of, 200–209
    classical, 209
    rational, 206
  direct sum of, 168
  elementary divisors of, 204
  invariant space of, 166–168
  minimum function of, 171, 202
  null space of, 151, 169–172
  nullity of, 151–152
Linear transformation, range of, 151
  rank of, 151–152
  representation of, 165–166
    diagonal, 180
    superdiagonal, 184, 229
Linear vector space, 32–33
  addition in, 32–33
  basis of, 41–42
  computation rules for, 34–35
  dimension of, 41–42
  scalar multiplication in, 33
  subspace of, 36–38
Linearly dependent set, 5, 38
Linearly independent set, 5, 38
Lorentz group, 145

M

Matrices, direct sum of, 168
  equality of, 71
  product of, 69–76
  sum of, 74
Matrix, 55
  column rank of, 61
  column space of, 61
  elementary column transformation of, 65, 81–82
  elementary divisors of, 204
  elementary row transformation of, 59, 77–78
  elementary transformation of, 65
  minor of, 98
  nullity of, 85
  partition of, 57
  polynomial, 191–197
  principal diagonal of, 56
  rank of, 62–63
  row rank of, 61
  row space of, 61
  scalar multiple of, 74
  (*See also* Adjoint matrix; Diagonal matrix; Matrices; etc.)
Matrix addition, 74
Matrix multiplication, 69–76
Maxima and minima, 127–128
Maximal linearly independent set, 13, 16
Minimum function, of linear transformation, 171, 202
  of square matrix, 171, 202
Minor, 98
  order of, 98
Monic polynomial, 158

N

Nonhomogeneous system of equations, 4
Nonsingular matrix, 64, 67, 80
Nonsingular substitution, 109
Nonsingular transformation, 137, 152
Norm, 217
Normal matrix, 234
  canonical form under unitary similarity, 251
Normal subgroup, 141
Normal transformation, 234–236, 241–242, 247
Normalized linear form, 8
Null space, 151, 169–172
Nullity, of matrix, 85
  of transformation, 151–152

O

One-one correspondence, 44–45
Order, of group, 144
  of minor, 98
Orthogonal complement, 221
Orthogonal group, 240
Orthogonal matrix, 240
Orthogonal projection, 219
Orthogonal set, 221
Orthogonal similar matrices, 251
Orthogonal similarity, 251
  canonical set under, 251
Orthogonal transformation, 240–241, 243–244
Orthogonal n-tuples, 21
Orthonormal basis, 222–224
Orthonormal set, 221

P

Parseval's equation, 224
Partition, 48
Permutation, 147
Permutation group, 147
Permutation matrix, 251
Polynomial matrix, 191–197
  elementary divisors of, 196
  invariant factors of, 194
Polynomials, 157–163
  equality of, 157
  product of, 158
  root of, 160
  substitution principal for, 159
  sum of, 157
Prime polynomial, 162
Principal axes, 250, 259
Principal diagonal, 56
Principal ideal, 161
Product set, xv
Projection, 153–154
Proper divisor, 160
Proper subspace, 36

Q

Quadratic forms, 111–112
  reduction of, 115–117, 119, 253–255
  theorems about, 115, 117, 119, 122–123, 251
  (*See also* Real quadratic form)
Quadratic function, 111–112
  representation of, 111
  (*See also* Real quadratic function)
Quadric surface, 70, 124, 252–255
Quotient group, 141–142
Quotient space, 53–54

R

Range, 151
Rank, of bilinear form, 108
  of bilinear function, 108
  of linear system, 13–14
  of matrix, 62–63
  of set of linear forms, 13
  of set of vectors, 40
  of transformation, 151–152
Rational canonical form, 206
Rational canonical representation, 206
Rayleigh quotient, 256
Real quadratic form, 119
  canonical set under equivalence, 120–121
  definite, 122
  indefinite, 122
  semidefinite, 122
  signature of, 120
Real quadratic function, 119
  canonical representation of, 119
  definite, 122
  indefinite, 122
  index of, 123
  semidefinite, 122
  signature of, 120
Real symmetric matrix, 121, 243
  canonical form, under congruence, 121
    under orthogonal similarity, 251

Real symmetric matrix, characteristic values of, 258
signature of, 121
Reducible polynomial, 162
Reflexivity, 48
Regular symmetric matrix, 119, 121
Relation matrix, 189
Relatively prime polynomials, 162
Representation, 106
of bilinear function, 107–109
of Hermitian function, 131, 250
of linear function, 105–107
of linear transformation, 165–166, 180, 184, 200–209
of quadratic function, 111
Residue class modulo p, 26
Residue class modulo S, 52
Rigid motion, 226–227
Ring, 155–157
with unit element, 155
Rotation, 226, 245
Row equivalence, 59–60, 81
canonical set under, 59–60
Row-equivalent matrices, 59
Row rank, 61
Row space, 61

S

Scalar, 32
Scalar matrix, 57, 74
Scalar multiple, 5, 33, 74
Schwarz's inequality, 216
Segre characteristic, 210
Self-adjoint transformation, 234
Set of linear forms, 5
linear combination of, 5
rank of, 13
theorems about, 13, 16
of vectors, Gramian of, 216
linear combination of, 37
rank of, 40
Sets, equality of, xiii
intersection of, xiii
notation for, xiv
subset of, xiii
Signature, 120
Similar matrices, 175–176
Similar transformations, 176
Similarity, 175, 201
canonical sets under, 200–209
Similarity factors, 199
Singular matrix, 64
Singular transformations, 151
Skew-symmetric matrix, 111, 118–119
Spanning set, 37
Square matrix, 56
minimum function of, 171, 202
order of, 56
Subfield, 24
Subgroup, 140–141
Submatrix, 57
Subspaces, 36–38
sum of, 37
direct, 51–52
Superdiagonal matrix, 56
Superdiagonal representation, 184, 229
Sylvester's law, of inertia, 119
of nullity, 85
Symmetric group, 147
Symmetric matrix, 111
theorems about, 114, 115, 117–121, 243, 251
(*See also* Real symmetric matrix)
Symmetric transformation, 240–243
Symmetry, 48, 245
System of linear equations, 3–22, 27–29, 42, 50, 62–63, 96
elementary operations on, 6–7
general solution of, 9–12
particular solution of, 18
rank of, 13–14

T

Transform, 175
Transitivity, 48
Translation, 145
Transpose matrix, 57–58
Transposed homogeneous system, 20
Triangular inequality, 216
Triangular matrix, 56

U

Unit element, of field, 23
of group, 139
Unit n-tuple, 32
Unitary group, 227
Unitary matrix, 228, 235
Unitary similarity, 250
canonical set under, 250
Unitary substitution, 229

Unitary transformation, 226–228, 234–235
Unitary vector space, 214

V

Vandermonde determinant, 102
Vectors, 31–32
  distance between, 217
  length of, 217
  (*See also* Set of Vectors)

Z

Zero element, 23, 140
Zero function, 5
Zero ideal, 161
Zero matrix, 56
Zero polynomial, 157
Zero transformation, 156
Zero n-tuple, 19
Zero vector, 34